AgriScience Explorations

Second Edition

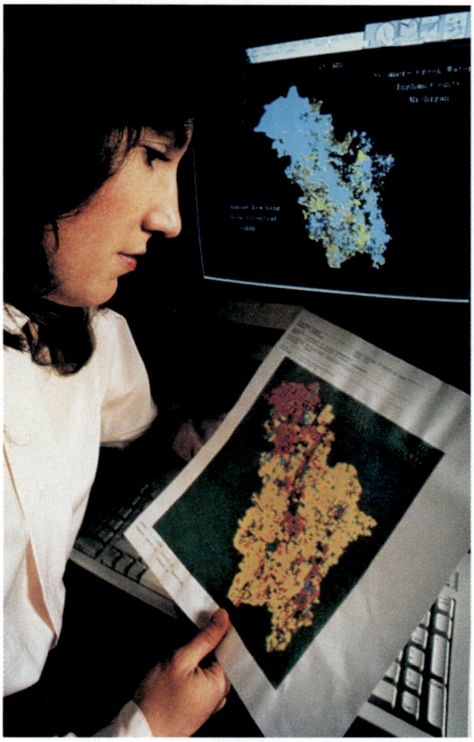

(Courtesy, Agricultural Research Service, USDA)

AGRISCIENCE EXPLORATIONS

IST

AgriScience and Technoology Series

Jasper S. Lee, Ph.D.
Series Editor

IPP

Interstate Publishers, Inc.
Danville, Illinois

ELIZABETH M. MORGAN

AgriScience Teacher
Presque Isle, Maine

RAY E. CHELEWSKI

AgriScience Teacher
Presque Isle, Maine

JASPER S. LEE

Agricultural Educator
Demorest, Georgia

ELIZABETH WILSON

Agricultural Education Coordinator
Apex, North Carolina

AGRISCIENCE EXPLORATIONS
Second Edition

Prior edition 1998.

Library of Congress Catalog No. 99-71694

ISBN 0-8134-3189-1

1 2 3 4 5 6 7 8 9 10 05 04 03

Order from

Interstate Publishers, Inc.

510 North Vermilion Street
P.O. Box 50
Danville, IL 61834-0050

Phone: (800) 843-4774
Fax. (217) 446-9706
Email: info-ipp@IPPINC.com
World Wide Web: http://www.IPPINC.com

PREFACE — Second Edition

This Second Edition ***AgriScience Explorations*** has been enhanced to make learning about our world more exciting. Just as all areas of life have changed, getting food, clothing, and housing has changed. Living and working are better when we know about agriculture and its products.

AgriScience Explorations, Second Edition is for you—a student who is beginning to study agriculture. It uses approaches and content for grades 7 through 9. Photographs have been used to make your study more exciting. Many of these feature young people. This will help you see how you can be a part of today's agricultural industry.

A modern approach to the agricultural industry is used in the book. Horticulture, forestry, natural resources, and the environment are included. Of course, the basics of plant and animal production are emphasized. The content is based on areas of science. The book has several chapters on being successful in agricultural education. It includes areas of personal development, such as leadership.

The book will help you become informed about agriculture. This is often known as agricultural literacy or awareness. The book also begins the process of career skill development. This is known as education in agriculture. It will help you get ready for the next step in learning. You can prepare for a career or go for additional education.

The Second Edition includes more emphasis on science and research. It contains the latest on FFA changes and how you can be involved. New, exciting photographs of students and scientists have been used.

Go ahead and get involved! The authors believe you will find ***AgriScience Explorations*** exciting. They have prepared the book with you in mind. Good luck!

ACKNOWLEDGEMENTS

The authors of ***AgriScience Explorations*** received assistance from many people in the preparation of the book. We say to everyone who helped, thank you!

Several individuals are acknowledged below for their special contributions. Without their efforts, this book would not have been possible.

Three persons were tireless in reviewing the manuscript for accuracy and relevance. They are Marvin Fadenrecht, Agriscience Instructor, Longmont, Colorado; John C. Cloran, Middle School Agriscience Instructor, Apopka, Florida; and Kirk Edney, Director, Agricultural Science and Technology, Texas Education Agency, Austin.

Several educators and agriculturalists in Maine are recognized for their assistance. The faculty and students of Maine School Administrative District #1 are acknowledged, with special appreciation going to Gehrig Johnson, Don Jordan, Debbie Martin, and Aaron Buzza. Greg Smith of Smith's Farm and Kathy Gunderman of the Natural Resource Conservation Service deserve special thanks. Paula, Dale, and Julie Chelewski are recognized for their support.

The staff and students of several schools were very helpful. These include Franklin County High School, Georgia; Newton High School, Mississippi; Paxton-Buckley-Loda High School, Illinois; and Leake County Vocational Center, Mississippi. Individuals include Lynn Wagner and Monte Ladner of Mississippi, Mike White and Doug Anderson of Illinois, and Wayne Randall of Georgia. Also acknowledged are the student technical models of Piedmont College, University of Georgia, Emory University, Mississippi State University, Georgia Tech, and North Carolina State University.

The authors would particularly like to thank the staff of Interstate Publishers, Inc. The enthusiasm and support of Interstate President Vernie Thomas and the dedicated assistance of Kimberly Romine, Jane Weller, Ron McDaniel, and Mary Carter are appreciated. Interstate Vice President Dan Pentony is recognized for his help in conceptualizing the book and analyzing educational needs served by the book.

Numerous other individuals, agencies, associations, and businesses are acknowledged throughout the book for their assistance.

CONTENTS

PART ONE—AgriScience and the Good Life

PART TWO—Agriculture and Science

PART THREE—AgriScience Applications

PART FOUR—Personal Growth and Leadership

1

LIFE HAS CHANGED!

Life is different today. We live in a modern world. Most Americans have plenty of food and clothing. We have good housing and health care. We did not just suddenly find ourselves in this situation. Much work was required to develop a modern system of providing for human needs.

Prehistoric people had to hunt for food and something to make into clothing. They had to seek shelter in caves or under thatched-together tree limbs. Imagine having to do that today! We wouldn't know where to begin. How would we go about trying to find food and clothing material? Some days we might not be able to find what we needed. The supply would not be dependable. We might be hungry and not protected from the weather.

Fortunately, today many products are available in easy-to-use forms. We have a wonderful agricultural industry!

1-1. New methods are used to study problems with plants. This shows probes being used to identify yellow stunt disorder in tomato and cucumber leaves. (Courtesy, Agricultural Research Service, USDA)

OBJECTIVES

This chapter explains the modern agricultural industry. It covers major events in the history of agriculture and the benefits of agriculture to the United States. The objectives of this chapter are:

1. Describe the modern agricultural industry
2. Trace major developments in the history of agriculture
3. Explain the rise of commercial agriculture
4. List the benefits of agriculture to the United States

TERMS

agribusiness
agricultural industry
agriculture
agriculture policy
aquaculture
commercial agriculture
farm
farming
forestry
marketing
natural resources
ornamental horticulture
self-sufficient farming
suburban farming

http://spacelink.msfc.nasa.gov/

(This site can be searched for information in many areas of agriculture.)

AGRICULTURE IS AN INDUSTRY

People have three basic needs to live: food, clothing, and shelter. If these needs were not met, we would be hungry, have little to wear, and live in poor housing. Meeting these needs is made possible by the agricultural industry.

1-2. How we get our clothing has changed. This shows a thick stack of cloth being cut. (Courtesy, Levi Strauss & Co.)

THE INDUSTRY

Agricultural industry is all of the activities needed to get food, clothing, and shelter to people for their use. Agricultural industry is big and diverse.

People often think that agriculture is only farming. Meeting the needs of people requires farming and activities not carried out on the farm. A large agricultural industry has resulted. Farming is a major part of the agricultural industry. But, it includes much more.

Agribusiness

Agribusiness is all of the nonfarm work in the agricultural industry. It includes many activities needed to meet the needs of people. Two main areas in agribusiness are:

1. Supplies and services—Many things are used in growing crops and raising animals. These are often known as inputs. These include seed, fertilizer, feed, and machinery. Without these inputs, farms would not be as productive as they are. Supplies and services also include some items used with lawns, flowers, trees, and related areas.
2. Marketing and processing—Once produced, products must be made into forms that people want. For example, wheat is milled into flour

1-3. Large manufacturing plants provide supplies for farming. This shows quality seed corn being bagged for distribution to growers. (Courtesy, DEKALB Plant Genetics)

and milk is pasteurized. Most people would find little use for unmilled wheat and raw milk! Marketing and processing include many activities.

1-4. Preparing food products is an important part of agribusiness. This shows wheat being checked at a flour mill. (Courtesy, Cargill)

Many people work in agribusiness. The work varies widely. Some people work in science laboratories studying new ways of farming. Others work in factories processing food, making clothing, or sawing lumber.

Ten times as many people work in agribusiness as on the farm. This does not mean that farm work is not important. It is! The work of one farm worker is far more important than ever before. We could not do without the farm worker!

AGRICULTURE

Agriculture is growing crops and raising animals to meet the needs of people. It involves science, such as how plants and animals grow. It includes how to conserve the soil, control pests, and use machinery. Managing the money used to carry out the work is equally important.

1-5. Animals are important for their products and for recreation. Many people enjoy horses! (Courtesy, DeShannon Davis, Mississippi)

Agriculture has changed in big ways! Early agriculture provided what a family needed to live. Most people had little extra. As time went by, developments were made. People began producing more than they

Career Profile

ANIMAL TECHNICIAN

Animal technicians work with animals. They care for animals by providing feed and water. They maintain facilities, herd animals, and change bedding. Most of the work is outdoors.

High school education in agriculture and science is important. College agriculture study in community colleges is nearly essential. Many get college degrees in animal science or a related area. Practical experience in working with animals is essential.

Jobs are found wherever animals are important in the lives of people. Most animal technicians work on ranches, in small animal facilities, or in zoos. This photo shows an animal research technician attaching a sensing unit to a steer. The unit records information so scientists know how much time the steer spends grazing each day. (Courtesy, Agricultural Research Service, USDA)

1-6. Powerful equipment is used to plant crops without tilling the soil. (Courtesy, New Holland North America, Inc.)

needed. The extra was exchanged with people who did not have crops and livestock.

1-7. Vegetables are common crops on suburban farms. Romaine lettuce is being harvested. (Courtesy, Agricultural Research Service, USDA)

Farming

Farming is using the land and other resources to grow crops and raise animals. These are used for food, clothing, and shelter.

Depending on what is produced, the nature of farming varies. Crop farming is using land, planting seed, and following good practices. Raising animals also varies. Some animals provide products, such as milk, wool, and eggs. Other animals become food.

A ***farm*** is a place where farming occurs. People often think of farms as large fields with crops or pastures with livestock. Not all farming is on the land. Some farming is in water.

Suburban Farming

Farming is not always a distance from towns and cities. Land areas in and near cities may be used for farming. This is known as suburban farming.

Suburban farming is using small areas of land in residential and business areas to produce crops and animals. The land areas are vacant lots and undeveloped tracts. The crops produced must conform to zoning requirements in these areas. Often, the land is quite valuable in the real estate market.

Common suburban farming includes vegetables, fruit, nut trees, horses, and exotic animals. Emu and ostrich may be raised on small acreage. Christmas trees and other forestry products may be on suburban farms.

Aquaculture

Farming in water is known as ***aquaculture***. Many kinds of crops grow in water. Some are animals, such as fish, shrimp, and oysters. Other "water" crops are plants, such as water cress and water chestnuts. Aquaculture helps meet the demands of people for fish and other foods.

AgriScience Connection

WATER FARMING

Some animals and plants grow on land; others grow in water. Those that live in water are known as aquatic organisms. Many aquatic organisms are popular foods, such as fish, shrimp, and oysters.

Natural supplies of some fish and other species have declined. The streams, lakes, and oceans cannot meet demand. This has resulted in water farming.

Water farming, known as aquaculture, is culturing crops in water environments. Most are in ponds. Others are in tanks and cages. Farmers need new kinds of skills to be successful. Aquatic species are a lot different from those grown on the land.

This shows fish being harvested. A seine moves all of the fish to one corner of a pond. A basket scoops up the fish and puts them into a tank for hauling. (Courtesy, James Lytle, Mississippi State University)

Ornamental fish are included in aquaculture crops. Fancy fish are raised for stocking aquariums. Many homes and offices keep ornamental fish in attractive aquariums.

Forestry

Forestry is the production of trees and manufacture of wood products. Some forests have native trees and receive little care. Other forests are on tree farms. Thousands of tree farms are planted to improved, fast growing kinds of trees. These farms take steps to provide good care for the trees. Much as agriculture, forestry is far more than on-farm production. It includes making products from the trees.

1-8. Trees produce many important products.

BEYOND FARMING

The agricultural industry is so large that it cannot be described in a few sentences. In fact, this entire book is about the agricultural industry.

Agriculture goes beyond food and clothing. It includes other areas that make the lives of people better. Three areas are ornamental horticulture, natural resources, and companion animals.

Ornamental Horticulture

Ornamental horticulture is producing plants for their beauty. It includes flowers and foliage plants as well as landscaping and landscape maintenance. Landscaping is using plants to make our environment more attractive. Shrubs, lawns, and flower beds are a part of landscaping.

Ornamental horticulture is often a part of suburban farming. Flowers, shrubs, sod, trees, and other plants may be raised in areas around cities.

1-9. Caring for flowers is a special hobby to many people.

Natural Resources and the Environment

Natural resources are all of the living and nonliving things found in nature. It includes plants, animals, and other organisms. Wildlife are popular with many people. Some people like to hunt. Others like to observe nature. Natural resources also include soil, water, air, and minerals. These are a part of environmental science.

Companion and Small Animals

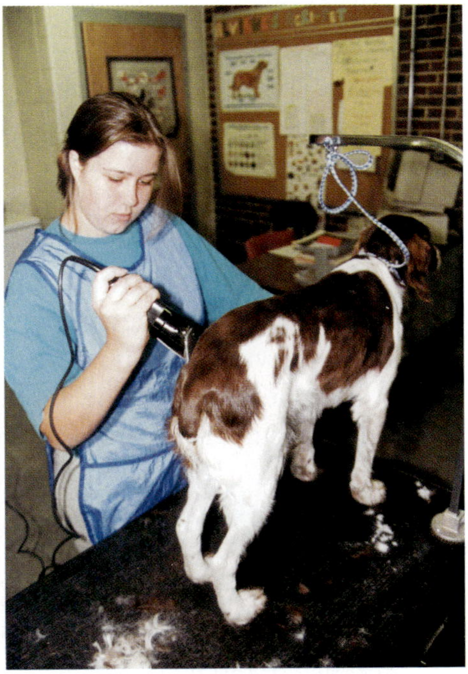

1-10. This New Hamshire student is learning skills in pet care.

People enjoy small animals. These animals are important in the lives of people. Sometimes known as pets, companion animals improve their owner's lives. Common companion animals are dogs, cats, hamsters, and birds. Others include snakes, African hedgehogs, tarantulas, iguanas, and pigs.

Keeping small animals involves knowing the needs of animals. Feed, health, and a good environment are important to their well-being.

HISTORY HELPS UNDERSTANDING

History helps us know "how we got here" and understand the way we live today. In the United States, agricultural history began with the Native Americans. Immigrants who moved here from many countries also had big impacts. Settlers brought crops and practices with them to the United States.

NATIVE AMERICAN AGRICULTURE

Early agriculture in what became the United States was practiced by the Native Americans. The Native Americans include Indians, Hawaiians, and Eskimos. Because of shortages, most of their time was used getting food. Major effort went into hunting and searching for wild foods.

Indians had wild berries, fruit, seeds, and animals for food. They used hides and feathers for clothing. Large leaves or woven stems were used for clothing. Their homes were made of wooden poles covered with hides or tree branches. Sometimes the homes were made of sod or, with Eskimos, ice and snow. The warm climate of Hawaii provided many fruits, berries, and vegetables. Fish were often captured in streams, lakes, or oceans.

Meeting daily needs by hunting was often difficult. The Indians learned to care for plants that provided food. This increased their food supply. They learned to collect seed for planting. They also learned about soil and planted seed in soil that was easy to dig (till). Using fertilizer in helping plants grow was a major step. The first fertilizer was a small fish placed beneath seed.

The Indians began simple farming some 7000 years B.C. Beans and squash had become major crops by 650 B.C. The early areas farmed were similar to small garden plots. By 1000 A.D., maize (corn) was grown in larger areas known as fields.

The practices of the Indians became the foundation on which settlers built agriculture in the United States.

COLONISTS

The early colonists tried to use European practices. These often failed to work well in the new land. Sometimes, the settlers learned practices from the Indians. Practices were changed little from the 1500s to nearly 1800.

Most early settlers used plants to meet their needs. Some seeds of plants were brought from Europe. Others learned to use the native plants. But, the settlers were not content with just plants—they wanted to raise animals! Also, animals could be used as power for plowing the land.

Hogs, cattle, sheep, goats, and horses were brought by the colonists. The first arrived in Jamestown in 1607. These were quickly consumed. None remained for production. A second shipment came in 1611. It was the foundation for raising livestock in Virginia and other colonies in the South.

Land was plentiful. Fields were farmed and abandoned. No effort was made to keep the soil fertile. New fields were available that had not been farmed.

At first, simple tools were used to till the soil. Broken tree branches, stones, or large animal bones were used to dig the soil. Later, wooden plows were pulled by oxen. This reduced the need for people to dig fields by hand. Regardless, farming required hard work.

1-11. A team of oxen is pulling a plow. (Courtesy, Jody Pollok, Michigan)

THE NEW NATION

The Declaration of Independence passed the Continental Congress on July 4, 1776. This created The United States of America. The 13 colonies became the first 13 states.

In the new nation, 90 percent of all people farmed. The early leaders for the nation were farmers. George Washington and Thomas Jefferson were large farmers who readily adopted new practices. Both also served as presidents of the United States. Today, we know that their zeal for new ways of farming led to the rise of American agriculture.

The homes of Washington and Jefferson have been preserved as historic sites. Mount Vernon in Alexandria, Virginia, is the former home of Washington. Monticello near Charlottesville, Virginia, was the home of Jefferson. Today, visitors to these homes can see how some farming was carried out. The new methods used by Jefferson are easy to see.

New crops were introduced and successfully grown in the new nation. Many kinds of vegetables and fruits plus cotton and tobacco were grown. These were used to meet the needs of the rapidly increasing population. Some crops, especially cotton and tobacco, were exported to Europe. Farmers often produced for a specific market, and this gave rise to commercial agriculture.

Trying to meet the demand for crops required bigger yields. The result was machinery and new ways for doing work. Later, crops were improved by new methods of selection. President Washington said that the government should promote agriculture.

1-12. President George Washington's Mount Vernon home and its surroundings are maintained for people to learn about the history of our Nation and Agriculture.

AGRICULTURE POLICIES

In the early 1800s, government leaders were slow to set up programs to encourage agriculture. Things began to change in the mid-1800s. The Federal government passed laws that helped. Laws of the government about agriculture form ***agriculture policy***.

The U.S. Congress set up a special committee on agriculture in 1825. With strong support from several groups, the U.S. Patent Office began agricultural research in 1852. Time was right ten years later for major agriculture policies.

With leadership from President Abraham Lincoln, three major acts were passed in 1862. The first was an act that set up an agency that later became the U.S. Department of Agriculture (USDA). The second was the Homestead Act. The third was the Morrill Act—an important education law.

The USDA has carried out many programs to promote agriculture. It has published bulletins and developed new crops. Programs were set up to help farmers receive better prices for the crops. Soil conservation, water use, and other laws helped conserve natural resources. Other programs helped farmers with loans and helped improve rural life.

In 1935, the Rural Electrification Administration was set up. This agency helped bring electricity to rural areas. City and rural people could have the same conveniences in life. Electricity changed farming practices, including how cows are milked and chickens are fed.

In recent years, policies have helped trade among nations of the world. One example is The North American Free Trade Agreement (NAFTA). Enacted in the 1990s, the goal was to open trade between the United States, Canada, and Mexico. The long-term effect on U.S. agriculture is not known. Many factors are involved. Countries that pay workers less can sell crops cheaper. Some people feel that importing crops will make it harder for U.S. farms to make a profit.

TODAY

Agriculture has had big changes since the Native Americans first began growing crops. Today, the United States has 1.9 million farms. These meet our needs for food, clothing, and shelter. One farm worker provides for the needs of about 125 people. This is possible because of the vast agricultural industry.

For every person who works on a farm, another 10 people work in the agricultural industry off the farm. These workers provide the supplies and services needed to grow crops and raise animals. Off-farm agricultural workers also market and process farm products.

How farming is done has changed. Farmers use the latest ways of farming. They use "smart" machinery and improved crops and animals. They

1-13. Today's farmers use modern equipment to grow crops. This shows a state-of-the-art Gleaner® Combine harvesting wheat. (Courtesy, AGCO Corporation, Georgia)

carefully manage the earth's natural resources. This helps assure food, clothing, and shelter for future generations. In short, they use technology!

COMMERCIAL AGRICULTURE

U.S. leaders recognize the importance of agriculture. They feel that prosperity depends on using the soil to meet the needs of people.

SELF-SUFFICIENT FARMING

Early farmers were self-sufficient. ***Self-sufficient farming*** means that the farmers used what they had available to produce for their families. Little extra was produced. If there was any extra, it was traded (bartered) with neighboring farmers. For example, a ham might be traded for a couple of chickens. Little from off the farm was used in farming. However, some trade with Europe developed.

As the United States developed, farming changed. Transportation was needed. Farms near rivers or oceans had advantages. Boats could travel on rivers and oceans carrying harvested crops. Roads were built to nearby boat docks. Farmers not on rivers could now produce crops for export.

Good inland roads and railroads were needed to haul crops. Some states set up "farm to market" roads that reached into the most rural areas. This road system was to make it easy for farmers to get supplies and to move products to market. These events led to a change in agriculture.

PRODUCING FOR OTHERS

Commercial agriculture is producing to sell to others. It is far more than meeting the needs of the farm family! Farmers find out what consumers want and go about meeting this demand. Specific crops and animals are grown. These are sold to buyers. The buyers process the crops and animals and deliver them to grocery stores and restaurants. Sometimes, several buyers may be involved in the process.

A vast marketing system is now used. ***Marketing*** is getting what people want to them in the desired form. Many steps may be involved. Marketing processes are required in commercial agriculture.

Many products are traded with other nations. In some cases, crops are planted specifically for meeting the demands of consumers in far away nations. An example is soybeans. Soybeans grown in Illinois may wind up in Europe.

The United States now imports from other nations. For example, most all coffee is imported. Little coffee is grown in the United States except Hawaii and Puerto Rico. Today, just about every food and clothing item may be imported and exported.

Many nations trade food, clothing, and shelter materials. This is known as international trade. Products may be sold and bought several times. Large companies are involved. Huge ports for oceangoing vessels are used, such as the Port of New Orleans for grain.

1-14. Large ships are used to transport grain. (Courtesy, Continental Grain Company)

BENEFITS OF AGRICULTURE

Agricultural production helps people enjoy life. The United States has good farms. Without them, the United States would not be strong. It would be a nation where the people might not have enough food or clothing.

Five benefits of agriculture in the United States are:

- Meets needs of people—The needs of people in the United States are met with plenty of food. Clothing and shelter are readily available. With these, prices are below what people in many other countries pay. People who have their needs met are happier and feel better about themselves.
- Makes country strong—A good agriculture helps make the United States a strong, secure place to live. The needs of people are met. Unrest is kept down. People are healthier.
- Provides jobs—The agricultural industry provides jobs for about 20 million people. These jobs provide money to help people enjoy life. People in the jobs must have the needed skills to do the work. Though most of the jobs are off the farm, they are very important in getting what we want to us.
- Provides trade—Agricultural products are traded with countries around the world. These countries often need food. They may have products needed in the United States, such as rare metals. The United States can keep trade nearly in balance among the other nations.
- Supports industry—A productive agriculture requires fewer farmers. This makes a labor force available to work in other jobs. These jobs manufacture things that people want, such as automobiles and computers.

1-15. Plenty of food allows people to enjoy life in many ways, such as vacationing at Hawaii's Waikiki Beach in the winter.

REVIEWING

MAIN IDEAS

Agriculture meets the basic needs of people for food, clothing, and shelter. It is more than farming. Agriculture has become a vast industry, known as the agricultural industry. Farming is the base for all of agriculture. It is using the land and other resources to grow crops and raise animals. In some cases, it involves aquaculture. Farming is also known as production agriculture.

U.S. agriculture history bagan with the Native Americans. It includes the practices brought by the settlers from Europe and other areas of the world. Early government leaders realized the importance of agricultural trade. Commercial agriculture developed to meet the needs of markets.

Agriculture has been important in the United States. Without a good agriculture, the United States would not have developed into a strong nation.

QUESTIONS

Answer the following questions. Use correct spelling and complete sentences.

1. What is farming?
2. What is the agricultural industry?
3. What role did Native Americans have in the early history of U.S. agriculture?
4. How did the settlers approach farming in the Colonies?
5. Name two early U.S. government leaders who were strong supporters of agriculture.
6. What is agriculture policy?
7. What is the nature of agriculture today?
8. What is the difference between self-sufficient agriculture and commercial agriculture?
9. What are the major benefits of agriculture to the U.S.?
10. What is the name of the town where the first livestock entered the United States?

EVALUATING

Match the term with the correct definition. Write the letter of the term in the blank provided.

a. agriculture
b. farming
c. farm
d. aquaculture
e. marketing
f. agribusiness
g. suburban farming
h. ornamental horticulture

____ 1. Farming on small acreage in residential and business areas.
____ 2. Growing plants and animals in water.
____ 3. Growing crops and raising livestock to meet the needs of people.
____ 4. Using the land to grow crops and raise livestock.
____ 5. A place where farming occurs.
____ 6. Getting what people want to them in the desired forms.
____ 7. Producing plants for their beauty.
____ 8. Nonfarm work in agriculture.

EXPLORING

1. Tour a museum that includes the history of agriculture for the local area. Prepare a written report on what you see. Give an oral report in class.

2. Write the dean of agriculture at a land-grant university in your state. Request information on the current agricultural situation for the state. Review the information and provide an oral report on your findings in class.

3. Explore the information available through the USDA World Wide Web site: http://www.usda.gov/. Give an oral report in class.

2

EDUCATION AND DEVELOPMENT

Education helps people have the knowledge and skills needed for daily living and work. No area has had more than the agricultural industry. People who work in agriculture must have good preparation.

Research provides new knowledge. Agricultural research has been a leading area of new knowledge and development. Improved methods and products have resulted. These improvements have changed what we need to know about agriculture.

People need to know more today than ever before. No area has a stronger science base than agriculture. Fortunately, you are enrolled in a class in agriculture. You are learning now. You will be ahead of other students. In order to stay ahead, you need to continue learning throughout life. If you do not, you will get behind!

2-1. A scientist is using an automatic DNA sequencer to analyze the genetic map of chickens. (Courtesy, Agricultural Research Service, USDA)

OBJECTIVES

This chapter explains agricultural education. It also introduces the important role of development. The objectives of this chapter are:

1. Describe agricultural education in the United States
2. Explain secondary agricultural education
3. Describe the importance of research and development

TERMS

- agricultural education
- agricultural experiment station
- CGIAR
- curriculum
- development
- FFA
- field plot
- Hatch Act
- Morrill Act
- prototype
- research
- secondary agricultural education
- Smith-Hughes Act
- Smith-Lever Act
- Squanto
- supervised agricultural experience

http://www.ed.gov/

(This site has information about many areas of education, including agricultural education.)

AGRICULTURAL EDUCATION

Agricultural education is education in and about agriculture. Both careers and literacy are included. The career focus helps students begin developing skills for job success. The literacy focus helps people be good citizens and users of agricultural products. All areas of the agricultural industry are included.

Pioneers learned the importance of education. The teachings of the Native American, ***Squanto***, to the early Plymouth Colony settlers are familiar. Some people consider Squanto the first agricultural educator in the United States. Many approaches have been used to provide education since the days of Squanto.

SOCIETIES AND FAIRS

Agricultural societies were set up in the late 1700s. The Philadelphia Society for the Promotion of Agriculture was the best known. Formed in 1785, its aim was to improve farm practices. Problems with crops, soil, livestock,

2-2. The National FFA Career Show features hundreds of new products.

and other areas were studied. Society members were informed about new practices. The number of societies grew to several hundred by the mid 1800s.

Agricultural fairs were sources of education. In the early 1800s, fairs put the latest ideas on display. Farmers could enter animals or crops in judging. By 1868, 1,367 fairs were held in the United States. Fairs are still held with less emphasis on education. Farm shows, forestry field days, and horticulture fairs are ways of showing new equipment and methods.

BOOKS AND MAGAZINES

Society efforts were made stronger by books and magazines. Some early farmers resisted "book knowledge." They relied on superstitions and practices of past generations. The "science" of agriculture was missing.

Career Profile

AGRICULTURE TEACHER

An agriculture teacher teaches students in agriculture classes. They write teaching plans and involve students in learning activities. They help students get experience and develop personally. Often they use outdoor labs in teaching.

Agriculture teachers need to know both their subject matter and how to be a teacher. They need education in the areas of agriculture being taught. They need to know how to prepare for teaching and use sound methods in the learning environment. Agriculture teachers help students with SAE and run an FFA chapter. A college degree in agricultural education with strong emphasis in agriculture is needed. Practical experience is very helpful.

Jobs for agriculture teachers are in schools where agriculture is taught. Salary depends on education, experience, and the scale used by the school district.

This shows a teacher pointing out the qualities of a pond-raised hybrid striped bass in a school lab. (Courtesy, Agricultural Research Service, USDA)

The first magazine about agriculture was *The American Farmer*. It began in 1819. As the editor, John Skinner is said to be the father of agricultural journalism. Other magazines were published in the 1800s. Few lasted beyond a year or so.

Several publishers had books on various subjects. Most agricultural book publishers did not last long. A few remain today, with Interstate Publishers, Inc., being the leader in books on agriculture. Students in the United States use exciting books that have a science-based approach to agriculture.

Times have changed. People now value the information in books and magazines. It is probably the best way to learn!

FORMAL CLASSES

In the early 1800s, a few schools had classes in agriculture. These often lacked good instruction. The instruction was often about tradition. In the mid 1800s, things began to change. People wanted more agricultural education.

2-3. Formal classes are used to teach many agricultural subjects.

College Level

With the leadership of President Lincoln, the ***Morrill Act*** was passed in 1862. This Act set up a system of colleges to teach agriculture and other subjects. Today, these are known as land-grant universities. Every state has at least one land-grant school.

2-4. The use of an electron microscope is being explained to students. (Courtesy, Agricultural Research Service, USDA)

Existing knowledge of agriculture was not enough. More information was needed. This resulted in Congress passing the ***Hatch Act*** in 1887. The Hatch Act set up a system of agricultural experiment stations at the land-grant colleges. The purpose was to generate new knowledge.

Members of the U.S. Congress realized that most people could not go to the campus of a school for education. This led to another law. In 1914, the ***Smith-Lever Act*** was passed. This Act created the Cooperative Extension Service. Education was provided on farming and home life. This work continues today in most local communities.

Secondary Schools

By the early 1900s, most local communities had grade schools. Some had secondary or high schools. More communities were setting up high schools. People were valuing education more. Classes in agriculture were sometimes offered.

In 1917, the U.S. Congress passed the ***Smith-Hughes Act***. This Act provided for agricultural education in the high schools. It was to be set up through state education offices in the states.

Today, more than 7,200 high schools have agriculture classes. Some 800,000 students are enrolled. These students have science-based agriculture classes. Students have supervised experience to gain firsthand skills. The FFA Organization has local chapters in these schools. The chapters de-

AgriScience Connection

THE MORROW PLOTS

The Morrow Plots at the University of Illinois are the oldest experiment fields in the United States. Set up in 1876, the plots have been used to study corn, soybeans, and other crops. The area is used for research each year.

The plots are in the middle of the campus of the University of Illinois in Urbana. The area is a highly desired site for buildings. A nearby library building was built underground to keep the plots available for research.

Maintaining the Morrow Plots shows how important research has been in the history of U.S. agriculture.

velop personal and leadership skills and encourage excellence in young people.

BROADCAST MEDIA

Several means of communication have been important in agricultural education. These help get new methods to farmers.

Radio was first. Many radio stations had farm-oriented programs. These programs gave the latest information on crops and livestock. Many local radio agricultural programs have been replaced with regional or national programs.

Television followed radio. Some TV stations had farm programs in the 1950s to 1970s. These have largely disappeared. Regional and national TV programs have replaced those of local stations. Some public TV stations carry agricultural programs.

2-5. Television is widely used to send agricultural information.

2-6. A teacher instructs a student in using the Internet to search for information. (Courtesy, National FFA Organization)

Today, closed video systems are being used. Satellites help share programs around the globe. Distance learning is used to help people gain knowledge and save travel to a college campus.

THE INTERNET AND WORLD WIDE WEB

In the 1990s, the Internet and World Wide Web became sources of information. Many people use the Internet to communicate. They use the World Wide Web to get information. The accuracy of the information must be carefully assessed. Using a site by a well-known provider is best.

Some sites are maintained by agriculture departments in land-grant universities. These often have accurate information based on research and the opinions of scientists. Some agribusinesses have excellent sites. The USDA maintains a site with much information. The address is: http://www.usda.gov/.

SECONDARY AGRICULTURAL EDUCATION

Secondary agricultural education is offered in many middle schools and high schools in the United States. Students enroll in the classes much as they would other classes in school. Once they have enrolled, students often discover a big difference! These classes extend beyond the classroom. They offer the opportunity to gain experience and develop personally.

Most classes in agricultural education include three areas: instruction, supervised experience, and student organization.

INSTRUCTION

The instruction is exciting! Classes are on useful subjects and include practical information. Agricultural education teachers use different approaches in teaching. The subjects in the classes are carefully planned.

2-7. An Illinois teacher is explaining plant care to a student.

Classes

Classes are offered in introductory or core subjects and specialized areas. Educators often refer to the "curriculum in agricultural education." The ***curriculum*** is all of the classes and learning experiences that students have. Curriculums vary from one school to another. Yet, many states have curriculum guides that provide an overall frame for agricultural education.

Students often take introductory classes in middle school and early high school. These courses often provide an overview of the agricultural industry. Many courses have a strong science base. Typical class names are "introduction to agriscience" and "agriscience explorations."

Specialized courses are offered in high school. These courses may be in horticulture, wildlife, agricultural mechanics, forestry, animal science, and other areas. In recent years, more schools have classes in aquaculture, environmental science, and biotechnology.

Teaching and Learning

Agricultural education classes use both classroom and laboratory instruction. Students learn from books, computers, and other resources in the classroom. In the lab, they develop skills and apply what they have learned in the classroom. Greenhouses, agriscience labs, nature areas, and shops are examples of laboratories. Many days, students learn skills they can go home and use that same day!

The classes are designed to help students in two ways. First, the classes prepare students to take advanced education. (These may be other courses in high school or courses after high school.) Secondly, the classes prepare students to enter and advance in careers. (Information about careers is often included. In addition, many students gain work experience as a part of the class.)

The Teacher

Instruction is provided by a trained agriculture teacher. Most agriculture teachers have college degrees in areas to prepare them to be teachers. They

2-8. Teachers often go to workshops for additional training such as this one in Georgia.

have taken courses in general education, agriculture and related areas, and teaching methods. Most agriculture teachers are excited about their profession.

Teaching is a big job. Agriculture teachers develop plans and organize facilities. They teach class and advise students. Their work includes serving as advisor of the FFA chapter and supervising experience programs. They are admired and respected in the community where they live.

SUPERVISED EXPERIENCE

Supervised agricultural experience (SAE) is the application of instruction in agriculture. Many kinds of experiences are possible. Students can explore areas to identify interests. They can practice specific job skills by working for another individual or by being an owner themselves. The agriculture teacher is closely involved in planning each student's SAE.

SAE is carried out after regular class time. Students are often busy on weekends, in the summer, and after the school day is over.

When students first enroll in agricultural education, they often do not know what their interests may be. This is a time of exploration. SAE that allows many kinds of activities may be planned.

2-9. An SAE placement student is discussing a customer's needs in an agribusiness. (Courtesy, National FFA Organization)

As students progress, their SAEs become more advanced. A student who began with one animal will expand. Another student who has a job for an SAE will do simple tasks at first. More advanced duties will follow.

Some students like to do research and experiments. These can be SAE if carefully planned. This type of SAE is carried out with assistance from an agricultural scientist in the local area or a science teacher.

More details on SAE are provided in Chapter 21, Developing Career Goals and Skills.

STUDENT ORGANIZATION

2-10. An FFA member is developing leadership skills in a meeting.

Nearly all schools with agricultural education have an organization for students—The FFA. The ***FFA*** is the organization for students enrolled in agricultural education.

Many exciting things are possible in the FFA. The emphasis is on learning new things and developing personal skills. Traveling to new places, meeting new people, and being recognized for what you do are a part of the FFA.

The FFA is an integral part of classes in agricultural education. The local FFA is known as a chapter because it is affiliated with a state and national FFA organization. Students pay a small amount for dues. They may hold special meetings and events, such as cookouts and banquets.

Most schools expect all agriculture students to join the FFA. Membership has so many benefits! You will want to join and keep your membership until graduation from high school.

More on the FFA is presented in Chapter 20, The FFA and You.

DEVELOPMENT

Development is creating something new. Many things may be involved. Most development begins with an idea. The idea is followed by research. Experiments are carried out to see how well an idea might work. If what was investigated shows promise, a new product may become available. This is the result of development work.

New products and methods do not just happen; someone has to plan and carry out the work. Many agricultural scientists are involved in research.

2-11. A sample of a fungal culture is being collected at the Agricultural Research Service Center in New Orleans. (Courtesy, Agricultural Research Service, USDA)

AGRICULTURAL RESEARCH

Research is using systematic methods to answer questions. The purpose of the research is to answer questions about the idea. Research gives facts that can be used to improve on what is being done.

Research is carried out in labs, fields, and other places. Field (or test) plots may be on farms or in other places. A ***field plot*** is a small area that has the new crop or whatever is being studied used on it. Careful steps are taken to have good test results. Once proven on a small scale, the product may be ready for use on a larger scale.

Agricultural research is done by experiment stations, private agribusinesses, the U.S. Department of Agriculture, and others.

Experiment Stations

The ***agricultural experiment stations*** are a part of the land-grant colleges and universities in each state. Scientists and students carry out research in many areas. The research is in labs on campus and at branch

2-12. Agricultural research is carried out on farms.

stations. The branches are in areas where climate or other conditions differ. Research on a campus in another climate or soil type might not get the same results.

Often, the research topics selected are based on suggestions from an advisory committee. The advisory committee is made of members who represent different areas of agriculture. They relay problems they are having with crops, livestock, or other areas. Members of the committees help the scientists design the research to solve the problems.

Most of the money for experiment station work is from government funds. It is up to Congress and the state legislatures to appropriate the money. Sometimes, government officials try to influence the research.

2-13. This research involves growing corn earworm pupae in a laboratory. (Courtesy, Agricultural Research Service, USDA)

Private Research

Agribusinesses do research on specific problems. Private funds are used. Well-trained scientists are often used to do the research. The major goal of the research is to improve how products are used or to make new products. Some agribusinesses have research facilities. Well-equipped labs and test farms are often used.

Overall, the research by agribusinesses is about immediate needs. The findings are used in manufacturing new products or helping users of products. Much of the important genetic engineering research has been carried out by private companies.

Farmers also do research. Sometimes, they help agribusinesses in doing research. Other times, farmers do research themselves. Ideas for new products often come from farmers. They see practical needs that others may overlook.

International Research

Some agricultural research deals with worldwide issues. Various government and private agencies are involved. Sometimes, religious groups support research efforts. International research often involves support from many sources. These efforts focus on solving world hunger.

The Consultative Group on International Agricultural Research (***CGIAR***) is an example. It was formed in 1971 by 53 public and private members. The mission of CGIAR is to promote sustainable agriculture in developing countries. It has 16 research centers in countries around the world.

One CGIAR research center is the International Rice Research Institute (IRRI) in Manila, Philippines. The IRRI was established to help increase rice production. Rice is an important food crop. Over half of the food eaten by one out of every three people in the world is rice. Some 1,400 people work for the IRRI. Besides research, the IRRI has education programs to train people in how to use the findings of research.

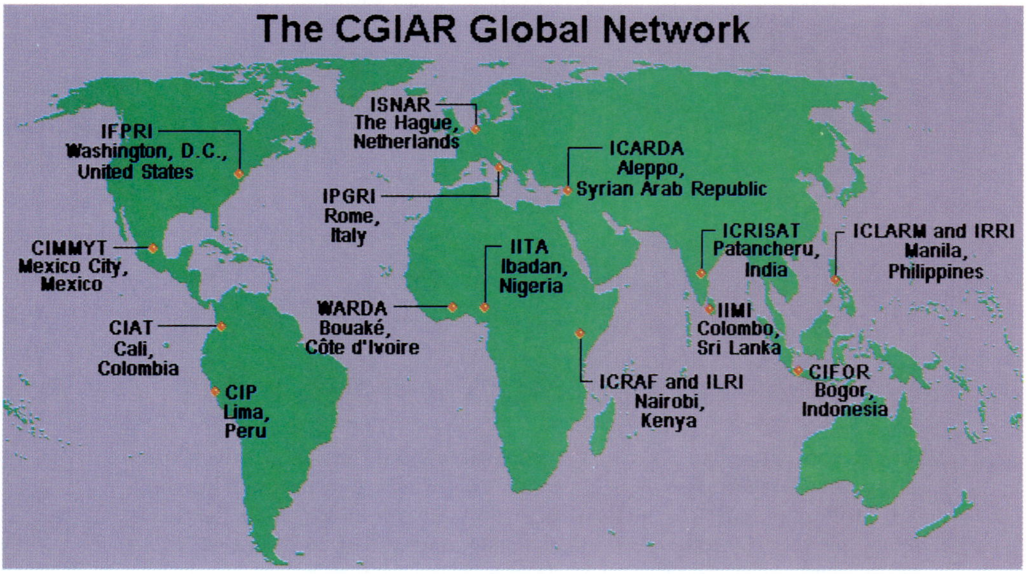

2-14. Map showing the locations of CGIAR international agricultural research centers.

PRODUCTS

New products often undergo a large amount of testing. How a product is tested depends on what it is or does. The product is carefully tested under controlled conditions.

New chemicals to use on crops must be tested over several years. Application to make the chemical for a particular use is submitted to the government. Gaining approval to use it is a long process. Much information is needed showing that the product will not harm the environment.

New machinery involves developing a prototype of the new item. A ***prototype*** is the original model tested. It becomes the pattern for making future models. The prototype is carefully tested under field conditions. Changes in design are made based on field test results. For example, if the steel frame for a new implement fails, a stronger steel frame is used.

After testing, a manufacturer begins production. Methods are used to efficiently make the product. People must know about and buy the new product. This is when marketing becomes important. Most agribusinesses use marketing specialists to help get people to use their products.

Getting farmers and others to buy and use a new product is often a challenge. Marketing is important. Education is needed. Books and magazines have information on the product. Meetings are held to promote the product. Field demonstrations are used.

2-14. The prototype of a planting system developed by design engineer Frank Faulring of New York is being field tested. This machine transplants growing plants through plastic. The machine burns small holes in the plastic, sets the seedling, presses the soil around the seedling, and waters the planted seedling—all in one operation. (Courtesy, Renaldo Sales and Service, Inc.)

REVIEWING

MAIN IDEAS

Agricultural education and development are important for advances in agriculture. People need good information. New products and methods need to be developed.

Agricultural education is education in and about agriculture. Education in all areas of the agricultural industry is included. Classes in agriculture are offered in colleges and secondary schools.

Secondary agricultural education classes are offered in many middle schools and high schools in the United States. This education begins important preparation for additional education and work in the agricultural industry. It includes instruction, supervised experience, and the FFA.

Development includes research and new products. These go together to improve agriculture. Research seeks answers to problems. New products or methods grow out of the research. Much testing is needed. Prototypes may be field tested to assure that a product works right.

QUESTIONS

Answer the following questions. Use correct spelling and complete sentences.

1. What is agricultural education?
2. How have societies and fairs been important?
3. What was one of the most important agricultural societies?
4. What was the title of the first magazine about agriculture?
5. What was the purpose of the Morrill Act?
6. How have the broadcast media been a part of agricultural education?
7. What are the three main areas of secondary agricultural education?
8. Why is development work important?
9. What is the purpose of research?
10. What group of colleges and universities carry out agricultural research?

EVALUATING

Match the term with the correct definition. Write the letter of the term in the blank provided.

a. Squanto
b. President Abraham Lincoln
c. Smith-Hughes Act
d. curriculum
e. SAE
f. field plot
g. CGIAR
h. prototype

____ 1. Promotes sustainable agriculture in developing countries.
____ 2. A way to get practical work experience.
____ 3. All of the classes and learning experiences students have under the direction of a school.
____ 4. A Federal law that provided for agricultural education in public schools.
____ 5. Provided leadership to pass a law setting up a system of colleges to teach agriculture.
____ 6. Considered to be the first agricultural educator in the United States.
____ 7. The model of a new product that undergoes considerable testing.
____ 8. Small area used to research a new crop.

EXPLORING

1. Use the Internet and World Wide Web to learn more about international agricultural research. Begin with the CGIAR and explore each of its research centers. The Web site for IRRI is http://www.cgiar.org/irri/

2. Tour an agricultural experiment station. Plan the tour well ahead of time and arrange for a scientist to explain the different research that is underway. (If a tour is impossible, invite an agricultural research scientist to serve as a resource person in school and discuss the research that is underway.)

3. Plan and carry out a research project. Identify a local problem in horticulture, forestry, farming, or other area. Select a problem that is small and can be done at the local level. Get a research scientist to advise you, if possible. Carefully study the background of your project. Control the variables so your findings will be valid. Enter your project in a local science fair.

3

TECHNOLOGY IN ACTION

Imagine life without the automobile? Automobiles have become widespread in less than a century. We all depend on them. They are important in our daily lives. This form of transportation technology has made big changes in the way we live.

Technology has also made big changes in our work. It has not happened in just the last few years. Technology has been around a long time. Every time a new tool has been developed, people made advances. In primitive times, technology changed how people lived much as it does today.

We live better today because of technology. Without it, our lives would be like those of people who lived years ago. Just think: Would you want to live the way people did 200 years ago? In fact, most of the things you use and enjoy would not be a part of your life. They were not around!

3-1. Remote sensing and microcomputers are being used to plot soil fertility information. (Courtesy, Agricultural Research Service, USDA)

OBJECTIVES

This chapter is about technology and how it has changed agriculture. Technology will be defined and examples given. The objectives of this chapter are:

1. Explain agricultural technology
2. List important inventions in agriculture
3. Describe technology with machinery, plants and animals, and pests
4. Describe the role of computers
5. Explain the meaning of site-specific farming

TERMS

agricultural technology
appropriate technology
food preservation
harvester
herbicide
insecticide
invention
mass selection
mechanical technology
parasite
pest
pesticide
planter
plow
site-specific farming
technology
transgenic

http://www.ag-com.se/
visit "Interesting Links"

(This site has information on a wide range of topics related to Global Positioning Systems and other areas.)

USING TECHNOLOGY

Technology is the use of inventions in working and living. The technology used in agriculture is often called ***agricultural technology***. Major changes in agriculture have resulted from the use of technology.

Most agricultural technology is based on science. Science methods have resulted in many new ways of producing food, clothing, and shelter. New technology results from the practical use of science. A good example is tissue culture. Science allows people to take tiny tissues or cells and grow another complete plant!

3-2. A hammer uses a simple machine—the lever.

Simple developments use science principles. For example, the handle in a hammer is a lever. A lever is a simple machine. It gives an advantage to the user in driving a nail. Just try to drive a nail without a handle in the hammer head!

Today, technology has given us the computer. Computers are widely used in our daily living. Nearly everything we do is touched by a computer!

APPROPRIATE TECHNOLOGY

Sometimes, people are not ready to use new things. The technology may cause problems for them. We may be unsure about the results. Most technology for agriculture is carefully studied. Research has proven that using it is safe. Are people sometimes afraid of new things without good reason?

Appropriate technology is the technology that people can use. People are not always ready for new things. The technology is said to be inappropriate.

AgriScience Connection

CHICKEN TECHNOLOGY

High technology is used in raising chickens! Both egg and meat production use many science-based inventions. This photograph shows White Leghorn chickens being observed. (White Leghorns are primarily for egg production.) Researchers carefully study chickens. The researchers learn the chickens' needs and how to provide a good environment. Differences in chickens for egg and meat production are carefully noted. New methods are developed that consider the well-being of chickens. Growers provide good feed and keep the poultry healthy. This helps consumers have top quality poultry products at reasonable prices.

Today, people around the world enjoy chicken. This is because new technology has made large-scale production possible. No area of agriculture has been changed by technology more than poultry production! (Courtesy, Agricultural Research Service, USDA)

People in lesser developed countries may not be ready for technology used in the United States. Large tractors might work well in the United States; however, they would not be appropriate where animal power is used to do work.

Decisions must be made about technology. Good information is needed in making decisions. Without good information, bad decisions are likely to result.

TECHNOLOGY BENEFITS

Technology helps people in many ways. Three benefits are:

- **Technology has made work easier.** Many jobs that required hard labor are now easier to do. Plowing is an example. By using a tractor, we can accomplish in a few hours what would have taken us days using a hand plow. Of course, we need training to learn how to use technology.

3-3. One person can plow several acres of land with a tractor-pulled implement. (Courtesy, New Holland North America, Inc.)

- **Technology has increased productivity.** We can do a lot more using technology. One hundred years ago, one farm worker produced enough for four people. Today, each worker supplies 125 or more people.
- **Technology has given a higher standard of living**. We have better food, clothing, and housing than ever before. This has resulted in our living longer. Life expectancy 100 years ago was about 47 years. Today it is about 75 years.

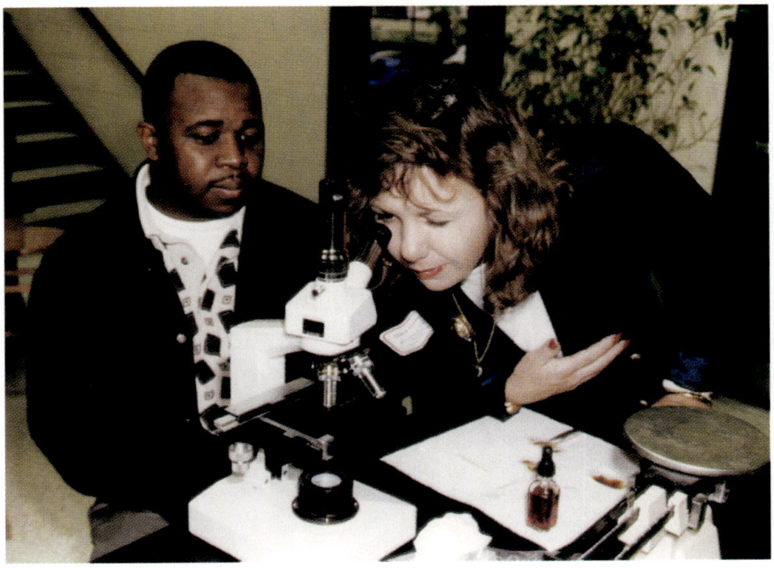

3-4. Much laboratory study goes into new technology.

INVENTIONS

An ***invention*** is a new device or product. Sometimes, inventions are new ways of doing work. An invention has not existed before. Something new has been created. You can be an inventor. How? Begin by identifying a task. Come up with a better way of doing the task.

There have been many inventions in agriculture. These inventions are new technology. Frequently, the benefits of inventions far outweigh the problems caused.

PREHISTORIC INVENTIONS

Some inventions were made before history was recorded. These are known as prehistoric inventions. The inventions were often primitive but helped move people toward bigger changes in the future.

The first plows and wheels were not as strong, smooth, nor polished as they are today. The first plow was developed from a tree branch about 4000 years B.C. The wheel came about around 3500 B.C. No one knows exactly how these inventions came about. Both have had major impacts on agriculture. For example, without wheels we would not have wagons, pickup trucks, nor tractors.

Some kinds of food preservation have been around a long time. Cheese was first made about 2000 years B.C. by nomadic Asian tribes. Yeast bread was first made by the Egyptians 2600 B.C. Many of today's food methods came as a result of making cheese and yeast bread.

AGRICULTURAL INVENTIONS

Most of the inventions in agriculture have been in the last 300 years. These inventions often began as simple, useful devices. They were improved and refined. Inventions have helped all areas of the agricultural industry: farming, supplies, and processing.

Farming

Farming is different! Inventions have made big changes. Three areas of significant change are machinery, plant and animal improvement, and pest

3-5. Automation and robotics are used in manufacturing equipment. (Courtesy, Automation International, Inc.)

control. One person can do a lot more work now—and have time left over to enjoy life!

Plows, planters, and harvesters have changed farming. These reduced the need for human power. At first, animal power replaced humans. In the last century, mechanical power has replaced animal power. Few farms operate today without tractors and machinery.

Many inventions are tied to the industrial revolution. Assembly line methods are used. Materials for making equipment, chemicals, and other inputs are being developed. New materials are replacing existing materials. A good example is plastic replacing metal in many uses.

Mechanical Technology. ***Mechanical technology*** is using physical force to do work. Devices are used to pull, push, or rotate objects. These form the machinery used in farming. Most farm machinery deals with plowing, planting, and harvesting crops. A power source is needed to operate the machinery.

A ***plow*** is a tool that loosens the soil. Finding a way to speed soil preparation was a major step in crop farming. Different soils require different shapes of plows. Thomas Jefferson used mathematics to study the design of plows. His findings were released in 1793. The heavy soils of the Midwest required a plow that would turn the soil and not quickly wear out. John Deere of Illinois developed the first steel plow in 1837. Improvements are continually being made in plow design.

A ***planter*** is a device that places seed in the soil. (Some special kinds of planters set out living plants, such as tomatoes.) Many planters operate at high speeds, covering the seeds the right depth with soil. Most lightly compact the soil over the seed. Some planters apply fertilizer and pesticide. Planter design varies with seed size and shape. Planters also vary by the arrangement of the seed in the soil. Some place seeds in rows. Others broadcast seed over the surface of the land. Prior to mechanical planters, seeds were planted by hand dropping and covering. Just think of all the bending and stooping people had to do!

3-6. High-speed planters cover much ground in a day. (Courtesy, The National Cotton Council of America)

A ***harvester*** gathers or picks crop products. A major advance in harvesting grain was made by Cyrus McCormick of Virginia. He patented the first

3-7. A combine can harvest several hundred bushels of soybeans in a day. (Courtesy, James Lytle, Mississippi State University)

Career Profile

AGRONOMIST

Agronomists study plants and soil. They learn how plants grow. Relationships between plants, soil, water, and growth are examined. New ways of helping plants grow are developed. Agronomists may work in fields, greenhouses, offices, or laboratories.

An agronomist needs a college degree in agronomy. Many get advanced degrees at the masters and doctoral levels. Their interest often develops while taking agriculture classes in high school. Science is important in the education of an agronomist. More agronomists are using site-specific farming. Some agronomists provide education for farmers and others in plant and soil growth. Practical experience working with crops and soils is important.

Agribusinesses, large farms, golf courses, and government agencies may hire agronomists. Salary and benefits are good.

The photo shows an agronomist examining a pod on a soybean plant. (Courtesy, Agricultural Research Service, USDA)

reaper in 1834. Many improvements have been made. Today, the reaper is a combine that can harvest many acres of grain a day. Other harvesters are cotton pickers, corn pickers, and hay mowers. All mechanical harvesters reduced the need for human labor in harvesting.

Early machines were powered first by humans and later by animals. Horses, oxen, and mules were often used as draft animals. Mechanical power used water wheels and other devices that took advantage of gravity. Later, steam power was used.

The internal combustion engine revolutionized power sources. Gasoline and diesel engines now power many machines. In 1904, Benjamin Holt invented the tractor for pulling implements. Holt's tractor also had a system of belts that could be used to operate other machines.

Large machinery companies often grew out of the initial inventions. One example is Deere and Company, which carries the name of John Deere. The company now produces a wide range of tractors and implements.

Plant and Animal Technology. Technology has been widely used to improve plant and animal production.

Plant improvement has focused on having plants that meet certain needs. A major goal has been to increase plant yields. Another goal has been to have plants with desired characteristics. A good example is the seedless grape.

Gregor Mendel studied the heredity of peas in an Austrian garden in the mid 1800s. He concluded that traits are passed from one generation to another. He reasoned that offspring receive traits from both parents. His work in genetics has remained important.

About the time of Mendel, Charles Darwin was at work. A British naturalist, Darwin developed the theory of evolution. Sometimes controversial, the evolution theory deals with the gradual change in plants and animals. The heredity of organisms may change. This causes future generations to vary from parents.

Mass selection was the method of plant improvement used for many years. ***Mass selection*** is saving seed from the best plants. In the mid 1900s, hybridization was started. This involved cross breeding plants, such as corn, wheat, and soybeans. Cloning and genetic engineering are now being used. With the latter, genes are transferred from one organism to another. The resulting plant is ***transgenic***. Transgenic means that the organism has genes from two different species of organisms.

Livestock and poultry have been improved through technology. In recent years, new developments with fish have created a new area of agriculture—fish farming.

3-8. Nearly all corn grown today is hybrid.

3-9. Turkeys have been improved to have more white meat. (Courtesy, Agricultural Research Service, USDA)

The heredity of animals has often been important. Animals were judged based on type or yield. Parents were selected to give offspring the desired traits. Embryo splitting, cloning, and other methods have been used. Hormones have been given to some animals to increase productivity, such as bST to improve milk production in dairy cattle. The use of animal technology will increase in the future.

Pest Control Technology. A ***pest*** is anything that causes injury to animals, plants, and property. Animals may be attacked by insects, worms, diseases, and other pests. Growing plants may be attacked by insects and diseases. Weeds cause big losses in some crops.

Some pests damage stored crops. For example, weevils and rats can ruin harvested corn. Gas-type pesticides may be used to control pests in harvested crops stored in enclosed places. Proper storage is important.

A few years ago, pesticides were developed for use on many crops and livestock. A ***pesticide*** is a substance that controls pests. Often, pesticides are developed for specific pests. Many pesticides are poisons. They can damage the environment and cause human health problems. Different pesticides are used for weeds, insects, and other pests.

A ***herbicide*** is used to control weeds. Many different herbicides are manufactured. Herbicides are chosen for specific weeds and crops. One kind of herbicide will not work in all crops. Some herbicides will kill crop plants along with the weeds. Herbicides must be carefully used to prevent damage to the environment.

3-10. Sprayers are often used to apply pesticides.

An ***insecticide*** is used to control insect pests. Proper identification of the insect is necessary. Chose an insecticide proven effective with the insect. Use insecticides according to directions. Always follow safety practices.

Other kinds of pesticides are used to control spiders, rats, mildew, and similar pests. Get information about which to use from an informed person.

3-11. An insect pest with harvested corn is the weevil (shown greatly enlarged). (Courtesy, Agricultural Research Service, USDA)

Some plants are being altered to resist pests. An example is Bt corn. A gene from *Bacillus thuringiensis* bacteria has been transferred into the corn. This causes the corn to resist damage by the corn borer. It makes its own pesticide!

Animals often receive medicine to help control disease. It may be in feed, drinking water, or given as a shot or in other ways. Animal medicines have reduced animal losses.

The work of Louis Pasteur is important. He pioneered disease prevention and treatment. He proved the value of vaccination in preventing anthrax in sheep in the 1870s. In 1881, he began studying rabies. By 1885, he had developed a vaccine that would prevent rabies.

Parasites are problems with animals. A ***parasite*** is a small animal that lives in or on another animal. Examples are round worms that live in the intestines and lice that live on the skin. Medications can be used to control parasites. Better pastures and feeds help animals resist parasites. More attention is now given to the well-being of animals. All animals need an appropriate environment.

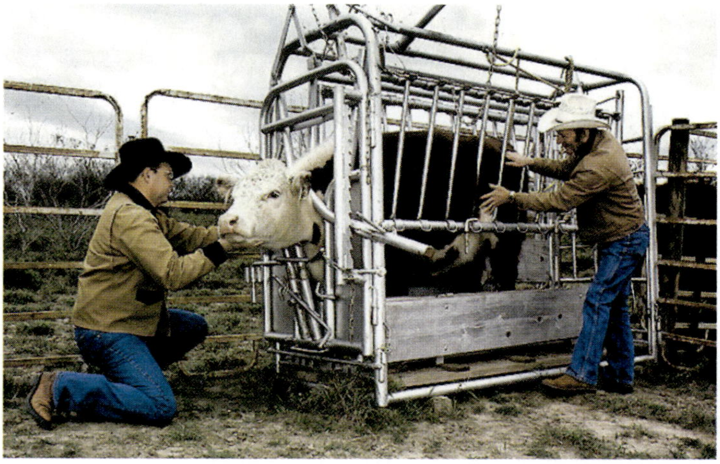

3-12. A beef animal is being inspected for ticks—a common external parasite. (Courtesy, Agricultural Research Service, USDA)

3-13. The work of Louis Pasteur continues to make our foods safe to eat.

Food Preservation

Food preservation is treating food to keep it from spoiling. Many changes have been made in this area. Drying, salt curing, and

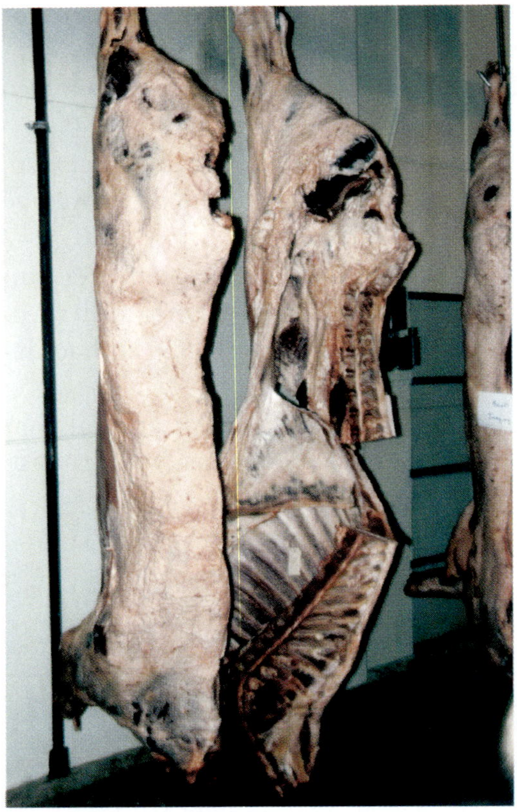

3-14. Beef carcasses are carefully prepared into meat products.

fermentation were first used. These are still important today. Canning and freezing are fairly recent in origin.

Canning was developed by Nicolas Appert in the early 1800s in France. In the mid 1800s, Louis Pasteur, a French chemist, found that heat could be used to prevent food spoilage. Today, we know the process as pasteurization. It is used with milk, juices, and other food products.

Ice making changed methods of keeping food. Low temperatures were known to keep food from spoiling. John Gorrie of the United States developed the first machine for making ice in 1851. Foods placed on ice could be kept several days. Clarence Birdseye of Massachusetts developed the process of quick-freezing food in 1925. This allowed fresh foods to be kept in frozen form. Refrigeration and freezing are widely used today.

COMPUTERS

Computers are used in many ways in agriculture. From 1950 to 1980, most computers were very large. They were used only by large businesses or agencies. Since the early 1980s, smaller computers have become widely used. These are having major impacts on the agricultural industry.

DECISION MAKING

Computers are now used as tools in decision making. Trying to decide what to do with a crop is not always easy. Programs are available that simulate crop production. Information about the crop, weather, pest problems, and other areas can be entered. The computer will provide suggested actions to take. It can predict what will happen if a certain decision is made. The computer does not make the decision—the producer does.

3-15. A computer and global positioning system on the tractor guide the application of fertilizer. (Courtesy, Top Soil Testing Service Company)

CONTROLLING EQUIPMENT

Computers are used in many ways with equipment. Both engines and implements powered by engines use computers. For example, rates of pesticide and fertilizer application may be computer controlled.

Using computer-controlled equipment makes more efficient work. It also helps us use resources wisely. For example, automatic feeders for poultry may be controlled by computers. Feeding no more than needed can save money.

KEEPING RECORDS

Computers are often used to keep records. The records are needed to prepare reports about a farm or agribusiness.

The records include expenses and income. Typical expenses include labor, supplies, equipment, and taxes. Income includes all proceeds from the sale of products. Money received from government programs is included as income.

Reports are prepared periodically. Many computers have record programs that make reporting easy. Some reports are sent directly over the Internet by computers.

SITE-SPECIFIC FARMING

Site-specific farming is using practices based on the specific needs of a location. Information helps us decide what to do. Computers direct equipment to assure that the work is done as needed. This is also known as precision farming.

Large areas are divided into smaller fields. Fertilizer and other materials are applied based on the needs of the smaller field areas. This prevents using inputs on larger areas than may be needed.

Site-specific farming uses grid maps of the land. Detailed information is collected on the grids. The information is entered into computers. Satellites help find the grids. Information sent to the equipment varies the rate of application. This is based on the need within each grid. The equipment uses controllers that direct how work is done. The controllers respond to computers.

Site-specific farming is continuing to be improved. Using satellites and global positioning is new in farming. Education is needed. Farmers are being trained in how to use it.

3-16. Site-specific farming uses several technologies, including global positioning. (The position of a location is determined using three satellites, with a fourth satellite used to correct for error.) (Courtesy, Spectra-Physics, Inc.)

REVIEWING

MAIN IDEAS

Agricultural technology is using inventions to provide food, clothing, and shelter. Most technology is based on science. Research has made many improvements in agriculture. With technology, we must select that which is best to use. Technology has made work easier and reduced labor and increased production. This has resulted in a higher standard of living.

An invention is a device or product that helps get work done. They are often based on technology. Some inventions have had major impacts on agriculture. Machinery, plant and animal improvement, and pest control are three major areas. Ways of preserving food helped make seasonal food items available year round.

Computers have made big changes in the last 25 years. They are used as aids in making decisions. Some computers control how equipment operates. Other computers are used for keeping records and preparing reports.

Since 1990, site-specific agriculture has emerged. This is a system that allows the use of inputs based on the unique needs of small parts of larger fields. A complex of technology is used. This technology includes computers, satellites, controllers, and maps.

QUESTIONS

Answer the following questions. Use correct spelling and complete sentences.

1. What is agricultural technology?
2. What are three benefits of technology?
3. What prehistoric inventions are important today in agriculture?
4. What is mechanical technology? What areas are included?
5. What sources have been used as power to operate machines?
6. What methods have been used to improve plants?
7. What is Bt corn?
8. How are computers used in agriculture?
9. What is site-specific farming?

EVALUATING

Match the term with the correct definition. Write the letter of the term in the blank provided.

a. technology	d. plow	g. pest
b. invention	e. harvester	h. insecticide
c. food preservation	f. transgenic	

____ 1. Anything that causes injury to animals, plants, and property.

____ 2. Used to control insect pests.

____ 3. A new device, product, or process.

____ 4. The use of inventions in working and living.

____ 5. Using methods that keep food from spoiling.

____ 6. A tool that loosens and pulverizes the soil.

____ 7. A plant or animal resulting from the transfer of a gene from another organism.

____ 8. A machine that gathers, picks, or reaps products.

EXPLORING

1. Prepare a report on a major agricultural invention. Examples of inventions include: plow, planter, combine, hybrid corn, Bt cotton, and bST hormone for dairy cattle. Give an oral report in class.

2. Interview a person who works on a farm or in an agribusiness about the role of machinery or other technology. Select one kind of technology and ask how it is used. Determine both advantages and disadvantages. Prepare a written report. Give an oral report in class.

4

AGRICULTURE MAKES LIFE GOOD

Agriculture helps us have good lives. It makes our lives better in so many ways. It provides the necessities for living—food, clothing, and housing. Every one needs these things. We could not live well without them!

We all have our favorite foods and clothing. You are probably thinking of pizza and jeans. Give thought to what it takes to have these products. How do you get the pizza? Jeans? A lot is required to get them to the restaurant or store where you buy them.

Comfortable housing helps us live well. Enjoying the outdoors is fun. Most all people have special outdoor things they enjoy—hiking, fishing, watching birds, and camping are favorites. Other people enjoy things at their homes such as pets, gardens, and butterflies. Our lives are better because we have these to enjoy!

4-1. People enjoy baby animals. This person is playing with tamworth piglets.

OBJECTIVES

This chapter explains how agriculture helps us have a good quality of life. The objectives of this chapter are:

1. Explain quality of life
2. Describe how to use the Food Guide Pyramid and list sources of nutrients
3. List agricultural products used to provide clothing and other materials
4. Identify agricultural products and natural resources used to provide human shelter
5. Discuss how the efficiency of modern agriculture benefits other aspects of society

TERMS

conifer
deciduous
domesticated
export
fluid milk
Food Guide Pyramid
hardwood tree
import
international trade
linen
natural fiber
nutrition facts panel
poverty
quality of life
softwood tree
vegetarian

http://www.usda.gov/

(This site has information on many areas of agriculture, including policy.)

AGRICULTURE: THE GOOD LIFE

Agriculture provides us with the food we eat, the clothes we wear, and the homes we live in. We are fortunate to have productive agriculture in the United States. It provides for our needs and for millions of people in other countries.

Quality of life is having a good environment for living. You may have nutritious food. Clean air and water are available. Having adequate housing and clothing is a part of quality of life. You may stay healthy and enjoy living. You may live in a safe, nurturing environment.

Agriculture contributes to quality of life in several ways. A few important ways are included here.

A STRONG NATION

The United States is a strong nation because of agriculture. We have an abundance of high quality foods. The foods are available at reasonable prices.

Americans spend less on food than do the people of any other nation. Only 11 percent of an American's total income is spent on food. Because our food cost is low, we have more money left over to spend on clothes, housing, and having fun. This allows us to have a better quality of life.

Agricultural scientists study ways of growing high quality foods. New types of plants and animals grow faster with less chemicals and fertilizers. This helps provide the abundance for Americans to have a better quality of life.

INTERNATIONAL TRADE

Agricultural products are often traded by nations. A system of international trade has been set up. ***International trade*** is buying and selling goods by two or more nations. The United States sells and buys many agricultural products.

Goods sold to or in another country are ***exports***. Rice, soybeans, wheat, poultry, and cotton are a few agricultural exports. Agriculture provides jobs for more people than any other industry in the United States. About 16 out of every 100 people in the United States work in agriculture. Our agriculture industry strengthens our country's well-being by providing jobs and by producing products to export.

4-2. Agricultural scientists study new technology to improve crop and animal production. (Courtesy, George Bostick, North Carolina State University)

The United States buys some products. Products bought from another nation are known as ***imports***. Examples of foods commonly imported include shrimp, coffee, and frog legs. The imports provide products that are in short supply in the United States.

POVERTY

Unfortunately, today some people live in poverty and are hungry. ***Poverty*** exists when people do not have enough money to buy food, clothing, and housing. This is usually due to a personal or societal problem, not because there is a lack of food to buy. Hunger is brought about when one does not have food to eat.

Poverty is not widespread in the United States, Europe, and other nations that have industry. Poverty is more common in some of the countries of Africa and Asia. As these countries develop, poverty will decline.

NATURAL RESOURCES

The United States has many natural resources. These make it possible to have a productive agriculture. Fertile soil helps crops grow. Good water supplies are needed by plants and animals. Fortunately, we are smart. We have developed ways of conserving our natural resources.

Natural resources are conserved by using them wisely. People who are involved with conservation are often known as stewards of the land. These stewards of the land are also increasing our quality of life. We have better quality air, water, and soil to use and enjoy.

4-3. The conservation of air, water, and soil is important for our future. (Courtesy, Agricultural Research Service, USDA)

AGRICULTURE PROVIDES NUTRITION

We have an abundance of food in the United States. This is due to agricultural technology. People in other countries are not as fortunate.

During the 1960's, more efficient fertilizers and crops were developed. These increased the amount of food grown. Today, many of these methods have been improved even more. Production is better than ever!

More types of food are now in our stores. Look in the produce section of your local grocery store. Many fresh fruits and vegetables are available even during the coldest weather.

How many different kinds of fresh vegetables are there for you to purchase? Today there are more than ever. This is good because we have choices to get the proper nutrition we need to grow healthy.

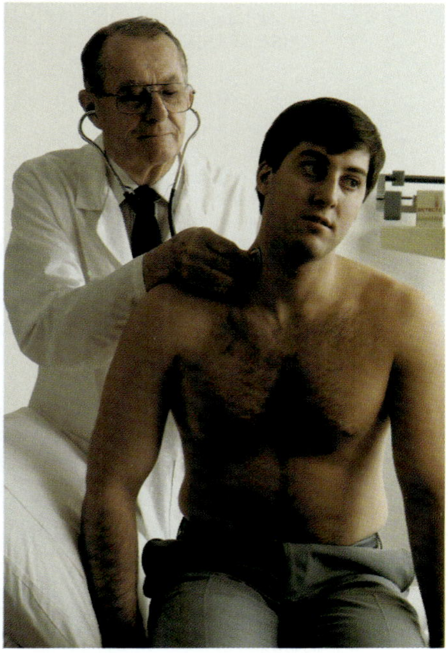

4-4. The food you eat is important to your health.

EATING RIGHT

Even with all the good food available, many people do not eat the most nutritional food. In many families, everyone works or goes to school. More people eat convenience foods. These foods are often higher in fat, salt, or sugar.

Much of our food is bought at a grocery store. This food is processed or manufactured in some way. The food was sold in raw form to processors who changed its form. People like easy-to-prepare and ready-to-eat forms, such as canned, frozen, and dried foods. Some foods are processed more than others. During this preparation, foods may lose part of their nutritional value. Fresh foods are the most nutritional if properly prepared.

The ***Food Guide Pyramid*** is a tool to help people eat the right things. It assures that people get the proper nutrients. The guide was created by the U.S. Department of Agriculture and the Department of Health and Human Services. Five main nutritional groups are used: bread, fruit, vegetable, milk, and meat.

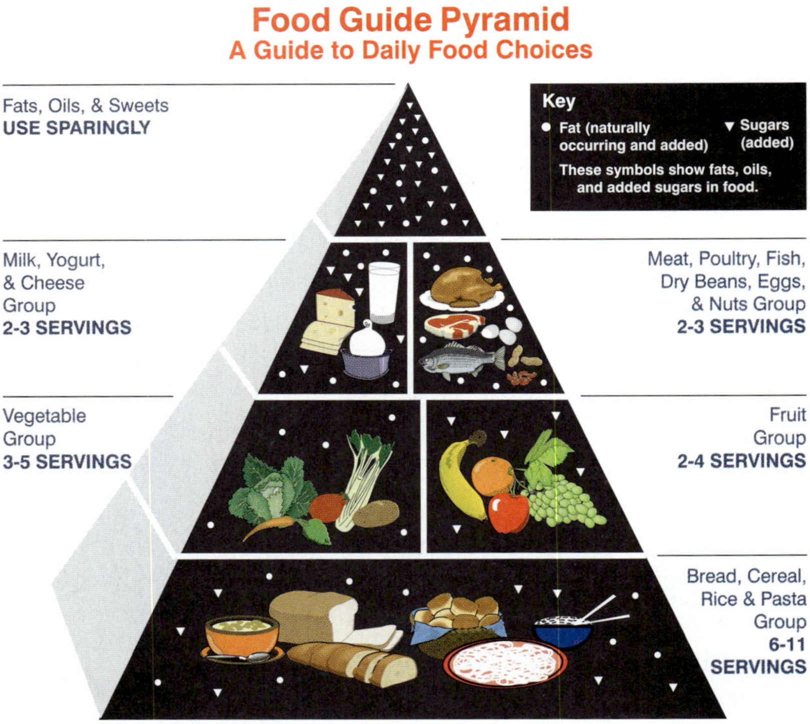

4-5. The Food Guide Pyramid is a guideline for healthy eating.

Table 4-1. A Sample Food Plan for a Day.

	1,600 Calories *For many sedentary women, and some older adults*	**2,200 Calories*** *Most children, teenage girls, active women, and many sedentary men*	**2,800 Calories** *Teenage boys, many active men, and some very active women*
Bread Group	6	9	11
Fruit Group	2	3	4
Vegetable Group	3	4	5
Milk Group	2–3**	2–3**	2–3**
Meat Group	5 ounces	6 ounces	7 ounces

*Pregnancy and breast feeding may require more calories.

**Women who are pregnant or breast feeding, teenagers, and young adults to age 24 need three servings.

(Source: U.S. Department of Agriculture)

Eating right depends upon your age, activity level, sex, and body size. Table 4-1 has information for several types of people. The most important guideline is to eat a variety of foods!

A diet should be balanced by eating foods from all five groups every day. Individuals should develop their own food pyramids by eating a good variety

4-6. A plate of nutritious and delicious food. (Courtesy, George Bostick, North Carolina State University)

of foods low in fat and cholesterol. Cholesterol is a waxy substance produced by the human body and found in animal products. Too much cholesterol will build up inside the arteries and cause health problems.

FOOD PRODUCTION

Where are all these nutritious foods grown? The majority are grown right here in the United States. We do import and export some of these products for economical reasons.

Bread

Foods in the bread group are made from grain. Wheat and rye are two important examples. These crops are grown throughout the United States. The greatest region of production is in the midwestern United States. Grain mills and bakeries are found in many locations.

4-7. Common ready-to-eat breads made with rye and wheat.

Fruit

Foods in the fruit group are grown throughout the United States. Some are better suited to certain climates than others. For example, oranges and grapefruit are grown in warm climates. California and Florida are well known for fruit production. We also import a large quantity of our fruit from

4-8. Apples are nutritious and popular fruits.

Mexico and South America. Some fruit crops also grow in the northern United States, such as blueberries in Michigan, apples in Washington, and cranberries in Massachusetts.

AgriScience Connection

UNITED STATES DEPARTMENT OF AGRICULTURE

The United States Department of Agriculture (USDA) is an agency of the federal government. Its headquarters are in Washington, D.C. The Department is administered by a Secretary of Agriculture. The Secretary is appointed by the President of the United States and is approved by the U.S. Senate.

The USDA has several programs to ensure a good food supply. Programs are offered to help produce healthy plants and animals. The USDA emphasizes proper care of the soil and water to prevent erosion and pollution.

Consumers benefit from research sponsored by the USDA. This research makes our food better. School lunches and food stamps are provided by the USDA to those in need to ensure proper nutrition for all. Meat and poultry products are inspected by the USDA for safety. Our national forest system is managed by the USDA.

USDA
United States Department of Agriculture

4-9. Broccoli is a vegetable high in vitamins. (Courtesy, Agricultural Research Service, USDA)

4-10. Fluid milk is a popular beverage.

Vegetables

Foods in the vegetable group are grown all over the United States. Cool weather crops, such as lettuce and broccoli, grow in northern states in the summer or in the southern states during the winter. Most warm weather vegetables are grown in California and Florida. Many of the vegetables we buy are also imported from South America and Mexico.

Dairy Foods (Milk)

Foods in the dairy group are made from milk. Most milk is from cattle on dairy farms. We also get milk from goats, and it is identified as goat milk. Dairy production is throughout the United States, with California and Wisconsin being the leading states. Wisconsin is the top producer of milk for making cheese. Milk in liquid form as sold in containers in supermarkets is known as ***fluid milk***.

Meats, Nuts, and Beans

Meat includes poultry, beef, pork, fish, and lamb. Nuts and beans sometimes substitute for meat in the Food Guide Pyramid. Poultry are any birds that have been domesticated and grown for food. An animal is ***domesticated*** when it is tamed, confined, and bred for human use. The most popular poultry is chicken followed by turkey. Chickens also produce the majority of eggs consumed in this group. Beef is the meat from cattle and is prepared into popular dishes, such as grilled steak and hamburger. Pork is the meat of swine. Swine is a plural term used to define hogs and pigs. Pork chops and bacon are two popular forms of pork.

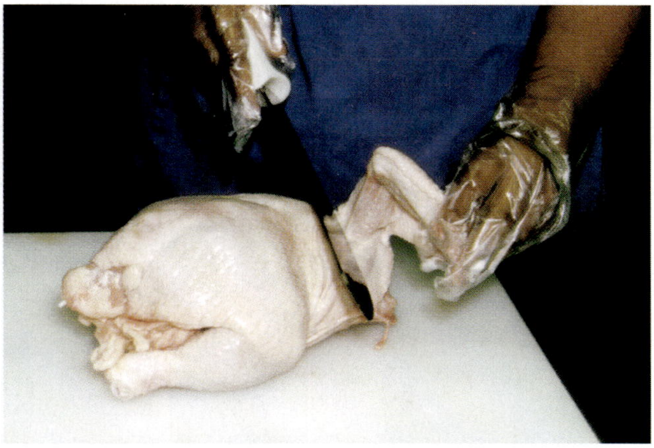

4-11. Chicken is a healthful meat when properly prepared.

Fish are in the meat group. More fish are being farmed. Oceans, streams, and lakes cannot provide enough fish to meet demand. Aquaculture is the production of aquatic plants and animals for food. Fish are grown in ponds or raceways and are fed daily. In the southern United States, catfish farming is an exciting new type of farming. The fish are fed

Career Profile

SOIL CONSERVATIONIST

A soil conservationist studies soil, water, and other natural resources. Work is often with land owners in using practices that conserve soil and water. A good knowledge of plants, especially row crops and pasture grasses, is needed. This photo shows a soil conservationist studying an erosion problem.

Soil conservationists need a college degree. In college, they study agronomy or related areas. Some have masters and doctors degrees. Soil conservationists should enjoy people and the outdoors. They usually work for state and federal government agencies that assist farmers and land developers.

To become a soil conservationist, you should take agriculture and science classes in high school. Participate in the FFA Soil or Land Judging Contests. You can also contact your local soil conservationist for more information. (Courtesy, Agricultural Research Service, USDA)

4-12. Agriscience provides many jobs in labs and fields. This shows a scientist studying pear blossoms. (Courtesy, Agricultural Research Service, USDA)

good feed, raised in clean water, and carefully harvested using nets. At the processing plant, they are prepared for the fresh or frozen market.

Nuts and beans are also agricultural products. There are a large variety of nuts and beans grown all over the United States. Peanuts and pecans are grown in the southern United States. Almonds and pistachios are grown in California. Beans, such as soybeans, are used as protein supplements in many prepared meat dishes. A ***vegetarian*** is a person who eats nuts, beans, and other foods for protein. Vegetarians do not eat meat.

FOOD LABELS

How do you know the nutritional value of prepared foods? A ***nutrition facts panel*** must be present on all food products except raw, single-ingredient products. This label was created by our government to help consumers follow a well-balanced diet. By reading this label, we know the nutritional value of a food. People can use this information when developing their own food pyramids.

Nutrition Facts

Serving Size 0 cup (000g)
Servings Per Container 0

Amount Per Serving

Calories 000	Calories from Fat 000
	% Daily Value*
Total Fat 00g	00%
Saturated Fat 0g	00%
Cholesterol 00mg	00%
Sodium 000mg	00%
Total Carbohydrate 00g	00%
Dietary Fiber 0g	0%
Sugars 00g	
Protein 00g	

Vitamin A 0% • Vitamin C 0%
Calcium 00% • Iron 0%

Percent Daily Values are based on a 2,000 calorie diet. Your daily values may be higher or lower depending on your calorie needs:

	Calories	2,000	2,500
Total Fat	Less than	65g	80g
* Sat Fat	Less than	20g	25g
Cholesterol	Less than	300mg	300mg
Sodium	Less than	2,400mg	2,400mg
Total Carbohydrate		300g	375g
Dietary Fiber		25g	30g

Calories per gram:
Fat 9 • Carbohydrate 4 • Protein 4

4-13. A Nutrition Facts Panel is required on all prepared foods. (Courtesy, U.S. Department of Agriculture)

FOOD SAFETY

Food safety is an important matter! Keeping food safe is a concern of everyone. Many steps are taken to be certain that food is safe to eat. It includes how foods are grown, harvested, and prepared for eating. Bacteria and other organisms can cause food to spoil. Chemical residues can make food unfit to eat.

The Food Safety and Inspection Service in the USDA has issued rules about meat food products. All raw or partially cooked meat and poultry must have a safe handling instruction label. This is to help prevent bacteria contamination. Some types of bacteria that might be on the meat can be harmful if eaten. When the meat is cooked properly, the bacteria are killed and the meat is safe to eat.

4-14. The Safe Food Handling Label gives instructions on how to store and prepare meat safely. (Courtesy, U.S. Department of Agriculture)

AGRICULTURE PROVIDES CLOTHING

Clothing is made from natural and synthetic fibers. A ***natural fiber*** is from a plant or animal. Synthetic fibers are manufactured from petroleum and other substances.

Agriculture provides not only food but also clothing for people. The most important plant fiber comes from the cotton plant. Yarn is spun from the fiber in the cotton boll. Cotton is widely grown in the southern United States.

4-15. Sheep are sheared for wool. (Courtesy, George Bostick, North Carolina State University)

Cotton is a versatile cloth. It is comfortable to wear and absorbs dye easily. Often, synthetic fibers are blended with plant fibers to make cloth more durable and to increase resistance to wrinkling.

The flax plant also provides clothing fibers. Bark fibers from the flax stem are used to make ***linen***. Linen is a popular fabric. Other plant fibers are kenaf, jute, hemp, and sisal. Kenaf is a new crop used to make cloth and paper. Jute is used to make burlap—a course cloth used to make bags. Hemp and sisal are plants that produce coarse fibers for rope. Sisal is a plant fiber taken from the stem of the agave plant.

The most common animal fibers used to make clothing are wool and fur. Sheep and goat fleece are two sources of wool. The sheep or goat's fleece is sheared, cleaned, dyed, and woven into thread. An Angora goat grows fleece that is woven into mohair. The mohair is then made into soft blankets or articles of clothing.

Fur is another source of animal fiber used to make hats, coats, and other clothing. Humans need clothing for warmth. Rabbits and mink are grown for their fur. They are grown in clean pens and fed balanced diets. Special hunting permits are required for animals to be captured from the wild for their fur.

FORESTRY PROVIDES SHELTER

Good quality of life requires proper housing or shelter. It protects us against nature's elements, such as cold and rain. It also provides a place for daily living. Humans are more comfortable in proper shelter.

Forestry is the science of planting, caring for, and harvesting trees in the forest. Agriculture produces products in the forestry industry that are used to build shelters, such as homes and office buildings. The furniture in these shelters may be made of forestry products also.

4-16. Plant scientists do research on how to grow bigger and better trees. (Courtesy, George Bostick, North Carolina State University)

The United States has 736.7 million acres of forest land. An acre is about the size of a standard football field. Can you imagine how many trees can be grown on 736.7 million football fields? Trees are growing faster, bigger, and stronger due to research. Foresters have learned how to use best management practices in caring for trees.

4-17. Trees are harvested for wood and fuel. This shows a large harvester that fells (cuts off) and moves whole trees. (Courtesy, George Bostick, North Carolina State University)

Forest products are grouped by the types of trees harvested. The two major types are hardwood and softwood. These groups are based on the kinds of trees grown and harvested. ***Hardwood trees*** are deciduous trees. A ***deciduous*** tree sheds its leaves in the winter. The wood has fine grains and is hard. Two common hardwood trees are oak and maple. Common hardwood products are veneer, furniture, and hardwood flooring for homes.

Softwood trees are conifers. ***Conifers*** are evergreen trees that usually have cones. Evergreen trees have needle-like leaves that do not shed all at once. The wood has a coarse grain. This is the largest group of trees used in shelter building construction. Common softwood trees are pine and fir. The wood from softwood trees can be made into lumber, plywood, composition board, and paper.

Forests provide shelter for animal wildlife. People often enjoy recreation such as camping in forests. Wood used as fuel produces 5 percent of the world's energy. Just over half of all the wood used in the world is for fuel. Developing countries often use wood for heat.

AGRICULTURE MAKES A GOOD ENVIRONMENT

Agriculture provides us many beautiful landscapes in the country and city. This beauty is often captured by artists and photographers.

Horticulture is the field of study in agriculture that deals with the skill and art of growing and using plants for their beauty. It also includes fruits and vegetables. Landscape horticulturists design, install, and maintain landscapes outside our homes and buildings. Horticulturists also grow plants to beautify the inside of our homes and buildings such as shopping malls. Flowers are grown by floriculturists and arranged by florist. These flowers are often given to others as gifts.

Agriculture makes our lives good by providing beauty in rural and urban places. An afternoon ride in the country or a stroll through a city park are often pleasant experiences. Plants make the scenery pleasing and relaxing!

4-18. A student and teacher review a landscape plan for the National FFA Center.

REVIEWING

MAIN IDEAS

Agriculture is important in the quality of life people enjoy. It provides food, clothing, and housing. Fortunately, the United States has an agriculture industry that produces an abundance of nutritious food, an ample supply of fibers for clothing, and wood for the construction of homes and offices.

Healthful foods in the Food Guide Pyramid include bread, fruit, vegetable, milk, and meat groups. Foods in all five food groups grow in the United States. The nutrition facts panel required on all prepared food products informs consumers of the nutritional value of the food product.

Cotton and flax are two popular plant fiber fabrics. Wool and fur are animal fibers used to make clothing also.

Forestry provides wood for shelter and heat. Softwood and hardwood are the two types of wood. Wood provides 5 percent of the world's energy used for heat. Forests are managed for recreational uses and wildlife habitat.

Horticulture makes the outside and inside of our homes and other buildings appealing. Flowers are important in many ways in the lives of people.

QUESTIONS

Answer the following questions. Use correct spelling and complete sentences.

1. What is quality of life?
2. What is the Food Guide Pyramid?
3. List the five main nutritional groups.
4. What is the difference between imports and exports?
5. What is the purpose of the nutrition facts panel?
6. What are two plant fibers used for clothing?
7. What are two animal fibers used for clothing?
8. How are trees used to increase the quality of life?
9. List the two major types of trees grown for wood products.
10. How does agriculture help make a good environment?

EVALUATING

Match the term with the correct definition. Write the letter of the term in the blank provided.

a. domesticated	e. deciduous	h. fluid milk
b. international trade	f. export	i. softwood
c. linen	g. poverty	j. nutrition facts panel
d. hardwood		

____ 1. A coniferous tree.
____ 2. To tame and breed for human use.
____ 3. Cloth woven from the stem of a flax plant.
____ 4. Buying and selling among two or more nations.
____ 5. A tree that sheds its leaves.
____ 6. A label on food that gives nutrition information.
____ 7. Milk as commonly sold.
____ 8. A deciduous tree.
____ 9. To send or carry goods to another country for sale.
____10. Exists when food and shelter are lacking.

EXPLORING

1. Look in the business section of your local newspaper. Find the Chicago Board of Trade report. Identify and list five animal and five plant commodities that are included.

2. Analyze the nutritional value of a typical school lunch by using the Food Guide Pyramid. Prepare a short written report on your findings.

3. List the first twenty agricultural products you touched after you woke this morning. Study the list and place a check by those grown near you. Also indicate if the product is from an animal or plant source by placing an "A" by those from animals and a "P" by those from plants.

4. Trace the agricultural production, processing, and marketing of your favorite food. Most food processing companies will provide this information if you write the company and request the information.

5. List the fibers used to make your five favorite pieces of clothing. These can be found on the tags inside the garments. Identify the natural fibers that came from plants and animals.

5

HOW WE GET WHAT WE WANT

Think of a favorite food. A hamburger is a good example. It has many ingredients—flour, meat, and tomato, to name a few. The ingredients came from several places. Many steps were needed to get the hamburger ready to eat. In a fast food restaurant, having a hamburger is easy. Place an order and pay for it! Much has to happen for this to be possible.

5-1. Pasta has increased in popularity in the United States. (Courtesy, ARA Services)

We use many things in daily living. We have these things because of an orderly way of doing business. How we do business is a part of our Nation's history. Many activities are involved.

In agriculture, crops and livestock are just the beginning. Transporting, processing, and packaging are needed. A method of paying is also required. These are part of "how we get what we want."

OBJECTIVES

This chapter is about the economic system used in the United States. It covers the ways of doing business and how consumers are important. The objectives of this chapter are:

1. Define economics and explain economic system
2. Describe free enterprise
3. List three ways of doing business
4. Describe the role of consumers and competition
5. Explain entrepreneurship

TERMS

business
competition
consumer
cooperative
corporation
demand
dividend
economic system
economics
entrepreneurship
free enterprise
money
monopoly
partnership
price
private ownership
property
risk
sole proprietorship
supply

http://aquanic.org

(This site has information on many subjects in aquaculture.)

ECONOMICS

Economics is a study of how people get what they want. It includes producing goods and services and getting them to people. Since most people want more than they can afford, they make other choices. The money to buy goods and services is scarce. The resources to provide goods and services are also scarce. There is never enough of everything!

Governments have systems to care for the needs of people. These systems demand some things of people. A common requirement is for people to work. Work is needed to produce goods and services. It allows people to have money to buy what they want. Their level of living is related to work and income.

Economics includes supply and demand, prices, and money.

5-2. Government laws make it possible to "do business."

SUPPLY AND DEMAND

The supply of and demand for a good (product) or service are both limited. Price varies with the supply of something as related to demand for it. This is the principle of supply and demand.

Supply is the amount of something that is available. As supply goes up, price goes down. Crop prices are lowest at harvest because supply is greatest then. Supplies can be reduced by floods, freezes, and other disasters. For example, a freeze in South Florida in the winter may destroy bean and tomato crops. This reduces the supply available. Prices for these foods will increase.

5-3. A "supply" of meat is being placed in a display case to meet the "demand" of consumers. (Courtesy, Mississippi State University)

Demand is the amount of something that will be bought at a given price. If prices go up, demand goes down. People will buy less of a product because their money is limited. They may buy other products—a substitute. If price goes down, more of the product will be sold. The price must cover the costs of production. If not, the producer will lose money and soon go out of business.

Supply and demand work together in setting price. Graphs are sometimes used to explain the principles of supply and demand. Two lines are plotted on a graph. One line represents supply. The other line represents demand. Wherever the two lines cross is the demand at a particular level of supply. This sets the price for a given amount of a product.

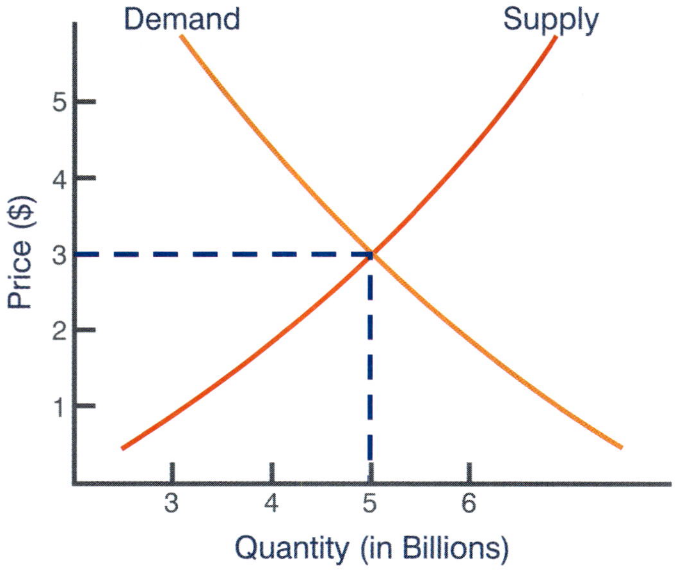

5-4. The supply and demand curve shows the amount sold at a given price.

PRICE

Price is the amount of money in buying or selling a good or service. Money is used as a medium of exchange. The price is the worth of the item in money. Supply and demand cause prices to vary.

Prices are sometimes stable. When this occurs, the price is said to be in equilibrium. At this price, buyers can buy all they want and sellers can sell all they want. Prices can easily get out of equilibrium. Minor weather changes can reduce supplies. The result is an increase in price.

MONEY

Money is anything exchanged for goods and services. The money unit may vary. Money used in a country is known as its currency. In the United States, the dollar is used. Other nations have different currency units. France uses the franc and England uses the pound. Canada uses the Canadian dollar, which has a different value from the U.S. dollar. Mexico uses the peso.

U.S. money consists of coins and paper bills. Buying and selling may involve using substitutes for coins and bills. Checks on bank accounts may be written or credit or debit cards used.

Career Profile

AGRICULTURAL ECONOMIST

Agricultural economists study how agriculture operates in the economic system. They collect information on trends that relate to agriculture. They analyze current situations and predict future supply and demand. Their work helps producers make decisions.

Agricultural economists have college degrees in agricultural economics or a related area. Many have masters and doctors degrees. Practical experience in agriculture is important. Many begin by taking agriculture classes in high school. High school courses in economics and government are helpful.

Most agricultural economists work for government agencies, agribusinesses, and universities. The photo shows an agricultural economist discussing cotton samples with a grower. (Courtesy, Agricultural Research Service, USDA)

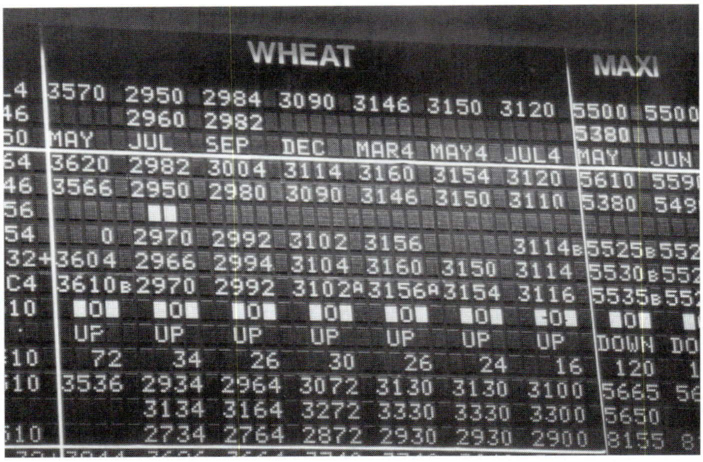

5-5. Electronic priceboards are used to keep traders informed at the Chicago Board of Trade. (This shows the futures prices of wheat for various months.) (Courtesy, Chicago Board of Trade)

The value of money is how much it will buy. Changes in prices cause the value of money to change. The value of money is known as purchasing power. We commonly hear that a dollar will not buy what it did just a few years ago. This means that a dollar has less purchasing power.

ECONOMIC SYSTEM

An ***economic system*** is how people go about doing business. It includes how things are created, owned, and exchanged. Economic systems include

5-6. The New York Stock Exchange is the largest stock exchange in the United States.

the agricultural industry. How people go about farming is related to the economic system. The returns (profit) they get for their work is important. Most people work harder to get a reward!

In any nation, creating goods and services is a major goal. These meet the basic needs of people. How people have what they want and need is based on the economic system.

Countries have different economic systems. Most are modified in some way to meet a particular goal. Examples of economic systems are capitalism, communism, and socialism. Many variations are found in each. None is found in its pure form.

Major differences in economic systems are based on property ownership and control. Government has an important role in the economic system of a nation. The United States has a form of capitalism known as free enterprise.

AgriScience Connection

COOPERATIVES

Cooperatives were first created to serve the needs of agriculture. In the early 1800s, farmers wanted more control over the prices they received for their products. They felt that joining together would help. One of the first was a dairy cooperative in Connecticut set up in 1810.

The U.S. government increased support for cooperatives in the early 1900s. Laws were passed that set up agencies to help farmers organize cooperatives. Today, cooperatives fill important roles in agriculture. Many farmers get supplies from cooperatives. Others sell their crops and livestock through marketing cooperatives.

The photo shows a rice and soybean facility of Riceland Foods, Inc.—a 10,000-member cooperative. (Courtesy, Agricultural Research Service, USDA)

FREE ENTERPRISE

Free enterprise is the way of doing business used in the United States. It allows people to go about business with a minimum of government control. Free enterprise allows people to work and make money. They may own property and get profit from what is produced. Free enterprise is also known as private enterprise.

PROPERTY OWNERSHIP

Owning property is a goal of many people. ***Private ownership*** is having things (property) that people can call their own. Some people are motivated by owning property. ***Property*** is anything that has value that can be exchanged. The two kinds of property are real and personal.

Real property is land and its natural features. Land includes buildings, landscaping, ponds, fences, and other permanent structures. Farm land with sheds and other improvements is real property. Real property is usually in a fixed place and not easily moved. Land and improvements may be known as real estate.

Personal property includes the things that people personally use. Most of the things we use in our daily lives are personal property. Clothing, personal computers, pickup trucks, and boats are examples.

Crops and livestock are property. Harvested crops are often known as commodities. Tractors and equipment are property. Flowering plants, pets,

5-7. Land and the permanently attached structures are real property. (Courtesy, U.S. Department of Agriculture)

and power tools are property. Logs, lumber, and trees are property. All property belongs to someone. It may be an individual, business, or government agency.

In the United States, people have respect for the property of others. Laws deal with taking the property of another without their permission. Trespassing is disrespect for another's property. Going onto another person's land without permission is trespassing. People should never trespass.

BUSINESS OWNERSHIP

A ***business*** is a person or group that produces and/or sells goods and services. A farm is a business. Manufacturers of equipment, fertilizer, and other inputs are businesses. Preparing food, clothing, and shelter for consumers is a business. Some businesses are large; others are small. Mowing lawns for others for pay is a small business.

In free enterprise, people own businesses. One person or two or more people may form a business. People are generally free to run their businesses as they please without interference. Of course, laws have been passed to protect businesses and consumers.

People who own a business take risk. ***Risk*** is the possibility of losing what has been invested. Risk is no more than the amount an individual has invested. Risk has benefits. Owners take risk in setting up a business. The owners get any profits from the business.

Most farms and other businesses are privately owned in the United States.

5-8. Advertisements are used to help consumers make choices.

CHOICES

Free enterprise involves freedom of choice. Freedom of choice means that people are free to choose what they want to produce, buy, and sell. Choice includes the seller and buyer agreeing upon a price.

Good decisions are needed. Poor decisions can result in no profit and failure of a business. Deciding requires information. People who set up businesses need to know what other people want and how much they are willing to pay for it. If other people do not want something, producing it is not smart.

CONTROL

Free enterprise has few government regulations. People can run their businesses as they wish. Unfortunately, people sometimes take advantage of others. Being dishonest is not acceptable.

Trying to do business without a few rules is nearly impossible. People benefit from some rules. Standards are kinds of rules. Many things in the United States must meet certain standards. It is important for products to be up to standard. Many examples are found in agriculture. These often involve protecting people from dishonest practices.

Standards help keep food clean and measured accurately. Most everyone agrees that rules to assure good food are important. People also want a full measure when they make purchases. For example, scales need to weigh accurately and containers need to be the right size. We want a full quart of milk when we pay for a quart! Inspection is used to assure compliance.

WAYS OF DOING BUSINESS

Free enterprise makes it possible to do business. All businesses are one of three types: sole proprietorship, partnership, and corporation. These are the ways of doing business.

SOLE PROPRIETORSHIP

A ***sole proprietorship*** is a business owned by one person. The owner is a proprietor. Proprietors are responsible for all areas of operation. They get profits, if any, and have losses, if any. Farms, garden centers, fruit stands, and fish hatcheries are examples of sole proprietorships.

A proprietor must have the money to start a sole proprietorship. They have many duties with the business. They manage it and see that all work is done. With small sole proprietorships, the owner may be the only worker.

Larger sole proprietorships may have many workers. Owners hire, buy, sell, supervise, submit government forms, and do other work.

Most of the time the name of a business shows if it is a proprietorship. Jones' Farm or Gonzales' Garden Center are examples of names. A business may have a name other than that of its owner. An example is Mountain Tree Farm.

5-9. This bee keeper is a sole proprietor. (A farm for bees is known as an apiary.) (Courtesy, Mississippi State University)

PARTNERSHIP

A ***partnership*** is a business owned by two or more people. The people are co-owners. They are bound as partners by a legal contract. Profit, if any, is divided among the partners. Losses are also divided among the partners. Each partner is liable for losses caused by another partner.

Partnerships vary in what each partner contributes and receives.

5-10. Two people are partners in a fish seining business. (Courtesy, U.S. Department of Agriculture)

Some partners may put up only money to get the business started. Other times, partners work side by side in the business. Good understanding is needed among the partners. Disputes can result in the business failing.

Farms, agribusinesses, logging operations, and many other businesses can be partnerships. The name of a business may say if it is a partnership. For example, Lee and O'Neal Pulpwood Company indicates a partnership. Names do not always reflect partnerships, such as Quick Cut Lawn Service.

CORPORATION

A ***corporation*** is an association of members for doing business. It is a legal entity with some of the same freedoms as a person would have in business. Corporations are chartered by state governments. They may sell stock or shares to many people. Selling stock provides money to start and operate the corporation.

The people who buy stock are known as stockholders. They hold meetings to vote on major decisions. The number of votes a person has depends on the number of shares of stock that they own. Stockholders elect a board of directors to oversee the corporation. Day to day operation of a corporation is by a hired manager and other workers.

5-11. A sign shows that a horse riding business is incorporated.

Businesses that are corporations are easy to identify. Their names usually end with "incorporated" or "inc." Large farms and agribusinesses are corporations.

Corporations may pay stockholders dividends. A ***dividend*** is part of the profit that the corporation makes. Dividends are returns to the person who took a risk by buying stock. Corporations that do not make a profit will not pay dividends. The dividends a stockholder receives are based on the number of shares owned.

Cooperatives are special kinds of corporations. A ***cooperative*** is an association to provide services to members. Cooperatives are widely used in agriculture. There are two major types: purchasing and marketing.

Purchasing cooperatives buy goods in large quantities to sell to members. Because of the large volume, the price is below what each person would have to pay if they bought individually. Purchasing cooperatives often handle seed, fertilizer, and other supplies. Some people refer to them as the local co-op or co-op store.

Marketing cooperatives help market farm products. They assemble larger amounts of produce than one farm would have. The larger quantity has appeal to buyers for canneries, supermarkets, and millers. Marketing cooperatives do business as grain elevators, packing sheds, and grower associations.

Members of cooperatives are known as stockholders. Members usually buy one share of stock. The number of votes of a member is one. A board is elected by the stockholders. The board oversees operation of the cooperative. Hired people manage and run it on a daily basis. If any profit is made, it is paid to the members as patronage dividends. Most dividends are based on how much business a member has done with a cooperative.

CONSUMERS AND COMPETITION

Businesses must have customers. Customers are people who buy the goods and services a business produces. This applies in agriculture just as in other areas. For example, the chemicals made by a manufacturer must be those that are in demand. If not, the business will not be successful.

CONSUMERS

Consumers determine what is produced. A ***consumer*** is a person, business, or agency that uses goods and services. Consumers of food and clothing are those who buy and eat or wear what they have bought. Most of the time we think of consumers as people.

Consumers make choices. All of them have limited money to spend. They decide how it will be spent. Consumers are free to buy as they choose. They can buy from whomever they wish at an agreed on price.

Most things consumers buy have a price set by the seller. This is known as the asking price. With small everyday items, the price is not negotiated. Prices of larger items may be negotiated. Many consumers refuse to pay the first price on automobiles, houses, and tractors.

Consumers usually pay the asking price on smaller items. If they do not wish to pay this amount, they can buy a different item or do without. A good example is a fast food hamburger restaurant. Several different hamburgers may be for sale. Some cost more than others. If money is low, the hamburger costing less will be bought. The consumer could go to a different restaurant where the prices are lower.

5-12. Consumers make food choices in a cafeteria. (Courtesy, ARA Services)

The preferences of consumers determine what is produced. If they do not buy a product, it would not be produced. Growers try to produce the kinds of crops consumers want. Agribusinesses are also driven by consumers. For example, the kinds of tractors bought influence the kind produced.

People who market products study what they should handle. They ask consumers what they want. They note the other products being bought and who is buying them. Important traits of consumers include age, income, education, and interests. Producers can use this information to decide what to produce.

COMPETITION

Consumers make choices. More than one product will meet their needs. They choose the one they like best compared with cost. They consider the money they have in terms of the money needed to buy it. Producers of goods and services try to attract consumers to what they have produced.

5-13. Consumers make choices in the clothes they wear. (Courtesy, Athletic Division, Russell Corporation)

Competition allows consumers to choose between similar products. A rivalry may exist between the producers. Each producer tries to attract buyers. Incentives may be given to get people to buy. Common incentives are discounts, coupons, and bonus or free items with a purchase. Competition does not exist if there is one source of a good or service.

A ***monopoly*** exists if one producer controls products and/or prices. The one producer can limit production and set the prices high. Monopolies do not have competition. Most monopolies are not good in a free enterprise system. In agriculture, monopolies are seldom a problem. Many farms produce products. After leaving the farm, the source of a product cannot be identified. Sometimes, only a few processing companies may be involved. These may develop a monopoly.

Advertising influences what people buy. Radio, television, newspapers, and other media are used to reach people. The purpose is to get people to select one product over a competing product. Honest advertising is an important part of free enterprise. Untruthful advertising should never be used.

ENTREPRENEURSHIP

Entrepreneurship is creating goods or services to meet unique needs. New "things" are created. Occasionally, new approaches in using existing "things" may be developed. Risk is involved.

An entrepreneur is a person who takes risk. In doing so, the entrepreneur must devise the product and develop ways of getting it to consumers. Entrepreneurs begin with dreams. They have a mental image of a product. They organize resources to achieve their dreams.

Everything we buy has involved entrepreneurship. Someone came up with the idea. Ways to make the idea a reality were sought. Not all entrepreneurs are successful. Some fail. Good planning and good decisions are essential.

Entrepreneurship is more than owning a business. Creativity is involved. They often manage the business that produces what they have developed.

5-14. An Arizona high school student is a proud entrepreneur producing koi—an ornamental fish.

REVIEWING

MAIN IDEAS

Economics is the production of goods and services for people to use. Governments use systems to make it possible for people to have what they need. This is an economic system.

The economic system used in the United States is free enterprise or capitalism. It deals with how things are created, owned, and exchanged. People privately own property. They may set up businesses, including farms, to produce what they want. Of course, risk is involved. Risk is the possibility of losing what has been invested. Free enterprise has few government controls. Some government controls are needed to protect both consumers and producers.

Business is done in three ways in the United States: sole proprietorship, partnership, and corporation. A sole proprietorship is one person owning a business. A partnership is two or more people owning a business. A corporation involves an association of people who form a business. In agriculture, a special kind of corporation is a cooperative. Cooperatives are intended to provide services to members.

Consumers are people who use goods and services. They create demand. Producers create goods and services. They compete for the scarce money of consumers. To do so, they use incentives to get people to buy their goods or services.

QUESTIONS

Answer the following questions. Use correct spelling and complete sentences.

1. What is economics?
2. How does supply and demand relate to the prices consumers pay?
3. What is currency? Give examples for three countries.
4. Why is creating goods and services a goal in any nation?
5. What are the principles of free enterprise? Briefly explain each.
6. What is the difference in real and personal property?
7. What are the ways of doing business? Briefly describe each.
8. Who is a consumer?
9. What is the difference between competition and monopoly?
10. What is entrepreneurship?

EVALUATING

Match the term with the correct definition. Write the letter of the term in the blank provided.

a. economics	e. money	h. partnership
b. supply	f. free enterprise	i. dividend
c. demand	g. risk	j. cooperative
d. price		

____ 1. Profit paid to stockholders by a corporation.

____ 2. Study of how people get what they want.

____ 3. The amount of product that will be bought at a given price.

____ 4. Anything exchanged for goods and services.

____ 5. The amount of money in buying or selling a good or service.

____ 6. Amount of a good or service that is available.

____ 7. An association to provide services to members.

____ 8. A business owned by two or more individuals.

____ 9. Freedom to do business with a minimum of government control.

____10. The possibility of losing something.

EXPLORING

1. Use a telephone directory to identify ways of doing business. List examples of businesses that are sole proprietorships, partnerships, and corporations. Identify at least one cooperative, if possible.

2. Attend a meeting of the patrons (stockholders) of a local cooperative. Prepare a report on your observations. Give an oral report in class.

3. Invite the owner/manager of a local agribusiness to serve as a resource person in class. Ask the person to discuss how the business is organized. Also, have the person describe what it takes to be successful in an agricultural business.

6

SAFETY IN AGRICULTURE

"Drive a little faster. We don't want to be late," said the passenger. "We are on a good highway. We can make it," he continued. The pickup gained speed going down the long hill. It was pulling a trailer with two horses—some 3,000 pounds of weight.

Suddenly, the pickup swerved from side-to-side. The trailer was taking control. What could be done? The driver struggled but it was too late now. The trailer instantly pushed the pickup onto the shoulder. Just as quickly, the trailer turned upside down. One horse was trapped inside the trailer. The other horse broke out of the trailer. Fortunately, neither horse was seriously injured. No people were hurt. The trailer was damaged but could be repaired.

Driving too fast was unsafe. Using a pickup and trailer to do more than it was designed to do resulted in an accident.

6-1. An upside down trailer along a highway made a dangerous condition.

OBJECTIVES

Some agricultural work is dangerous. We can make it safer by going about the work properly. This chapter covers several areas of safety in agricultural work. The objectives of this chapter are:

1. Explain the importance of safety
2. Define personal protective equipment and give examples
3. Explain important areas of safety with machinery and tools
4. List safety practices in laboratory work

TERMS

- accident
- earplug
- ground fault circuit interrupter
- hazard
- material safety data sheet
- operator's manual
- particulate mask
- personal protective equipment
- respirator
- safe
- safety
- safety glasses

http://home.earthlink.net/~zinkd/index.html

(This site has information and links on food safety.)

BEING SAFE

Being safe is an important part of working and living. ***Safe*** means that a person is free from harm and danger. The risk of being injured has been reduced as much possible. People can take steps to be safe.

Knowing about accidents and safety helps prevent them.

ACCIDENTS

An ***accident*** is something that occurs unintentionally. People do not know an accident is going to happen. If they did, they would avoid it. Accidents are more likely to occur in some situations. These situations are hazardous.

A ***hazard*** is a danger. Risk is present. Injury to people and property may occur. A hazardous situation is filled with risk. The hazard may not cause injury every time. Fortunately, people can help reduce the dangers in hazardous situations.

Accidents are the leading cause of death in the United States. Nearly 100,000 people are killed in accidents each year. About nine million people are injured. Highway and road accidents account for half of the deaths and over a third of the injuries. Accidents happen in homes, places of work, and public areas, such as beaches and parks.

6-2. A tractor that was "totaled" in an accident.

AgriScience Connection

WORKER PROTECTION STANDARD

The Worker Protection Standard is a federal law on safety in using pesticides. It is designed to protect people who use them. The law was issued by the Environmental Protection Agency.

An employer must take steps to prevent injury to workers. Workers must be trained in how to safely use pesticides. Information must be provided on specific chemicals being used. Equipment must be provided to protect workers on farms. Forests, nurseries, and greenhouses are also covered by the law.

The Worker Protection Standard deals with health hazards. People who use pesticides should know the signs of exposure. They should know the steps in case of an emergency. The goal is to protect people when using pesticides.

This photo shows an equipment operator applying a pesticide. Note the personal protection equipment worn. (Courtesy, Agricultural Research Service, USDA)

Accidents happen on farms and in agribusinesses. Accidents also occur in forestry, horticulture, and related areas. The following agriculture work may have particular hazards:

- using power tools, such as saws, welders, and nailers
- using tractors and equipment
- working with animals that can attack, such as dogs or cattle
- applying chemicals, such as pesticides, fertilizers, and hormones
- working with plants that cause allergies, such as poison ivy
- using materials that may cause fires and explosions, such as fuel
- doing construction work, such as fencing and building repair
- working on stairs or roofs

Some injuries are serious; others are minor. Serious injuries may leave a person injured the remainder of his or her life. Blindness, deafness, loss of limbs, and brain damage may result from accidents. Some people are injured so they are unable to work.

Knowing and following safety rules helps make working in agriculture safer.

SAFETY

Safety is preventing injury and loss. It refers to the precautions that people take to be safe. These help people enjoy good health. People must be informed about dangers and safety.

Table 6-1. General Safety Rules

- Think safety.
- Learn how to be safe.
- Always follow safe practices.
- Anticipate dangers and take steps to avoid them.
- Read and follow instructions.
- Use personal protection equipment, such as safety glasses and ear plugs.
- Keep equipment in good condition.
- Remove hazards from the environment.
- Alert other people to hazards.
- Never take unnecessary risk.

Dangers are present in our living and working activities. Steps can be taken to prevent injuries. Begin by knowing general safety procedures. Safety is the responsibility of all people. It begins with you!

Using safe practices protects both you and other people. Both adults and children are protected. Following general safety rules will help prevent injury.

PERSONAL PROTECTIVE EQUIPMENT

Personal protective equipment (PPE) are devices to protect people from injury. It is each individual's responsibility to have and use PPE. Having and not using PPE provides no protection. Using PPE the wrong way does not offer the protection needed. For example, safety goggles on the forehead provide no eye protection.

Each person must have the needed PPE. Sometimes people have to share the devices. If so, keep them clean. Goggles, helmets, gloves, and aprons are often shared. Ear plugs should not be shared.

EYE PROTECTION

You may work where eye injury can occur. Eye protection is needed in using chemicals, doing shop work, or where bright light may cause injury. Fly-

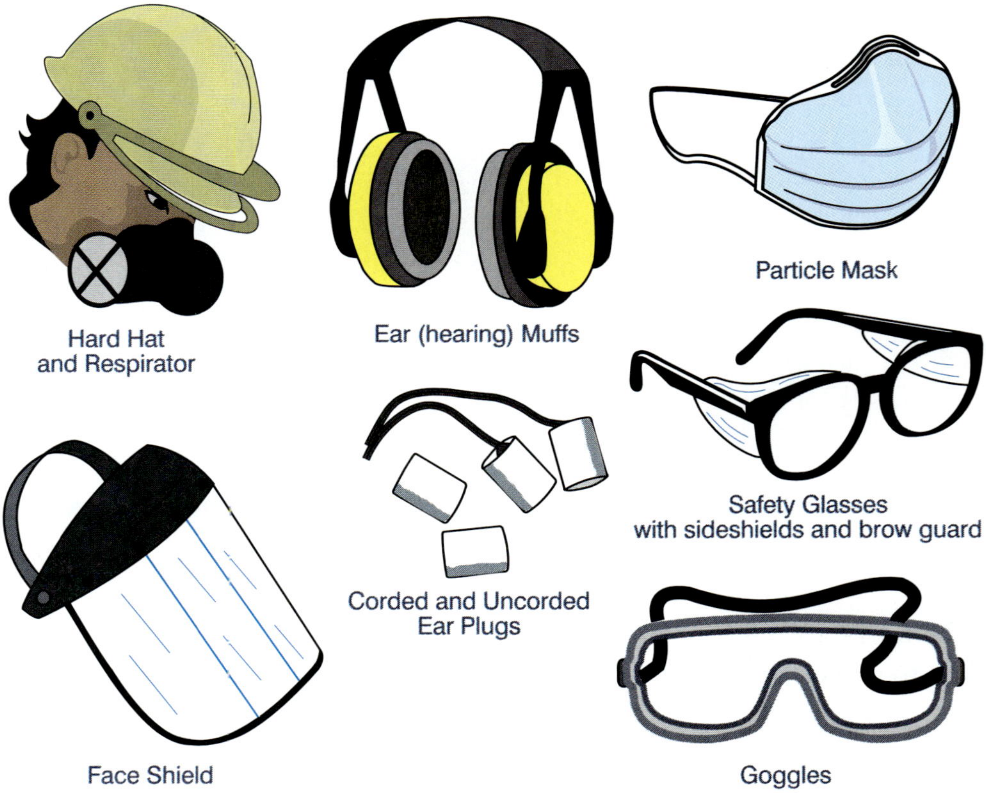

6-3. Examples of common personal protective equipment used in shop work.

ing objects may get into your eyes. In shops, small pieces of metal or wood can fly from materials being worked on. Bright lights may burn the eyes and create pain and damage to vision. Always use proper eye protection.

Common eye PPE includes:

- Safety glasses—***Safety glasses*** protect eyes from injury. They keep tiny particles out of the eyes. The lens should be shatterproof (not break when struck). The glasses should also have side panels and brow bars. These provide the best protection from small pieces of metal, plastic, glass, or wood. Tinted glasses may be needed if harmful light rays are present.

6-4. Safety glasses should be worn when using most tools.

- Safety goggles—Goggles are larger than safety glasses and cover more area. Most goggles are made of flexible plastic material. The lens area should re-

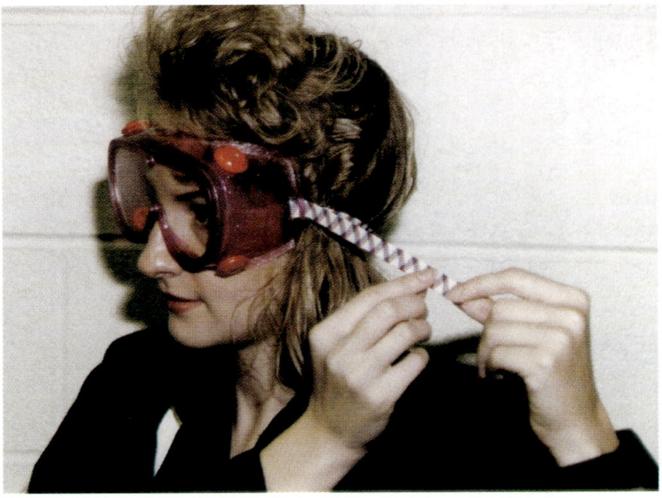

6-5. The strap on safety goggles should be adjusted for good fit.

sist impact. An elastic strap around the back of the head should hold the goggles snugly in place. Some goggles have air vents. These goggles seal tightly around the face. They are used when liquids are being handled. Goggles may have tinted lens for use around flames and bright lights.

- Safety shields and helmets—Face shields and helmets protect the sides and front of the face. They also provide some eye protection. Normally, safety glasses are worn underneath shields and helmets. Shields are used in chemical laboratories and for shop work. Helmets with approved lens are used in welding. Always select a helmet with the right lens for welding.
- Eye wash—Eye wash is needed if hazardous substances, such as chemicals, get into the eyes. A stream of water is used to wash substances from the eye. Eye wash systems may be mounted on the wall or a table; others are portable.

6-6. A safety shield and safety glasses should be worn when using a grinder.

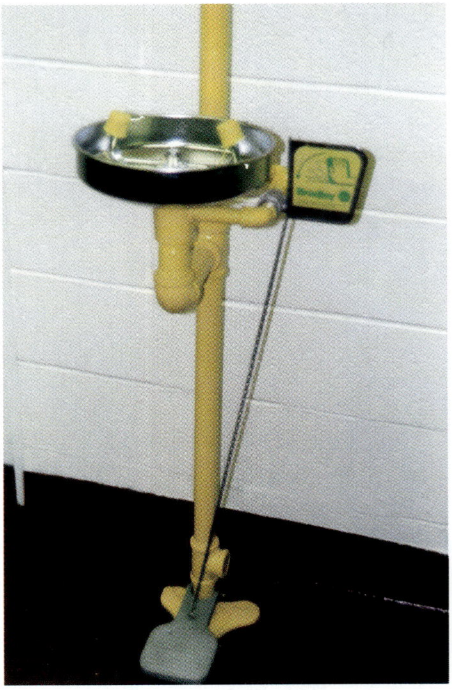

6-7. An eye wash station that can be turned on with a foot or hand.

HEARING PROTECTION

Hearing protection is worn to prevent hearing loss. Damage to delicate nerves in the ears is a common problem. Many power tools and activities make loud noise. The noise from engines and blades may damage hearing. Sudden loud noise can break an ear drum.

If possible, reduce noise level. Use materials that absorb sound inside shops. Do noisy work outside rather than inside. Select equipment that makes less noise. Use hearing PPE.

Common hearing protection includes:

6-8. Ear muffs and safety glasses are being worn while using the table saw.

- Earplugs—An ***earplug*** is a soft pliable device that fits into the ear canal. Plugs are shaped for good fit. Plugs reduce the level of sound that enters the ear. Some plugs have cords that connect the two individual plugs. Hearing bands are sometimes used. They are flexible plastic or metal straps that hold earplugs in place.
- Earmuffs—Earmuffs fit over the entire ear. A plastic or steel band may go over or around the head to hold the muffs in place. Some earmuffs are attached to hard hats. The ear covering is a soft, foam material. It conforms to the ear and head.

RESPIRATORY PROTECTION

Materials in the air may be inhaled. Some cause damage to the nasal passages and lungs. Common forms are dust, mist, fumes, and vapors. The hazard posed depends on the kind of material. Gas vapors are more hazardous than wood dust. Select the kind of protection based on the conditions of the air. Dust can be removed by filter masks. Air with harmful fumes may need to be purified. In extreme cases, masks will need to be connected to a supply of clean air.

Select PPE to meet your needs. Common respiratory protection devices are:

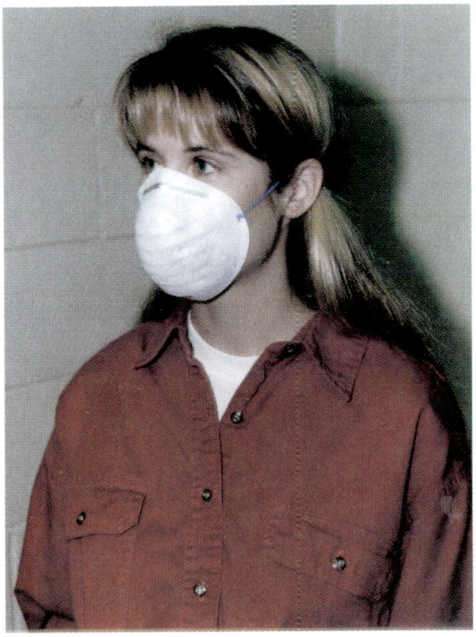

6-9. A particulate mask is used to filter dust particles from the air.

- Masks—A ***particulate mask*** is placed over the nose and mouth to remove dust particles from the air before it is inhaled. These tiny particles result from sawing, grinding, sweeping, and other work. Particulate masks are made of layers of fibers. An elastic strap is attached. Many masks have a nosepiece that helps the mask stay snug and fit properly. Be sure the elastic strap holds the mask tightly. No unfiltered air should enter around the sides of the mask.
- Respirators—A ***respirator*** purifies the air that passes through it. Some cover the full face; others cover only the nose and mouth. The strap behind the head should hold the respirator snugly in position. The air is filtered by cartridges of filter material. Select cartridges based on what is to be filtered from the air. Cartridges should be replaced as needed.
- Supplied air systems—Some substances are not readily removed from air by respirators. These substances are especially dangerous. Hoods may cover the entire head and be connected by a hose with a source of air. Air systems should be carefully selected to meet the needs of a situation.

6-10. Rubber gloves, long sleeves, a hat, and goggles are worn to protect the body from tiny droplets of bleach solution sprayed to control mildew.

SKIN AND BODY PROTECTION

The skin may be exposed to toxic substances (poisons). Cover the skin, such as hands, arms, the face, and other areas, for protection. Gloves, aprons, hats, long sleeves, and long pants offer protection from many substances. Protective shoes or boots may be needed. Depending on the kind of hazard, gloves, aprons, and boots may be made of rubber, plastic, or similar material.

Rubber gloves are needed in handling many materials. Gloves protect the skin from chemicals. These include pesticides as well as paint, solvents, and fuel or oil. Wear rubber or plastic gloves when handling automobile or tractor batteries. Gloves should be worn in handling animal tissues and materials, such as when cleaning a cage or testing blood.

Hard hats may be needed in places where objects may fall from above. Shoes with steel toe protection are used where heavy objects may fall on the feet. Workers in construction and forestry should use hard hats and high-top shoes with steel toes.

Clothing exposed to hazardous materials should be washed. Do not wash it with other clothing. Wash clothing that has been worn while applying pesticides in machines used only for that purpose. Always follow instructions on pesticide containers.

Emergency showers are needed where chemicals are used. In case of an accident, stand underneath the shower and pull the handle. Do not take the time to remove clothing. Quickly washing the hazardous substance off is more important than protecting clothing.

Body protection should be observed when eating and drinking. Always wash hands before eating. Never eat around pesticides or other chemicals. Food and beverages should never go into a laboratory. Traces of harmful substances can get into the food or drink and injure the body when eaten.

SAFETY WITH MACHINERY AND TOOLS

Many kinds of machinery are used in agriculture. These often pose safety hazards. Machinery that has power and is made for cutting is especially hazardous.

POWER AND ENGINES

Engines are used to provide power for operating agricultural equipment. The common engines use fuel that creates heat. The fact that engines burn fuel signals danger.

An operator's manual accompanies every engine and piece of power equipment. An ***operator's manual*** is a written document that gives informa-

6-11. Gasoline has been kept in an approved safety can.

tion about how to safely operate the engine. Always study the operator's manual before using equipment.

Here are some safety hazards and how to deal with them.

- Fuels—Keep fuels in approved containers. Avoid fires around fuel. Most fuels are readily ignited by a spark or flame. Avoid breathing the fumes from fuel. Avoid letting fuel come into contact with the skin. Avoid spilling fuel on the ground, equipment, or other property.
- Batteries—Most engines have a storage battery that provides electricity to start the engine. The processes in the battery create hydrogen and oxygen. Hydrogen is highly flammable in the presence of oxygen. Follow directions in the operator's manual if jump-starting an engine. Avoid flames around batteries.
- Electricity—Most engines have electrical systems. Electrical systems may be connected to lights and motors in other locations on the equipment. The systems can cause shock if improperly used. Avoid contact with uninsulated electrical wires.

6-12. Always use care in servicing an engine, such as the air filter on this tractor engine.

- PTO and pulleys—Engines often have power take off (PTO) shafts or shafts with belts and pulleys. These turn at high rates of speed. Safety shields should be in place. Avoid any contact with moving parts. Always turn off the engine before attempting to work on moving parts.
- Cooling systems—Cooling systems absorb heat from the engine. These systems often get quite warm. Most are under pressure and can reach temperatures above the boiling point. Serious burns can result. Avoid contact with a hot cooling system. Never open an overheated cooling system. Steam and water spray can cause serious burns.
- Moving parts—Engines have parts that move. Some have fans that turn at high speeds to cool the engine. Avoid contacting any moving part. Turn off an engine before attempting to service or repair it.

TRACTORS AND EQUIPMENT

All equipment should be properly maintained and operated. Follow the routines printed in the operator's manual. Never push tractors and equipment beyond what they are designed to do.

- Speed—Most equipment is designed to operate within a speed range. Do not exceed this speed. Going too fast can cause tractors and equipment to turn over. Excessive speed increases wear and causes parts to fail.
- Slopes—Tractors and equipment may need to be operated on inclines or sloping land. Excessive slope may cause a tractor or equipment to roll over. Never drive equipment onto slopes that could cause it to roll over. Driving across a

6-13. Always read and follow warning decals on equipment.

slope is more likely to cause roll over than going up and down a slope. Tractors have high clearance and are easier to turn over.

- Hitch properly—A hitch is used to attach equipment to a tractor. Plows, harrows, and wagons are often attached with hitches. Hitching too high raises the center of gravity. This increases the likelihood of a tractor turning over. Hitches should be used so the tractor keeps its balance.
- Condition—Equipment in good condition is more likely to be safe. Keep bolts and nuts tight. Keep all operating parts properly adjusted. Safety shields should always be in place.
- Safety features—Most equipment is manufactured with safety devices. Examples include roll bars and seat belts. Keep these devices in good condition. Use them properly when operating tractors and equipment.

6-14. Two ways of protecting operators in case a tractor turns over: cab (on left) and roll bar.

POWER TOOLS

Power tools include saws, sanders, planers, air compressors, and welding machines. These may be portable (moved about) or stationary (stay in one place). The tools may be used in a shop or on a job site. Some power tools use electricity; others have engines. Always read the operator's manual and follow the safety instructions. Power tools are dangerous!

With electricity, be sure all cords and plugs are in good condition. Use only plug-ins that are properly grounded. A ***ground fault circuit interrupter*** (GFCI) should be used. These protect people from shock. If a GFCI is not made into the electric wiring, use one on the cord of the tool being used. Do not stand in water or damp places when using electricity. Turn off all electrical equipment when finished.

Hazards with power tools include eye injury, hearing damage, and cuts. Some cuts may be severe. Cuts can include small skin lacerations to cutting off fingers, hands, arms, feet, and legs. In some cases, deep cuts or injury may be made to the body. With cuts, stop the bleeding and get medical help immediately. Severed fingers, hands, and other body parts can sometimes be saved with prompt medical help.

Properly use personal protection equipment. This includes safety glasses, helmets or hats,

6-15. Safety in welding includes a helmet, welding gloves, long sleeves and pants, a hat, and leather shoes.

AgriScience Connection

SMV EMBLEMS

Agricultural equipment should have Slow Moving Vehicle (SMV) emblems. The emblems are mounted on the backs of tractors and other equipment that travel slowly on public roads. Emblems should be clearly visible for several hundred feet.

SMV emblems are shaped like a triangle. The emblems are made of a thin metal painted with red and orange paint. The paint reflects light and is easy to see at night. A bracket is used to mount the emblem with small bolts. SMV emblems should never be used for other purposes.

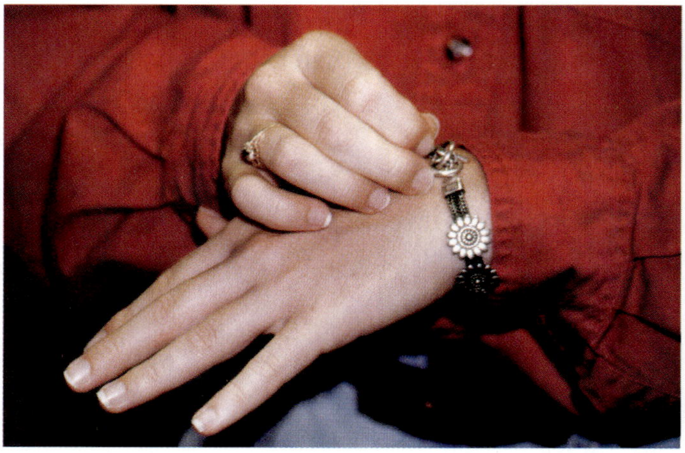

6-16. Always remove jewelry before working in a shop.

hearing protection, and protective clothing. Each power tool has unique hazards. For example, loud operating noise may cause gradual hearing loss. Just because the loss is not immediate does not mean that hearing protection should not be used! (A reference with safety information on common power tools is *AgriScience Mechanics*, available from the Interstate Publishers, Inc.)

6-17. Wear eye protection when using wrenches.

Dress properly when using tools. Remove jewelry. Wear clothes that fit and are not loose. Tie long hair back. Dress appropriately for the job to be done. Improper dress can contribute to risks.

SAFETY WITH HAND TOOLS

Hand tools are small powerless tools. They help do work that cannot be done by hand. Try driving a nail without a hammer! Many kinds of tools are used in mechanics. Common hand tools include hammers, screwdrivers, pliers, wrenches, and saws.

Here are several safety rules in using hand tools:

- Wear eye protection.
- Select the right tool for a job. (Use a hand tool only for its intended purpose.)

- Keep tools in good condition. (Do not use tools with cracked handles or other unsafe conditions.)
- Use tools properly. (Hold the handle properly.)
- Store tools after use.
- Carefully handle tools with sharp edges to avoid cuts. (A chisel is a good example.)
- Never throw tools. (Carefully hand a tool to another person.)

SAFETY IN AGRISCIENCE LABORATORY WORK

Agriscience laboratories may have chemicals, living organisms, and devices that can cause injury. Learn how to do an activity before you begin. Always follow the rules. Know where safety devices are located and how to use them. Know where the telephone is located and the emergency number to call.

Some common laboratory safety practices are given here.

- Use chemicals safely—Some chemicals are hazardous. Use these carefully. Learn as much as you can about a chemical before using it. Never touch or taste a chemical. Never smell substances directly. Conduct experiments in ventilation hoods. Avoid inhaling fumes. Do not return unused chemicals to bottles in the storage area. Refer to the ***Material Safety Data Sheet*** (MSDS) for details. These sheets are sent with each shipment of chemicals. Information about safe use of the chemical is included.

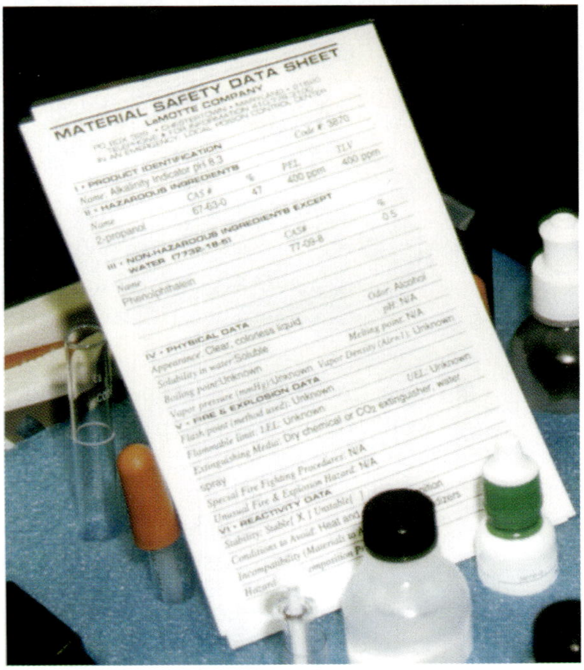

6-18. Chemicals must have a Material Safety Data Sheet (MSDS).

- Use equipment safely—Several laboratory devices are regularly used. Some are made of glass. Broken glass poses particular dangers. Never use chipped or broken glassware. Do not force parts together, such as a glass tube through a rubber stopper. Be careful with scalpels and scissors. Never allow direct sunlight on a microscope mirror. Think carefully about safety hazards in using equipment.
- Handle organisms carefully—Plants, animals, and other organisms are often used in laboratories. Some of these are live specimens. Others have been carefully preserved. Know the characteristics of an organism before working with it. Some plants can cause allergic reactions, such as poison ivy. Living animals should be handled properly. They can bite, scratch, and cause other injury if handled wrong. Any discarded specimens should be properly disposed of. With live animals, always practice animal well-being.

6-19. Know the plants being used in a lab.

- Use fire and heat properly—Lab work often involves heating materials to study reactions. Be careful with hot plates and open flames. Use only Pyrex glassware for heating. Be sure the table or bench is level. Place containers on burners so they are secure and will not turn over. Use tongs and heat-resistant gloves to handle hot or warm materials. Always turn hot plates and burners off after use. Never leave a heat source without someone to watch it.
- Know the lab—Locate all safety devices in the lab. Know how to use them in case of an emergency.
- Know the procedures—Always follow the rules in the lab. Never joke around. Tell your teacher or leader when an accident occurs. Take steps to control accidents.

REVIEWING

MAIN IDEAS

Being safe is being free from harm and danger. Safety is taking steps to prevent injury and loss. Accidents sometimes happen. Good planning and following safety rules can prevent most accidents. Safety is everyone's responsibility!

Personal protective equipment is used by people to prevent personal injury. Specific equipment is available for eyes, hearing, the respiratory system, and the skin and body. Protective equipment must be properly used to offer good protection. It is best if people have their own safety glass or goggles, earplugs, and masks or respirators.

Machinery and tools pose special hazards. Always refer to the operator's manual for information on proper use and safety with machinery. Tractors and equipment involve powerful engines and moving parts. These can be used safely by knowing how to operate them and following good procedures. Power tools can quickly cause injury. If not used properly, hand tools can also cause injury.

Agriscience laboratories may have special hazards. Chemicals, equipment, organisms, and fire and heat all pose dangers. Know where the safety equipment is located. Know how to use the equipment properly.

QUESTIONS

Answer the following questions. Use correct spelling and complete sentences.

1. What is an accident?
2. What is safety? Who is responsible for safety?
3. What is personal protective equipment?
4. What eye protection may be used?
5. What hearing protection may be used?
6. When is respiratory protection needed?
7. Why are skin and body protection important?
8. What are the major safety hazards from power and engines?
9. What should be considered in safely operating tractors and equipment?
10. What are the hazards with power tools?

EVALUATION

Match the term with the correct definition. Write the letter of the term in the blank provided.

a. safety
b. earplug
c. accident
d. operator's manual
e. particulate mask
f. respirator
g. safety glasses
h. personal protective equipment

____ 1. An unintentional event.
____ 2. Preventing loss and injury.
____ 3. Devices designed to protect people from injury.
____ 4. Eye protection.
____ 5. Hearing protection.
____ 6. Protect from dust.
____ 7. Printed material that gives information on equipment.
____ 8. Device that purifies the air.

EXPLORING

1. Invite a safety specialist to serve as a resource person in class. Ask the person to discuss the work of a safety specialist and how hazards in agricultural work can be avoided. (Many agricultural organizations, such as the Farm Bureau, have safety specialists. If not, contact the Cooperative Extension Service at the land-grant university in your state.)

2. Take a guided tour of the agricultural mechanics and/or horticultural laboratories at your school. Note the possible hazards that you see and the safety devices that are used. Ask the teacher to explain safe procedures while in the laboratory. Prepare a brief oral or written report on your observations.

7

SCIENCE IN ACTION

Agriculture is science in action! Understanding plants and animals involves science. Think about living plants and animals. What do they need to live and grow? All of what they need are science-based.

A neat thing about studying agriculture is that it shows the importance of science. It gives science meaning. Science principles are easy to understand when used in agriculture. In fact, many scientists have strong interests in agriculture.

Science is a part of all of the agricultural industry. Processing food involves science. Designing new machinery involves science. Forestry, horticulture, and wildlife are science-based. This does not mean that learning about them is hard. It means that science is easier because of the practical use it has.

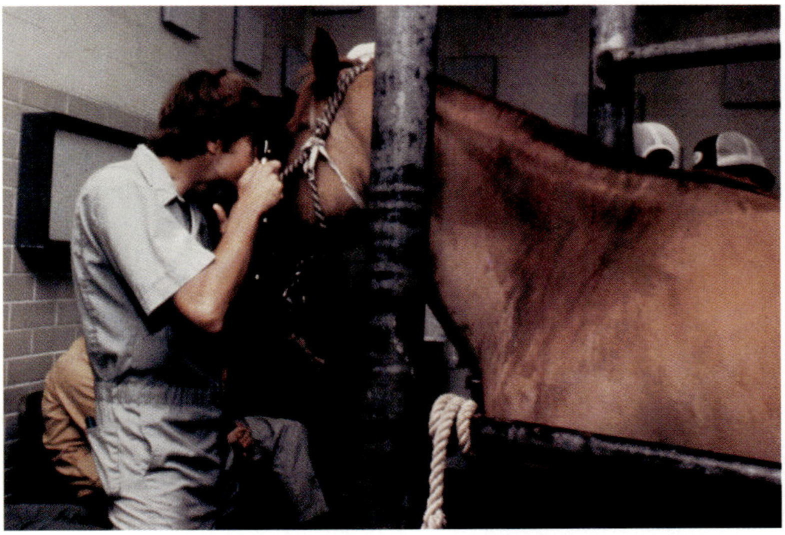

7-1. Treating a horse involves science in action. (Courtesy, Veterinary Medical Center, Texas A&M University)

OBJECTIVES

All areas of science are a part of agriculture. This includes the environment and keeping the earth a good place to live. The objectives of this chapter are:

1. Explain agriscience
2. Relate agriscience to important areas of science
3. Describe the relationship of agriscience to the environment

TERMS

agriscience
agronomy
animal
animal science
botany
chemistry
ecosystem
entomology
environment
environmental science
horticulture
organism
physics
plant
science
soil science
sustainable agriculture
zoology

http://www.carolina.com/

(This site has information on a wide range of science materials.)

AGRISCIENCE

Science is learning about the world in which we live. Careful study is made of living and nonliving things. Truths or laws are developed to explain what is happening and why. Today, we use the term "agriscience."

Agriscience is the application of science in producing food, clothing, and shelter materials. It deals with both "why" and "how" things are done. Knowledge helps people with their work. Studying agriscience gives knowledge about useful areas of science. It helps meet the needs of people.

All areas of science are in agriscience. The most important relate to plants, animals, soil, water, and machinery. Areas of environmental science are more important. This is because many agricultural activities affect the quality of the environment.

Technology is a part of agriscience. It is the application of science. This leads to new devices and methods. These make work easier and increase yields.

Scientific principles form the base on which agriscience knowledge and work are built. For example, people who know plant science are better at growing plants. The same is true about animals and animal science. Science-based agriculture is the current state-of-the-art.

7-2. A genetically engineered vaccine has been used to protect these pigs from African swine fever. (Courtesy, Agricultural Research Service, USDA)

SCIENCE AREAS

Animals, plants, chemicals, and machinery involve science. All work in the agricultural industry uses science in some way. The areas are related. None is isolated from the others. For example, designing a plow requires knowledge of plants, soils, and plow design—biological science, earth science, and physical science, respectively.

Science is a big field of study. It is divided into several specific areas that often overlap with other areas. The major areas of science are life science, physical science, mathematics, and social science.

7-3. Growing rice involves scouting for pests. (Courtesy, Agricultural Research Service, USDA)

LIFE SCIENCE

Life science is about living things. Plants, animals, and other organisms are included. Life science is often known as biology.

Living things are ***organisms***. All organisms carry out life processes. If the life processes stop, the organism is no longer living. Life science helps explain the needs of organisms to live and grow.

All known living organisms have been classified. Scientists study the similarities and differences. They place the organisms in groups based on what they see. Today, five groups are used: plants, animals, fungi, protista, and monera. All organisms fit into one of the groups, known as kingdoms. Plants and animals are most important in agriscience.

Plants

A ***plant*** is an organism that makes its own food. Plants use nutrients in a water solution to make food. Usually, the water solution is taken in by plant

7-4. Cotton is the most important fiber crop. Bolls containing cotton and seed will form following the flower. (Courtesy, Mississippi State University)

roots from the soil. Plants must have light to make food. Plants are in the plant kingdom.

Plants are important to people. They are used for food and to make clothing and housing. Plants are also used as food by many animals, such as cattle and sheep. Other plants have personal appeal. These are used as flowers and shrubs to make life more attractive.

The study of plants is known as ***botany***. Some plant scientists specialize in food plants or plants used for beauty. The study of plants for crops is known as ***agronomy***. In agronomy, new kinds of plants are developed. Developing new ways of growing plants and using the soil is also in agronomy. Common crops in agronomy are corn, wheat, soybeans, cotton, and rice. Many of these are made into high-quality food products. Even the seeds of cotton are used to make oil!

7-5. Marigolds are popular flowering plants.

Horticulture deals with using plants for personal appeal or food. Those plants used for beauty and appeal are known as ornamental plants. Examples of plants used for beauty are roses, petunias, and azaleas. The food horticultural plants are vegetables,

7-6. Redwood trees grow quite large.

fruits, and nuts. Most of these are used in human food. Some are used for other products, such as the oil from a tung nut used in manufacturing paint.

Forestry involves plant science. It includes growing trees for lumber, paper, and other purposes. Just think of all the wood products we use! Furniture, baseball bats, and picture frames are made of wood. In addition, we enjoy the beauty of trees. Scientists know that trees have a positive impact on the environment. We bring trees into our homes. To some people, the holiday season would not be complete without a decorated tree!

Animals

An ***animal*** is an organism that must obtain its food, often from plants. The nutrients in the food are digested and used by the body. Many animals can move about and all have circulation systems. Common animals include dogs, fish, horses, insects, and birds.

7-7. Caribous are magnificent wildlife animals. (Courtesy, U.S. Fish and Wildlife Service, Dean Biggins)

The study of animals is ***zoology***. Several specialized areas of zoology are important. Some of these areas are studied in agriscience and are called animal science.

Animal science is the study of animals used for food and other purposes. Poultry science deals with chickens and turkeys. Aquaculture deals with fish farming and related areas. Beef cattle are raised meat. Dairy cattle are kept for milk. Horses are used for pleasure and, sometimes, work.

Many animals use plants for food. Those that do not eat plants consume them indirectly. For example, a fox may prey on a rabbit. The rabbit has used plant materials for food.

Animals are important as pets. Another term used to describe a pet is companion animal. You or someone you know has a pet that is a part of their lives. Common pets include dogs, cats, and fish. Other animals kept in homes include gerbils, mice, snakes, and African hedge hogs. Always know how to care for a pet before you get one. Consider the animal's well-being!

7-8. Some people enjoy keeping African hedge hogs.

Insects are important animals. Some are beneficial; others are pests. The beneficial insects have useful roles. They pollinate flowers and help assure good crops. Insect pests cause damage. They may attack plants, our homes, or us directly—such as the mosquito. The area of insect study is ***entomology***. Sometimes, other small animals are included in entomology, such as spiders and mites.

PHYSICAL SCIENCE

Physical science deals with nonliving things. All of the things in our environment that are not living are included. Close relationships exist between living and nonliving things. This will be obvious as you study this chapter.

There are three large areas in physical science: earth science, chemistry, and physics.

Earth Science

Earth science is the environment in which we live. It includes air, water, soil, and other areas.

Soil is a part of earth science. Learning about the soil is often called soil science. ***Soil science*** is the study of the soil. It includes soil structure and nutrient content. Soil science includes ways of keeping the soil fertile and preventing the loss of soil.

Close relationships exist between the soil and living organisms. Plants gather nutrients from the soil. Plenty of nutrients must be present. If not, the plants will not grow well.

7-9. Soil samples are taken for testing to learn the fertilizer needed.

Chemistry

Chemistry deals with the makeup of materials. Materials are known as matter. This means that they occupy space and can be weighed. All matter is made of elements arranged in varying combinations. These are known as chemical elements. Oxygen, hydrogen, carbon, and iron are examples of chemical elements. Scientists have found 92 natural elements on the earth. Artificial elements have been made in laboratories. Seventeen artificial elements have been named. Elements attach to each other to form compounds.

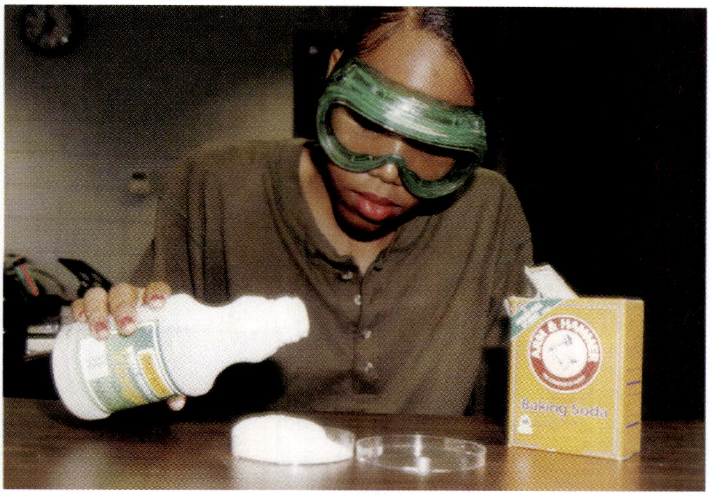

7-10. A chemical reaction occurs when vinegar is added to baking soda.

For example, water is a compound formed of oxygen and hydrogen. This is easily shown in the formula for water: H_2O.

Career Profile

BIOCHEMIST

Biochemists study relationships between living organisms and chemical processes. They investigate what makes plants and animals unique. Some study the nature of life processes and how these can be improved.

Biochemists usually have advanced degrees in biochemistry or a related area. Most have bachelor's degrees in biochemistry, biology, or chemistry. Many begin with high school courses in biology and chemistry. Those who work in agriscience areas need practical experience with agriculture. Training in research and laboratory methods is essential.

Jobs are mostly in biochemical businesses, universities, and government agencies. Some are with biotechnology firms. The photo shows a biochemist using a micropipet with a disposable tip for safety and sanitation.

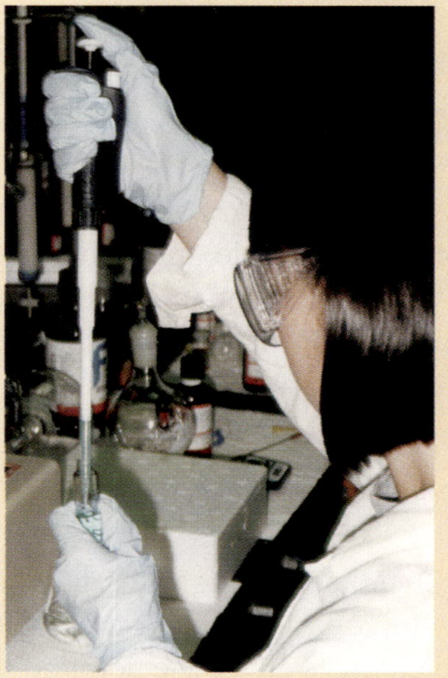

Physics

Physics is about the physical nature of objects. This includes heat, light, electricity, and mechanics. Simple machines are used to help do work. These include levers, pulleys, and inclined planes. Tractors, lawn mowers, and other equipment are made of many simple tools.

Physics principles are used in our everyday lives. The internal combustion engine is a good example. Heat from the burning of fuel causes gases to expand. This expansion moves engine parts that create mechanical power.

7-11. A pistachio harvester involves many principles of physical science. The harvester shakes the tree, causing the nuts to fall onto a catching frame. (Courtesy, California Pistachio Commission)

MATHEMATICS

Mathematics is the science of numbers. It is used in making observations. For example, mathematics helps measure objects. Without mathematics we would be unable to buy and sell things. We would not be able to build structures or measure land. Standards are used to assure uniform measures. All of us want a gallon of milk always to be a gallon!

SOCIAL SCIENCE

Social science is the study of human society. Many different human behaviors are included. Some areas relate to agriculture. For example, population changes affect the demand for food. With social science, more is learned about the nature of human actions.

THE ENVIRONMENT

Living things are not alone. They live in a world with other living things and nonliving things. Important relationships exist. Most of the relationships are easy to understand.

Where we live is our environment. ***Environment*** is the surroundings of an organism. It includes the air we breathe, the water we drink, and much more. Many people use science to explain the environment.

ENVIRONMENTAL SCIENCE

Environmental science is the study of the environment. Emphasis is on areas of science that help us understand the environment. Areas of life science and physical science are important.

Ecosystems

Plants and animals grow under certain conditions. These conditions form their ecosystem. An ***ecosystem*** is made of all the parts of an organism's environment that help it live and grow. In some ways, it is a community of living and nonliving things. Look under a log. Here small worms, beetles, and other things interact. This tiny area is an ecosystem. Other organisms have larger ecosystems. An example is a beef cattle herd. The cattle may have several hundred acres on a ranch for their pasture. They interact with the grasses, insects, water, and other things that are present.

7-12. Red foxes thrive in the proper ecosystem. (Courtesy, U.S. Fish and Wildlife Service, J. Stutzman)

Ecosystems sometimes lose their ability to support life. The log that was moved to see worms and beetles disrupted the ecosystem. These organisms no longer found it a good place to live. The same can be true with cattle. A drought can result in the pasture not growing. Polluted

7-13. Air samples are collected to detect pollution.

water can cause disease. Many things can change an ecosystem. Some changes make it a better place to live. Other changes make it less desirable.

Human Ecosystems

Humans need certain things to live in their ecosystem. People want to breathe good air, drink good water, and have other needs met. A good environment is essential. If not, life is threatened. Humans may suffer disease and die early. Fortunately, science is helping people learn more about the environment. It helps in keeping a good place to live.

Some people are not concerned about the environment. They do things that damage it. Polluting the air and water are two examples. Should we be concerned? Yes!

Pollution changes the environment. It may create an unhealthy place to live. All of us have a role. We must prevent pollution. We must work to have an earth that will support future generations.

7-14. Humans live together in a complex ecosystem.

AGRISCIENCE RELATIONSHIPS

Agriscience helps in understanding the environment. How plants, animals, air, soil, and other areas relate is important. The way people go about growing plants and raising animals affects the environment. Using the best practices is essential. Select those that do not damage the water, air, or soil.

Damaging Practices

Some practices that damage the environment are easy to see; others are not. Careful study will show things that cause damage.

Look at the water in a stream. Is it clear or muddy? Muddy water is caused by tiny, suspended soil particles. These wash into a stream when it rains. Muddy water is polluted!

Some pollution is not as easy to see as muddy water. Chemicals may get into water without changing its color. Laboratory tests are needed to know if they are present. You can do some tests yourself using simple water test kits.

AgriScience Connection

WATER TESTING

Water may need to be tested. Testing determines what makes it impure. All water from the earth contains some substances. Testing shows what the substances are. This allows a person to know if using the water is a hazard.

Where testing is done varies. Testing may be done on site at a stream, lake, or other source. Sometimes, water samples are collected and tested in a laboratory. A good sample is needed. The reliability of a test can be no better than the sample tested. A poor sample can provide results that are wrong.

This photo shows wastewater being tested for phosphorous. Washing detergents are major sources of phosphorous in streams and lakes. Phosphorous changes the water. High levels result in more algae and plants in the water. Fish may be killed off. (Courtesy, Agricultural Research Service, USDA)

Here are a few examples of activities that can damage the environment:

- Clearing land—Removing trees and grass from land leaves the soil bare. Rain and wind can wash or blow it away. This is often a problem on construction sites. Building roads disturbs the protection provided by plants. The same is true on sites for buildings. Water runoff can carry soil into streams. Wind can pick up tiny soil particles and pollute the air. No one wants to breathe dusty air! Exposed soil can be covered with straw or other materials. Silt fences can be used to catch any soil before reaching a stream.
- Plowing the wrong way—Plowing disturbs the soil. Proper plowing helps keep soil in place. Some land can be farmed without plowing at all. The remains of previous plants can be left on the land to prevent soil loss. The direction of rows is important. Rows should go across the slope of a hill rather than up and down the hill. Rows are laid out on a contour to allow very gradual slopes.

7-15. This land has eroded because of improper care.

- Using chemicals incorrectly—Chemicals have many good uses. Using them the wrong way can create pollution. Some chemicals used in the home are hazardous. Washing detergents damage the water. Insect sprays foul the air. Crop and livestock production often requires chemicals. When used improperly, pesticides and fertilizers can get into the air and water. Sometimes, chemical buildups occur in soil. Always use approved chemicals only as needed.
- Manufacturing—Factories may produce chemical byproducts in manufacturing. Smoke and wastewater are examples. These can damage the environment. Fortunately, many factories take steps to prevent pollution. The wastes are disposed of properly. This protects the air, soil, and land. It makes the environment better for all people.
- Dumping wastes—People create waste. The animals they keep create wastes. Manufacturing results in waste materials. Getting rid of these wastes can be

done carelessly. Fortunately, steps are being taken to dispose of wastes properly. Animal wastes may go into lagoons for treatment and then be spread on the land. Many materials can be recycled. Materials made from plants or animals can be composted. This allows the materials to decompose before making good use of them as fertilizer.

- Using engines—Engines use fuel and create exhaust. The exhaust can contain harmful substances. Much of the air pollution around big cities is due to the exhaust from engines in trucks and cars. Keeping engines in good condition reduces pollution. Using engines no more than needed reduces exhaust and saves energy.

7-16. The dark area near the ground behind the Washington Monument and U.S. Capitol is air pollution.

The Good News: Sustainable Agriculture

The environment can be protected. Many people are being good citizens. They are taking steps to reduce pollution. Everyone has a role in keeping a good environment. This assures food, clothing, and shelter in the future. Today, sustainable agriculture is used.

Sustainable agriculture is using practices that maintain the ability to grow crops and raise livestock. This assures that the needs of future generations should be met. The environment is protected. Chemicals are not wasted. Natural processes are encouraged. Scarce fuel is saved.

People view crop and livestock production as a system. No practice is done alone. One practice has an impact on other practices. Using more than one practice in combination improves efficiency. Also, the activities of one person influence another person. What happens on one farm can affect a neighboring farm. Practices are not carried out in isolation.

Four major areas in sustainable agriculture are:

- Diversify—This means that land is planted to different crops in succeeding years. Often, the kinds of crops alternate. Corn may be planted one year and soybeans the next. This helps maintain the soil and reduce pest buildup. For example, soybeans are legumes. Legumes naturally produce nitrogen on their roots in the soil. Corn uses much nitrogen. The nitrogen formed on the roots of soybeans is used by corn the next year. Less fertilizer is needed. The chance of excess fertilizer being washed into streams is reduced. Water pollution is reduced.
- Use biology—This involves knowing how organisms grow and relate to each other. It is sometimes known as biological pest control. Many practices can be used. Beneficial insects can be used to control pests. Crop plants can be spaced to shade the ground and keep weeds from growing. Varieties of crops can be planted that resist insects or disease. Using biology saves the use of chemicals. This protects soil and water. It helps make future production possible.

7-17. A beneficial insect has attacked pests—the pirate bug is feeding on whitefly nymphs. (Courtesy, Agricultural Research Service, USDA)

- Prevent disease—Both animals and plants may get diseases. Taking steps to prevent disease saves the environment. Vaccinating animals keeps them healthy. Antibiotics are not needed to treat disease. Raising animals that resist disease is another practice. Some animals have a natural resistance to disease. One example is cattle with dark pigment around their eyes (cattle with dark pigment are less likely to have eye cancer).
- Use improved crops—Several new crop varieties have been developed using biological engineering. These crops have traits that prevent disease and protect the environment. Hazardous pesticides are not needed. Sometimes, the crops have been developed to withstand drought—this saves irrigation water. So many things are taking place with crop plants. A smart grower keeps up to date on developments.

REVIEWING

MAIN IDEAS

Agriscience is the use of science in growing crops and raising animals. It includes important areas of science: life science, physical science, mathematics, and social science. Each of these have special uses in agriculture.

Life science deals with plants, animals, and other organisms. Plants and animals are particularly important. Some are used for food, while others are used to make clothing and shelter. Plants and animals also make life better in other ways. Some plants are grown for beauty. Some animals are kept as pets. Both plants and animals have important roles in the lives of people.

Physical science is important because it involves nonliving things on the earth. Earth science includes soil and related areas. Chemistry and physics are increasingly used in agriscience.

The environment is important to all living organisms. The parts of an organism's environment that help it live and grow form an ecosystem. Agricultural practices are related to ecosystems and the environment. Some practices cause damage. Fortunately, steps can be taken to reduce damage. Sustainable agriculture is now a major goal in agriscience.

QUESTIONS

Answer the following questions. Use correct spelling and complete sentences.

1. What is agriscience? Why is it important?
2. What are the four major areas of science? Briefly explain each.
3. What is the major difference in how plants and animals have food?
4. Why is physical science important in agriculture?
5. What is the environment?
6. What is an ecosystem? How do changes in an ecosystem influence life?
7. What are examples of activities that can damage the environment?
8. What is sustainable agriculture?

EVALUATING

Match the term with the correct definition. Write the letter of the term in the blank provided.

a. science	d. entomology	g. ecosystem
b. organism	e. chemistry	h. sustainable agriculture
c. agronomy	f. environment	

____ 1. A living thing that carries out life processes.

____ 2. The study of crop plants and soils.

____ 3. The study of insects.

____ 4. Surroundings of an organism.

____ 5. Parts of an organism's environment that help it live and grow.

____ 6. Using practices that maintain the ability to grow crops and raise animals.

____ 7. Study of the world by observation.

____ 8. Study of the makeup of materials on the earth.

EXPLORING

1. Prepare a report on the important plants and animals in the local area. Identify their use—food, clothing, or housing. Sometimes, the plants may be ornamentals and used for making the environment more attractive. Include these on your list.

2. Make a field trip to a farm that uses sustainable agriculture. Have the owner or manager explain the practices used. Make photographs that depict what you see. Prepare a report on your observations. Also, assess what you saw and include in your report suggestions for improving the farm.

3. Prepare a report on the sources of pollution in your community. Include farms, factories, homes, businesses, and other sources of pollution. Offer suggestions in your report on how the pollution could be controlled.

8

BIOTECHNOLOGY

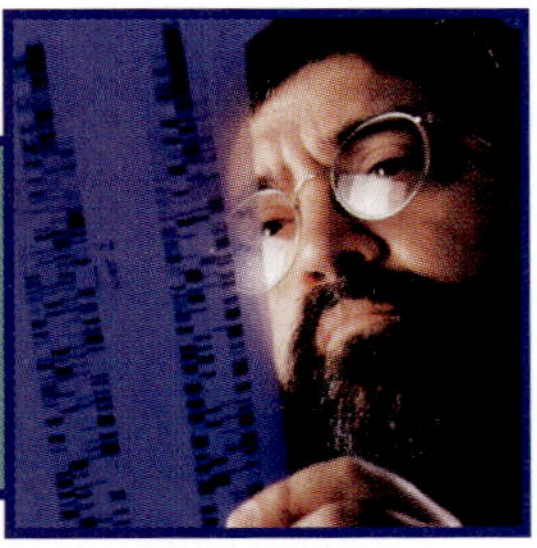

Can plants and animals be improved? Yes! Scientists have learned ways of making plants and animals better. Some methods use simple skills that you can do. Others require specialized training and equipment.

Should plants and animals be improved? Yes! But, not everyone agrees. Some people feel that living things should not be changed. People who are informed feel that making improvements is needed. Plants and animals can be made to serve the needs of people better.

How do you feel about making plants and animals better? Learn about biotechnology before taking a stand. Study the benefits, risks, and issues. Use science to help make your decision.

8-1. Gennie the pig has been altered to produce a human blood factor needed in surgery. (Courtesy, Virginia Tech)

OBJECTIVES

This chapter provides an introduction to biotechnology. The emphasis is on using biotechnology in agriculture. The objectives of this chapter are:

1. Define biotechnology and list examples in agriculture
2. Explain the meaning of recombinant DNA
3. List the benefits of biotechnology
4. Identify issues associated with biotechnology

TERMS

biotechnology
chromosome
cloning
cuttings
gene
gene mapping
gene splicing
genetic code
genetic engineering
genome
grafting
growth implant
hormone
issue
particle gun
recombinant DNA
tissue culture
transgenic organism

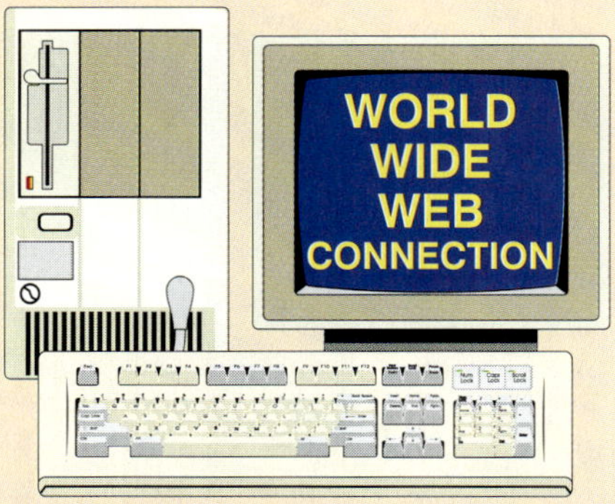

http://www.cato.com/interweb/cato/biotech/

(This site has information on many areas of biotechnology.)

BIOTECHNOLOGY: WHAT IT IS

Biotechnology is using science to change living organisms. New products may be obtained. New ways of making products may be developed. Much of the work involves the tiny building blocks of organisms known as cells.

The word, biotechnology, tells a lot about what is involved. First, is "bio" or biology. This includes life processes and the structure and functions of cells in living organisms. Secondly, "technology" is used to change organisms. Technology may be used to make the environment for the organism more favorable. Other times, the genetic material of an organism may be altered. Changing genetic material is a complex process. It is the area that is sometimes controversial.

8-2. Plastic mulches of different color are being tested. (The purpose is to determine the effect of color on tomato yield.) (Courtesy, Agricultural Research Service, USDA)

ORGANISMIC BIOTECHNOLOGY

Organismic biotechnology is simple. It is using living things—organisms—as they are. They are not altered. Their heredity material is not changed except for that which occurs naturally. The organisms are used as they are with life made better. New breeding methods may be used to improve the traits of offspring.

This kind of biotechnology has been used a long time. Plant and animal producers use it nearly every day. It is as simple as using fertilizer to help plants grow. Other organismic biotechnology uses specific skills.

Cloning

Cloning is making two or more organisms out of one. The pieces are separated and grow into complete organisms. A new organism has only one parent and is identical to it. Cloning is often used with plants.

Tissue culture is a kind of cloning. It is using cells or small clusters of cells from a parent to produce a new living thing. The tiny cell cluster is known as an explant. Tissue culture uses specialized skills and equipment. The work is done in an aseptic environment. Aseptic means that it is free of

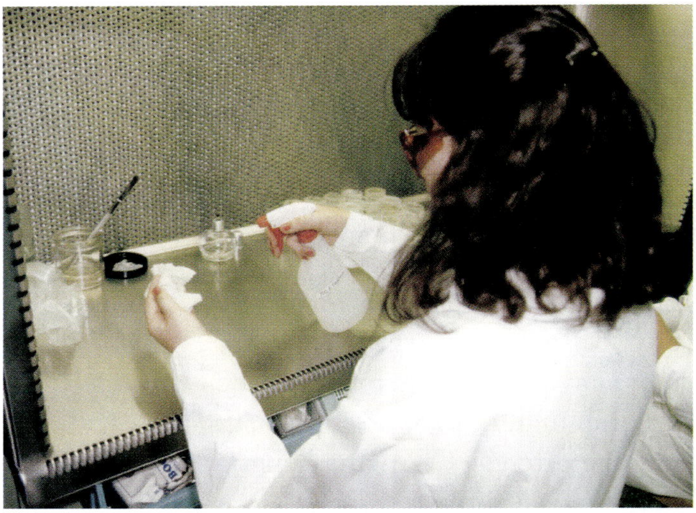

8-3. Preparing an aseptic work area in a flow hood to do tissue culture work. (The area is sprayed and cleaned with a mild bleach solution.)

8-4. Cutting an explant from a parent for a tissue culture. (The parent is a Venus fly trap.)

8-5. Placing an explant in a jar with agar (agar is the medium in the bottom of the jar).

8-6. Checking the growth of tissues after a few weeks of growth. (Features of the Venus fly trap are evident.)

germs. Laminar flow hoods help keep the proper environment. The surfaces and tools used are sprayed with a bleach solution to kill any germs. A plastic bag is sometimes substituted for a hood. Care is used to avoid getting germs into the work area.

Cuttings may be used to clone plants. This involves removing sections of stems or leaves from plants. The cuttings are placed in a growing medium, such as sand or soil. A damp, warm environment is provided. Roots are formed and the stem or leaf grows. Soon a new plant has

8-7. A plastic bag can be used in place of a laminar flow hood for small scale tissue culture work.

8-8. Air layering is used to clone some plants. (The plant grows roots inside the plastic, which contains soil or rooting material. The top is cut off and the roots placed in soil.)

formed. Cuttings work only with certain plants, such as roses and azaleas. The beautiful holiday season poinsettia is produced from cuttings.

Grafting is moving a small section of one plant onto another plant. The small section is a scion; the other plant is the stock. The two parts grow together. The scion bonds with the stock. The growth from the scion has the characteristics of its parent, not of the stock. Several methods of grafting are used. Whip-and-tongue, bark, cleft, and budding are most common. Many fruit and nut trees are grafted. This allows large fruits or nuts to grow on a stock that was inferior.

More advanced methods of cloning are possible. Embryo splitting is often used with livestock. The embryo is cut in half while only a few cells in size. It is about six days old. Each half can then

8-9. How a cleft graft is made.

AgriScience Connection

TISSUE CULTURE

Tissue culture uses a few cells from a plant to produce another plant. Much of the early work is done in an aseptic environment. Bacteria and fungi can readily grow on tissue cultures.

Follow these steps in an aseptic environment:

1. Remove small pieces of plant material from the parent plant. Use a sharp scalpel. (The small pieces are called explants.) (Be safe! Scalpels can cause injury.)

2. Clean the explant to remove bacteria or fungi. Use a small amount of the mild bleach solution for this purpose.
3. Place the explant on an agar medium in a glass test tube or bottle. Use an air tight cover on the jar or tube, as shown in the photo.

Place the jar or tube in a warm place with light. Observe daily for growth or changes in the explant. As growth begins, it will need to be transplanted into another medium. Hormones are used to encourage root growth. After a time, the plants will have roots and be large enough to set into soil.

become a separate animal. Other methods are being used experimentally with animals, such as sheep and monkeys.

Altering Growth

Many approaches are used to help plants and animals grow better. Some are common; others are experimental.

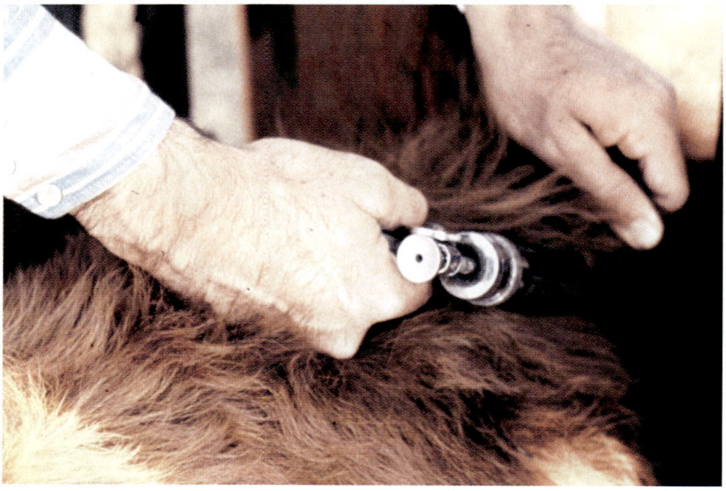

8-10. Placing a growth implant in a growing beef animal.

Growth hormones are used with both plants and animals. A ***hormone*** is a substance produced by an organism that has a specific effect. Giving an organism more of these hormones changes the rate of growth or production. One example is bST, which stands for bovine somatotropin. This is a hormone naturally present in cows. It influences milk production. By raising the amount of bST in a cow, milk production may increase 25 percent or more. Today, many dairy farms use bST with cows. Similar hormones are used to get more lean meat from hogs.

A ***growth implant*** is a small pellet made of hormones or hormone-like substances. It is placed just under the skin of animals—often behind the ear. The implant helps animals make better use of feed. Implanting beef cattle has been used for several years. It increases rate of growth.

Other methods are used to alter growth rate. Female fish can be injected with a hormone to get them to spawn (lay eggs). Other hormones can be used to make all of the fish in a tank the same sex. This helps raise them to market size in less time. Applying fertilizer to crops is another approach in organismic biotechnology.

MOLECULAR BIOTECHNOLOGY

Molecular biotechnology is changing the genetic makeup of an organism. It is often called ***genetic engineering***. Cell structures are changed. Genetic information is modified by moving genetic material from one organism to

Career Profile

GENETIC ENGINEER

A genetic engineer studies and creates transgenic organisms. The work is usually in laboratories with highly technical equipment. Patience is needed to reduce errors and wait for results.

Most genetic engineers have advanced graduate degrees in molecular biology, biochemistry, or a related area. They have a good background in biology, chemistry, and genetics. Begin preparing to study genetic engineering while in high school. If your interests are in crops, get practical work experience on a farm. In other cases, working in horticulture or with animals will be helpful.

Jobs are with biotechnology businesses, universities, and government agencies.

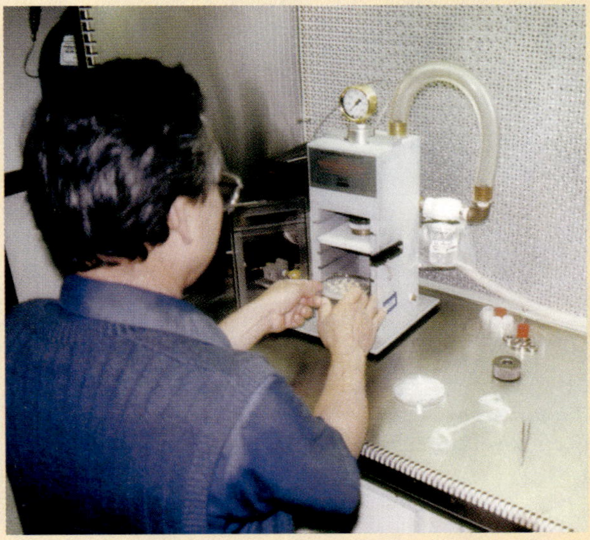

This shows a particle gun being used in genetic engineering. (Courtesy, Ronald J. Biondo, Illinois)

another. This is done in the tiny DNA of a cell. An organism with new traits results. These new traits have useful benefits.

The new organism is said to be transgenic. A ***transgenic organism*** is one that has new genetic material combined from two widely varying organisms. Its heredity has been changed. This is done to get an organism, such as a crop plant, that has desired traits. Several transgenic plants are now grown on farms.

One example is tomatoes. Fresh tomatoes do not keep very long in a refrigerator. A tomato that could be kept in a refrigerator longer would be more valuable. The MacGregor tomato was developed for this purpose. A gene from a flounder (a fish) was moved into the genetic material of a tomato. The resulting MacGregor tomato can be kept in a refrigerator longer without spoiling. The benefits have proven to be good.

8-11. The tomato was the first transgenic food plant. (Courtesy, Agricultural Research Service, USDA)

Another example is cotton. Cotton is a crop with many insect pests. Toxic pesticides have been used to control the pests. Some pesticides damaged the environment. Scientists have developed transgenic cotton that resists bollworms. No pesticides are needed to eliminate the bollworm. Here is how they did it: They knew that the bacterium, *Bacillus thuringinesis*, attacks and kills bollworms. The gene in the bacteria responsible for killing the bollworms was moved into the cotton gene. The resulting cotton plants are known as Bt cotton. (Bt is the initials of the kind of bacteria.) The Bt gene helps cotton plants repel attacks from bollworms.

8-12. Thousands of acres are now planted to Bt cotton. (Courtesy, Mississippi State University)

Some people are concerned about genetic engineering. They fear the results may bring future problems. Understanding the processes helps people make better decisions.

RECOMBINANT DNA

Genetic engineering is used to change the traits of an organism. Only those traits needing to be changed are altered. The work begins with the nature of cells and genetics.

HEREDITY MATERIAL

Cells are the building blocks of living things. They divide and grow and direct life processes. Cells contain important structures, including the genome.

The ***genome*** holds the heredity material. Heredity material passes the traits of parents to offspring. It determines the species of an organism and its unique traits, such as eye and hair color. Organisms can be changed by altering the material in the genome.

Genomes are made of chromosomes. A ***chromosome*** is a threadlike structure that contains genetic material and protein. Chromosomes are in pairs. The genetic material in chromosomes is in genes.

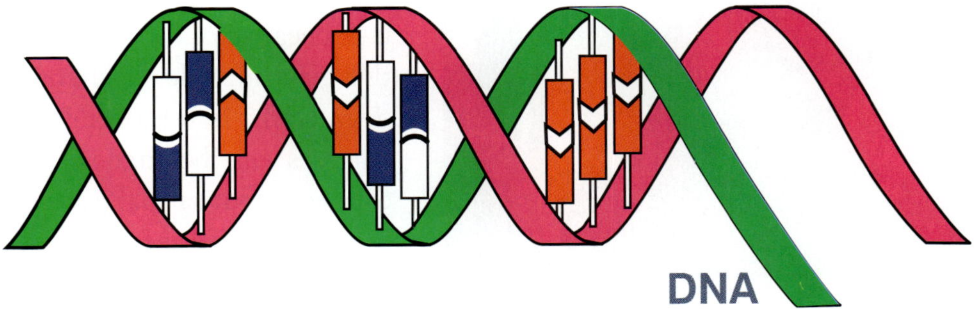

8-13. The general structure of DNA.

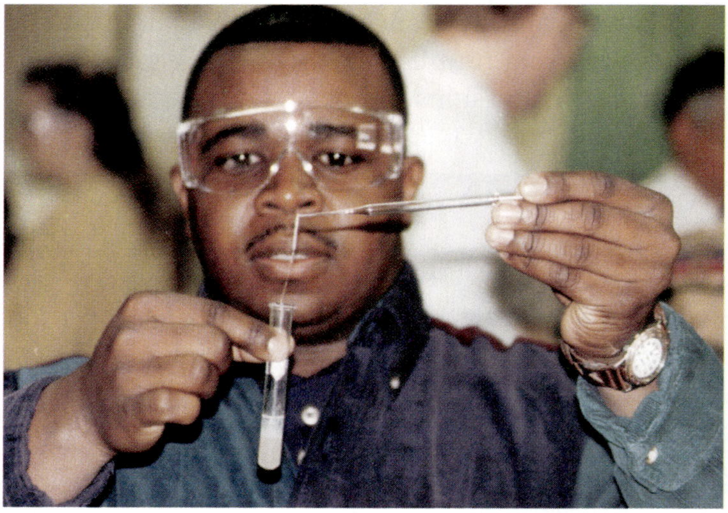

8-14. DNA is easily extracted from some cells. (This shows DNA from the thymus gland of a cow.)

8-15. The DNA sequence of a beef animal is being read. (Courtesy, Agricultural Research Service, USDA)

A ***gene*** is the unit of heredity on a chromosome. Genes contain the heredity traits of an organism. A gene is made of DNA (deoxyribonucleic acid). Many genes are found on a chromosome. Most genes have specific places on the chromosomes. Genes influence chemical and physical processes in an organism.

The information in all the genes in an organism forms its ***genetic code***. An organism is what it is because of its genetic code. If the genetic code is changed, the organism is changed. Before changes can be made, the genes that control a particular trait must be identified.

The process of locating and identifying genes by trait is known as ***gene mapping***. It is useful in replacing tiny segments with other segments. A replaced segment may change one or more traits of an organism. A good understanding of genome content is needed for genetic engineering.

RECOMBINANT TECHNOLOGY

Genetic engineering creates a new DNA form. It is known as recombinant DNA. ***Recombinant DNA*** has genetic traits that differ from the original DNA. This results from removing and replacing one or more genes in a chromosome.

Gene splicing is joining DNA from one organism with the DNA in another. The gene added is intended to give a desired trait, such as resistance to disease. Cells with new DNA are known as transformed cells. Getting the new DNA to join with the DNA in a cell is not easy. The quantities are very small. Many tries are usually made to be successful.

Genetic engineers use a device known as a ***particle gun*** or gene gun. The particle gun "shoots" tiny particles coated with DNA into cells. This is time consuming, expensive work. The particles are made of gold. Most of the attempts are not successful in making a gene splice. (The equipment is also known as a micro projectile unit.)

Considerable training is needed to create transgenic organisms. An in-depth knowledge of biology is essential.

BENEFITS OF BIOTECHNOLOGY

Biotechnology has many benefits. Of course, opinions differ. Not everyone agrees on the specific benefits. Most people do agree in some areas.

Five general benefits are listed below.

- Reduce Pollution—Plants that resist insects would not require the use of pesticides. This would stop the release of these materials into the environment. Pollution of the air, soil, and water would be reduced. Beneficial insects would not be killed by pesticides.
- Improve Food—Food from crops grown without pesticides would not have residues. The food would be more wholesome. Some people feel that pesticides cause disease. Not using pesticides should get rid of any concern.

8-16. Wind has blown a pesticide onto these leaves. Pesticides would not be needed with transgenic plants.

- Conserve Resources—Fuel, equipment, and labor are needed to apply pesticides to crops. Growing crops that do not require pesticides would save money. It would also help conserve limited supplies of fuel and other natural resources. Pollution of the air by equipment exhaust emissions would be reduced.
- Reduce Hunger—Better crops should result in more food. This will help solve food shortages. People who do not get enough to eat might have the food they need.
- Improve Health—Some biotechnology processes make substances that promote good health. Enzymes and other substances from animals can be used to promote human health.

ISSUES WITH BIOTECHNOLOGY

Using biotechnology is acceptable to some people. It has useful benefits. Other people are concerned about using it. One approach is to study the issues associated with biotechnology.

An ***issue*** is a question or problem that has more than one possible answer. Several issues surround some uses of biotechnology. Good information is needed about issues. The findings of research are often best. People are sometimes emotional and fail to think clearly. Deciding an issue without good information may result in the wrong decision.

8-17. People make better decisions if they are informed about biotechnology. (Courtesy, Agricultural Research Service, USDA)

ISSUE ONE: DO NEW LIFE FORMS POSE DANGERS?

Some people fear what new life forms will do. Scientists carefully study new forms before suggesting their use. A drawback is that the long-term dangers are not known. People who fear biotechnology think that "monsters" may be released. You can decide for yourself. Get the facts first!

ISSUE TWO: WILL FOOD FROM NEW LIFE FORMS BE HARMFUL?

Some people fear that food from transgenic organisms is not healthy. They fear that those who eat it and prepare it are in danger. Scientists do not know what the source of the danger could be. If handled properly, is not a slice of tomato a slice of tomato even if it is from a transgenic tomato plant? You can decide for yourself. Get the facts first!

ISSUE THREE: SHOULD ORGANISMS BE KEPT NATURAL?

Some people are concerned that transgenic organisms are not natural. They do not think that people should alter organisms. Their thoughts may be based on religious beliefs. Other people do not see problems with transgenic organisms. They think it is okay if the resulting plants and animals are made better. You can decide for yourself. Get the facts first!

ISSUE FOUR: IS IT RIGHT TO JOIN PLANTS AND ANIMALS?

Some people are concerned about "moving" animal "parts" into plants and vice versa. They feel that a plant should be "all plant" and that an animal should be "all animal." Genetic engineering may move selected genetic material from one to the other. With care, only the desired trait is moved. Lab testing is done to assure that the "move" was okay. You can decide for yourself. Get the facts first!

ISSUE FIVE: SHOULD TRANSGENIC PRODUCTS BE LABELED?

Some people feel that they have a right to know if products are from transgenic sources. They want such products labeled. Other people are not concerned. They want good products no matter the source. You can decide for yourself. Get the facts first!

ISSUE SIX: SHOULD PEOPLE BE FRIGHTENED BY BIOTECHNOLOGY?

Some people fear new things. They are not comfortable in changing to new ways of doing things. Biotechnology is a big concern to some people. Other people readily use new things. They are the first to try something new. Part of this issue deals with being well informed. Reading and learning help in making good decisions. You can decide on this issue for yourself. Get the facts first!

REVIEWING

MAIN IDEAS

Biotechnology is using science to change living things. Biology is the most important area of science. Some biotechnology uses organisms without changing their genetic material. This is known as organismic biotechnology. Cloning, tissue culture, and grafting are examples. Using growth hormones to increase production is another use.

Molecular biotechnology changes the genetic material in an organism. It is sometimes known as genetic engineering. A good knowledge of cell structure and genetics is needed. The structure of chromosomes and genes is important. Gene mapping is used to locate genes by the trait they influence. Once this is done, gene splicing is used to insert a small segment of new DNA. Particle guns are used in gene splicing.

Biotechnology is controversial. Several major benefits exist. Some people are concerned about how it is used. Several issues have arisen. These are best solved when people get the facts. Good decisions are made when people have the needed information. The best decisions are made by not getting emotional.

QUESTIONS

Answer the following questions. Use correct spelling and complete sentences.

1. What is biotechnology?
2. What is cloning? What are three ways cloning is used with plants?
3. What is a hormone? Give one example of how hormones have been used to increase production.
4. What is a transgenic organism?
5. What are two examples of transgenic crops? Why are these useful?
6. Where is heredity material?
7. What does a genome contain?
8. What is gene splicing?
9. What are the benefits of biotechnology?
10. What issues are associated with biotechnology?

EVALUATING

Match the term with the correct definition. Write the letter of the term in the blank provided.

a. cloning	d. growth implant	g. gene mapping
b. grafting	e. chromosome	h. gene splicing
c. tissue culture	f. gene	

____ 1. Segment or unit on a chromosome.

____ 2. Transferring a small section of one plant onto another.

____ 3. A small pellet containing hormones that is placed under the skin of an animal.

____ 4. Making two or more organisms out of one.

____ 5. Joining the DNA of one organism with that of another organism.

____ 6. Locating and identifying genes by trait.

____ 7. A form of cloning that uses cells or small clusters of cells.

____ 8. Threadlike structure that contains genetic material.

EXPLORING

1. Use tissue culture to produce a new plant. Follow the proper procedures in an aseptic environment. Have your instructor demonstrate how to make a tissue culture. Several weeks may be needed to determine your success. For more information, use references listed in the bibliography on horticulture or plant and soil science. Be sure to follow all safety rules in making a tissue culture.

2. Make a graft. Before you begin, study the different kinds of grafts and how they are done. Select a plant that can be successfully grafted. Use any method of grafting that is appropriate for the scion and stock you have selected. Have your instructor demonstrate methods of grafting. Be sure to follow all safety rules in grafting. For more information, refer to the bibliography for books on horticulture. (Note: If you cannot do a graft, prepare a poster that illustrates one method of grafting.)

3. Prepare a report that addresses an issue in using biotechnology. Examine all sides of the issue and take a stand. Defend your stand with facts. Make an oral report in class.

9

LIVING BETTER THROUGH AGRISCIENCE

The United States has a major role in world agriculture. New technology has been developed and shared worldwide. Trade among nations is at an all-time high. Export markets create jobs and income from agriculture. People around the world are living better than ever before!

Agricultural development began with the first Pilgrims. Fertile soil, good water supplies, and other natural resources were important. The Pilgrims had to learn how to grow crops and raise animals in their new land. They brought seed and farming practices with them from their old land. They discovered new crops and practices in their new land. Together, the old and the new, resulted in a highly productive agriculture.

The early settlers endured hardships and worked long hours to overcome adversity. The work ethic was an important part of survival. People worked hard and gained personal satisfaction from their work.

9-1. Horses have long been important for work and pleasure. This is a jumping event at a horse show.

OBJECTIVES

This chapter describes the importance of agriculture to the people on the earth. It covers world population trends and trade. The objectives of this chapter are:

1. Identify global nutrition and health trends
2. Describe global differences among nations
3. Explain world population trends
4. Describe trends in the global food industry
5. Explain the role of agriculture in international relations

TERMS

Codex Alimentarius
commodity
culture
customs
developed country
developing country
FDA
GATT
global economy
hormone residues
hunger
malnutrition
nutritional deficiency
pesticide residue
plantains
sub-Saharan Africa
tariff
trade balance
trade barriers
WTO

http://www.IPPINC.com/

(This site has information on books and other materials in agriculture, environmental science, and other areas.)

GLOBAL NUTRITION AND HEALTH

Important questions need to be answered to feed the world population in the future. Who is responsible for having enough food? How can we predict

9-2. People need food to enjoy life.

the amount of food that will be needed? How do we ensure that nutritional foods will be available to all people?

FOOD DEFICIENCIES

Agriculture is the basis for the nutritional health of people. The main food groups are bread, fruit, vegetables, milk, and meat. All involve agricultural products. The absence of these foods will result in deficiencies. A ***nutritional deficiency*** is a lack of proper food nutrients.

Food deficiencies can cause ***malnutrition***. Disease can result from malnutrition. One example is scurvy.

9-3. Vegetables are good sources of nutrients.

9-4. Fresh fruit is promoted around the world as a good source of Vitamin C. (Courtesy, John Richardson, North Carolina State University)

Scurvy is a disease that results from a lack of Vitamin C in the diet. Vitamin C is found in many fruits and vegetables.

Hunger is discomfort caused by a need for food. Many people in the world are hungry. They cannot get enough food to eat. A big concern in the 1960s was the human population was increasing too fast. People feared starvation.

AgriScience Connection

FFA MAKES A WORLD OF DIFFERENCE

The National FFA has international programs for students in high school. FFA members can travel to other countries to live and learn about agriculture.

A Summer Homestay to England or Australia is popular. It involves living three weeks with a farm family. The FFA Explorers Program is a six-week, three-month, or six-month experience. Some 25 countries are available. Students learn the language and culture. Many have youth projects while living with host families.

Other opportunities exist for travel in the FFA. The winners of national FFA contests are awarded trips. Special touring trips are also available. Contact the National FFA Center for more information. This shows students attending the Green Week agricultural fair in Germany. (Courtesy, Kristen Effle, North Carolina FFA Association)

THE GREEN REVOLUTION

New crop production methods made it easier to grow food. Better ways of farming were discovered. This led to increased production. All of these developments were called the Green Revolution. Because of the Green Revolution, there was less hunger by the 1980s.

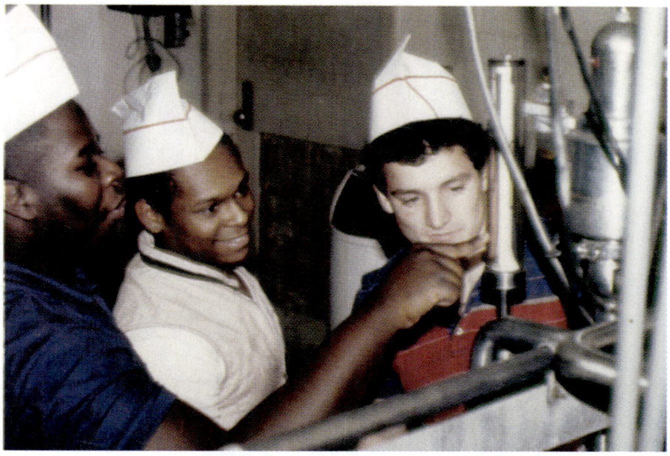

9-5. People must have jobs and income to earn money to buy food. (Courtesy, George Bostick, North Carolina State University)

9-6. Rice is a major food in many countries. (Courtesy, Agricultural Research Service, USDA)

Today, there is enough food grown for everyone. Unfortunately, hungry people cannot always get the food they need. This is usually in developing countries. These countries have primitive agricultural methods and poor transportation. This limits the available food. These countries often have climates and soils that are not good for farming. Many people do not work. Without jobs and income, people do not have money to buy food.

INCOME AND FOOD CHOICES

More than one billion people in the world earn less than one dollar a day. People with low incomes buy low

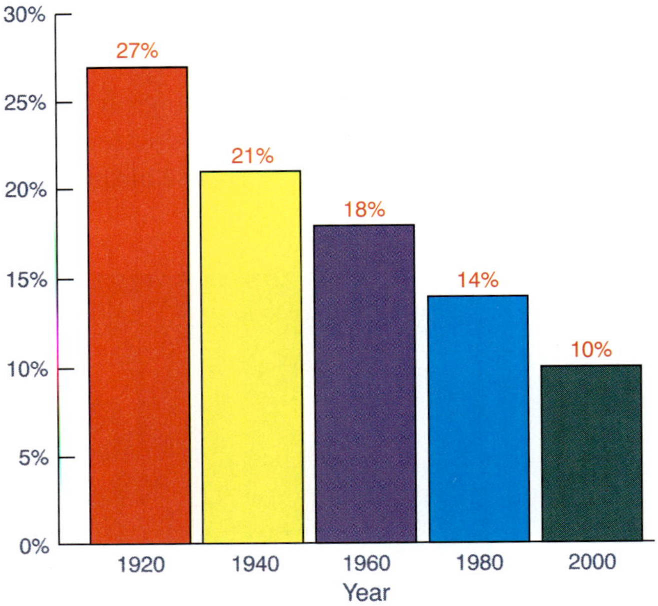

9-7. Percent of income spent on food in the United States.

9-8. Research finds better ways to grow crops in different climates and soils. (Courtesy, Agricultural Research Service, USDA)

cost foods. Examples of low cost foods are rice and potatoes. These foods are nutritious but some variety is needed.

Food choices are based on what is available. The most available foods are grown locally and are cheaper. For example, roots, tubers, and plantains are low cost foods in the sub-Sahara. ***Plantains*** are fruits in the banana family. They are usually cooked before eating. ***Sub-Saharan Africa*** is part of the African continent south of the Sahara desert.

When people have more money for food, their food choices change. They will usually eat more variety. Meat and other nutritional foods are eaten. In some countries, people must spend most of their limited income on food.

Countries with advanced food systems can produce more food cheaply. Therefore, their citizens spend less money on food and have more food choices.

Human nutrition is usually improved when more food is available. To have more food, developing countries need to put more resources into agriculture. They must invest in research, mechanization, and technology to increase food production.

GLOBAL DIFFERENCES

Differences among nations are often described as developed and developing. A ***developed country*** has jobs so people can work and earn money. The country is said to be industrialized. North America, Europe, and Australia are developed areas.

A ***developing country*** is not industrialized. Many people have low incomes. There are few jobs. People barely get by. Many are sick due to a lack of food. Modern conveniences are lacking. Good drinking water is not widely available.

For every developed country, there are three developing countries. Forty-five countries are said to be developed; 125 are developing.

9-9. People in all nations need food.

DEVELOPED COUNTRIES

Developed countries have advanced food systems that are efficient and productive. They are more successful in agriculture for many reasons. Here are a few:

- Technology—Science and research provide new supplies, methods, and products for agriculture.

- Knowledge and Education—Farmers and others in agriculture have a broad knowledge of many subjects. They know and use science in their work.
- Marketing—Good transportation, storage facilities, and processing plants exist to get food to the consumer.
- Mechanization—Advanced machinery is used to cultivate, plant, and harvest crops.
- Improved Crops and Animals—Plants and animals with superior qualities, such as vigor and disease resistance, are produced.
- Environmental Protection—Soil, water, and other natural resources are conserved.
- Soil and Climate—Desirable soil and climate are available for producing plants and animals.

9-10. Developed countries use power machinery to do work. (This shows a self-propelled rotary-disc windrower in an alfalfa field.) (Courtesy, AGCO Corporation)

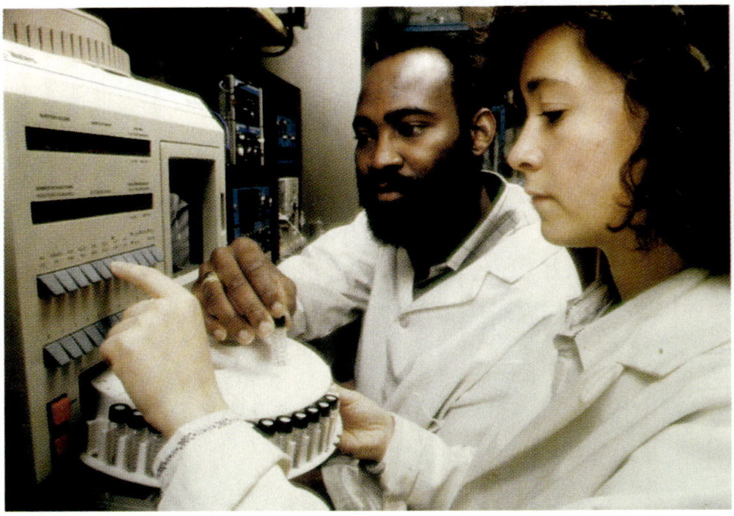

9-11. Technology is the result of research. (Courtesy, George Bostick, North Carolina State University)

CUSTOMS AND CULTURES

Food and clothing vary from one country to another. These differences are a part of the cultures and customs. ***Customs*** are long established ways of doing things. Customs, which create the ***culture***, are passed from one generation to the next. Many customs, such as celebrations, are a result of agricultural successes. Fall festivals and Thanksgiving in the United States are such customs.

The foods we eat are also part of our customs and cultures. In the past, people ate locally grown foods that became part of their culture. For example,

Career Profile

INTERNATIONAL AGRICULTURAL MARKETING SPECIALIST

International agricultural marketing specialists are involved in trading agricultural products by countries. They need to know the laws dealing with international trade. They should have a background in agriculture and business.

An international agricultural marketing specialist travels to other countries. He or she must speak foreign languages and know the customs. A college degree in economics, agribusiness, international relations, or similar area is needed.

Careers in this field are usually in large corporations that export or import food. Some jobs are also with federal agencies that regulate trade, such as the FDA. Good communication skills are essential in international agricultural marketing. (Courtesy of Marshall Stewart, North Carolina State University)

people in Asia often eat more rice and duck. These foods are associated with their culture.

Customs teach us food preferences at an early age. These preferences affect our food choices. Food choices have an influence on the marketing of foods.

9-12. A local Japanese market sells customary foods. (Courtesy of Marshall Stewart, North Carolina State University)

In developed countries, people have more food choices due to higher incomes. Their incomes allow them to try different cultural foods, such as Chinese and Mexican food. This creates new import markets.

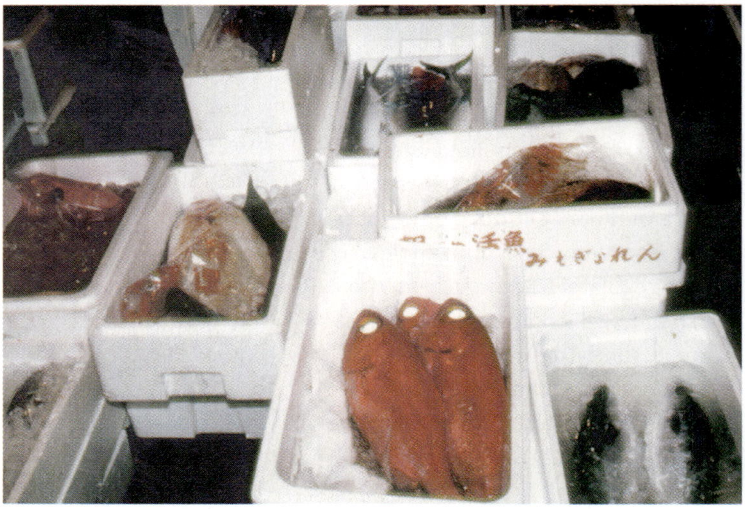

9-13. Seafood is popular in Asia. Many types are imported into other countries. (Courtesy of Marshall Stewart, North Carolina State University)

WORLD POPULATION

Today, the world's population has many cultural, racial, educational, and religious groups. If there were only 100 people on the earth with all existing cultures and races in the same proportion:

- 57 would be Asians
- 21 would be Europeans
- 14 would be Western Hemisphere people (North and South America)
- 8 would be Africans
- 70 would be non-white; 30 white
- 70 would be non-Christian; 30 Christian
- 50 percent of the entire world wealth would be in the hands of only 6 people
- All 6 would be citizens of the United States
- 70 would be unable to read
- 50 would suffer from malnutrition
- 80 would live in substandard housing
- 1 would have a college education
- 51 would live in a city
- 79 would live in a developing country

9-14. Over half the people in the world are of Asian descent. (Courtesy, Marshall Stewart, North Carolina State University)

World Population Milestones
World Population reached:
1 billion in 1804
2 billion in 1927 (123 years later)
3 billion in 1960 (33 years later)
4 billion in 1974 (14 years later)
5 billion in 1987 (13 years later)
6 billion in 1998 (11 years later)
World Population may reach:
7 billion in 2009 (11 years later)
8 billion in 2021 (12 years later)
9 billion in 2035 (14 years later)
10 billion in 2054 (19 years later)
11 billion in 2093 (39 years later)

9-15. World population milestones. (Source: United Nations, New York)

9-16. Large cities have many people in a small area.

This view of our world today may be surprising. Often, we view the world only by what we see in our own neighborhoods. Agriculturists must think about how many people will live in the future. They must consider the number of people who will need food.

Today, 5.6 billion people live on the earth. By the year 2093, the population is predicted to increase to 11 billion. Most population increases will occur in the developing countries, such as China, sub-Saharan Africa, India, and Central America. The developing countries will account for 93 percent of the total population increase.

Fewer children are being born in the world today than a few years ago. More are surviving to adulthood. More people are now in the childbearing ages. These people may have fewer children but more people are having children. Advancements in nutrition and health care are also allowing people to live longer.

France, Italy, and Japan have experienced a population decline. By the year 2040, the number of people in the United States will also begin to decline. However, California, Texas, and Florida are expected to continue to increase in population as people move to these states.

China is the country with the most people. It is followed by India and the United States. Japan is one of the most populated nations in relation to land area. Japan imports much of its food. This is because land is limited, which restricts the ability to grow food. When a country has a natural disaster or a political event, such as war or revolution, food production may be stopped or slowed. In the 1990s, Russia experienced this phenomenon and had to rely more on imported food.

Table 9-1. Twenty-five Countries with the Largest Population

Rank	Country	Population July 1996 (estimated)
1	China	1,215,609,480
2	India	953,122,675
3	United States	266,504,935
4	Indonesia	206,759,795
5	Brazil	162,698,486
6	Russia	150,208,907
7	Pakistan	133,225,657
8	Bangladesh	131,066,751
9	Japan	125,908,113
10	Nigeria	104,431,190
11	Mexico	95,771,579
12	Germany	81,549,019
13	Vietnam	75,665,450
14	Philippines	74,899,407
15	Iran	66,105,378
16	Turkey	64,654,615
17	Egypt	63,575,636
18	Thailand	61,018,664
19	United Kingdom	58,452,516
20	Italy	58,384,321
21	France	58,376,462
22	Ethiopia	57,708,770
23	Ukraine	51,888,575
24	South Africa	46,272,450
25	South Korea	46,027,642

(Source: World Factbook, 1996)

THE GLOBAL FOOD INDUSTRY

The United States exports more agricultural goods than any country. Products that are hard to produce are imported. The United States is the third largest importer of goods.

9-17. Large ships are used to transport food products.

The United States exported $28 billion more in agricultural goods than it imported in a recent year. The difference between the amount a country ex-

9-18. Barges are often used to transport grain. (Courtesy, Cargill)

ports and imports is ***trade balance***. A positive trade balance is good for a country's economy. "Positive" means that more is exported than imported. The United States sometimes has a negative trade balance for non-agricultural goods. In a recent year, $180 billion more non-agricultural goods were imported than exported. The rate of increase in agricultural exports declined in the late 1990s. This was due to poor economic conditions in Asia.

The ***global economy*** results from the trade between countries. Various events influence agricultural trade. Drought in one country can create a demand for food because less is available. High energy costs can cause prices to go up because the use of machinery and transportation would cost more.

9-19. Cultural traditions influence food preferences. This shows Polynesian men attending to a luau barbecue pit.

Asian countries are now consuming more beef, beverages, and prepackaged foods. This demand creates an opportunity for the United States to export more of these products. For example, poultry is often exported. The processed food products are prepared to meet specific cultural demands.

Agreements influence trade because they set the rules for trade, including taxes to be imposed. These factors influence the global economy.

INTERNATIONAL RELATIONS

Agriculture is important in international relations. International relations are relations among countries. Many countries rely on each other for most of their food and food choices.

TRADE BARRIERS

Trade agreements between countries are often a result of political harmony. Sometimes countries will have trade barriers. ***Trade barriers*** are government policies of a country that slow or stop trade with other countries.

The United States has used trade barriers with countries that did not respect human rights—not treating people fairly. Human rights are basic freedoms all people should have. Our country has expressed belief in human rights by creating trade barriers against countries who were oppressing groups of people.

Trade barriers are also used when countries do not agree on health and sanitary rules. Food sanitation is important for the safety of those who eat the food.

Trade barriers are used by a country to protect the profits of farmers in their own country. By not allowing their country to import products, people in the country must buy products from their own farmers. This may only be good in the short run. Free trade promotes competition. Without competi-

9-20. Proper food packaging and processing are important. (Courtesy, George Bostick, North Carolina State University)

tion, consumers may not get good value for their money. Free trade is usually good for the economy.

FOOD SAFETY

An international group called ***Codex Alimentarius***, meaning food code, sets food and commodity standards. A ***commodity*** is an agricultural product, such as grain and fiber. The standards are rules about such things as levels of ***pesticide residue***. Pesticide residue is a small trace of a chemical that remains on a product when it is sold. These chemicals are used to kill insects and weeds in growing the crop. If eaten in high levels, pesticides can be dangerous to health.

9-21. Food products must meet standards to be traded. (This shows lemons being inspected.) (Courtesy, U.S. Department of Agriculture)

The U.S. Food and Drug Administration (***FDA***) is responsible for preventing high levels of pesticides and hormones in imported foods. ***Hormone residues*** are traces of chemicals that regulate activities in the body. Some animals are given synthetic hormones to make them grow faster. The FDA monitors imported foods by testing samples of food. If high levels are found, the food is not imported. Often, the FDA standards are stricter than the Codex standards.

These rules or standards have a big influence on food safety and global trade. Countries must follow the rules if other countries buy their products. This makes trade fairer among countries. Products must be up to the standards of quality.

9-22. Foods are tested for levels of pesticides and other substances. (Courtesy, Agricultural Research Service, USDA)

TRADE AGREEMENTS

9-23. Coffee, a traded commodity, is being graded. (Courtesy, Coffee, Sugar, and Cocoa Exchange, New York)

In the past, trade barriers and tariffs have been used for many reasons. A ***tariff*** is a tax placed on imports or exports. Countries may create barriers and tariffs to encourage their people to buy their own goods. The price of imported goods will be higher because of the tariff.

The General Agreement on Tariffs and Trade (***GATT***) and the World Trade Organization (***WTO***) Agreement are agreements between countries to promote trade. These agreements encourage trade and discourage trade restrictions. All laws are applied equally to all countries. The WTO Agreement can be used to settle trade disputes.

Removing trade barriers benefits consumers. More choices are available at competitive prices. Fair competition encourages trade around the world. U.S. agricultural exports will likely continue to increase. This will create more farm income and jobs.

REVIEWING

MAIN IDEAS

Agriculture is important to the nutritional health of people. Unfortunately, many people in developing countries have malnutrition. The people may not have income to buy food. Developing countries often lack technology and transportation. These are necessary for good commercial agriculture.

The world population is made of a variety of cultures, races, and religious groups. The majority of people in the world are of Asian descent, are non-white, are unable to read, live in cities, and suffer from malnutrition. World population is expected to double by the year 2093 or before. Most of this growth will be in developing countries.

Agriculture is a vital part of the global economy. Countries import and export agricultural commodities. The United States is the largest exporter of agricultural goods and the third largest importer of goods.

The global economy is the result of trade between countries. Trade agreements may set the standards and rules for trade. These rules have a big influence on the global economy. They often eliminate trade barriers and unfair practices. This allows more free trade. Free trade promotes competition. Competition promotes the production of quality goods at their cheapest price.

QUESTIONS

Answer the following questions. Use correct spelling and complete sentences.

1. What is the result of poor human nutrition?
2. How does income influence what people buy?
3. List seven reasons why developed countries are more successful in agriculture.
4. How do customs and cultures affect consumer food choices?
5. What is the predicted world population trend in the twenty-first century?
6. Why are people living longer?
7. Why does the United States have a positive agricultural trade balance?
8. Explain two reasons trade barriers are imposed by countries.
9. What can be done to be sure that food imported into the United States is safe?
10. What is GATT?

EVALUATING

Match the term with the correct definition. Write the letter of the term in the blank provided.

a. commodity	d. sub-Saharan	g. GATT
b. trade barriers	e. malnutrition	h. FDA
c. tariff	f. global economy	i. Codex Alimentarius

____ 1. An international body, meaning food code, that has developed food and commodity standards.
____ 2. A product that is bought and sold.
____ 3. A federal agency called the Food and Drug Administration.
____ 4. The General Agreement on Tariffs and Trade, which is a multinational agreement.
____ 5. A system of trade between countries.
____ 6. The state of receiving poor nutrition.
____ 7. A part of the African continent south of the Sahara desert.
____ 8. A tax placed on imports and exports.
____ 9. Can slow or stop trade between countries.

EXPLORING

1. Research the agriculture system of a developing country. Use materials available in the school library. Interview a person who has been involved in international work. Use the Internet and World Wide Web for information. Prepare a written report on your findings.

2. Explore the culture and customs of a country. What are the favorite food dishes? Trace them back to the agriculture industry. Use resources in the school library. Interview people who have traveled in the country.

3. Make a copy of a world map and highlight the developing and developed countries. What are their geographical similarities? Use resources in the school library to help with your work.

4. Read the newspaper and search for current events dealing with importing and exporting trade. Prepare a news review to give to your class in the form of a radio announcement.

5. Ask your teacher and guidance counselor for information about jobs dealing with global trade. What skills do these jobs have in common?

10

NATURAL RESOURCES

Agriscience uses many natural resources. Plants and animals require these resources to grow. Sometimes we take natural resources for granted. We should not! We need to know about these resources and how to use them wisely. Wasting them may cause shortages in the future. No one wants to run out!

We can take steps to sustain our resources. Every person has a role. All of us can be smart users of resources!

10-1. Many people enjoy watching animal wildlife. (Courtesy, U.S. Fish and Wildlife Service)

OBJECTIVES

This chapter is about the meaning and importance of natural resources. It includes how to sustain the resources and compares conservation and preservation. The objectives of this chapter are:

1. List and describe major kinds of natural resources
2. Classify natural resources as renewable or nonrenewable
3. Explain sustainable resource use
4. Compare and contrast conservation and preservation

TERMS

- air
- conservation
- endangered species
- erosion
- extinction
- feral animal
- fossil fuel
- game
- habitat
- mineral
- nonrenewable natural resource
- preservation
- recycling
- renewable natural resource
- soil
- soil conservation
- sustainable resource use
- water conservation
- wildlife

http://www.ncg.nrcs.usda.gov/

(This site has a wealth of information on the conservation of natural resources.)

KINDS OF NATURAL RESOURCES

Natural resources are the things found in nature that humans use for their benefit. Humans cannot make these resources. They are available to meet human needs. Wise use is essential. If wasted, the resources will be gone and none available in the future.

All natural resources fit into seven groups. These groups are a part of agriscience in some way.

10-2. The clear water in the Soque River of Northeast Georgia is refreshing.

SOIL

Soil is the outer layer of the earth's surface. It supports terrestrial plant life. The contents of soil vary. Soil high in the nutrients that plants need is

10-3. The fertile soil in this California field helps lettuce grow.

most useful for crops. Materials known as fertilizers are added to soil to increase fertility. Fertile soil has more nutrients than infertile soil.

Soil is made of sand, silt, and clay. The amounts of these vary. The amounts determine how a soil is best used. Soil also contains decaying remains from plants and animals. Many organisms, such as earthworms and bacteria, live in soil. The presence of these organisms in soil determines how well plants will grow.

WATER

Water is found in nature in three forms: liquid, ice (solid), and gas (vapor). The liquid form is found in streams, lakes, oceans, and aquifers deep in the earth. Much of the ice is in large areas around the North and South Poles. These areas are known as ice caps. Some of this ice is thousands of years old.

Water vapor is in the air in all places on the earth. It is measured as humidity. The air near streams and lakes has more water vapor than in desert areas. Steam is a kind of water vapor that has special uses. It is important in cooking food and providing power for manufacturing.

10-4. Lake Tahoe is a freshwater lake. (This shows Fannette Island in the Emerald Bay area of Lake Tahoe.)

Much of the earth's water contains salt. In fact, 97 percent of it is saltwater. Only 3 percent is freshwater, with two-thirds of the freshwater being frozen in the ice caps or deep in the earth. This leaves only 1 percent of the earth's water in a form that humans can readily use! Saltwater cannot be used to drink, irrigate crops, or in most manufacturing.

All living things need water. Without plenty of good water, plants and animals may fail to grow. If no water is available, death is certain. Most drinking water is from streams, collected runoff, and deep in the earth. Purifying large amounts of saltwater is too costly.

The earth has a way of cleaning water after it has been used. If water is too dirty, the earth may be unable to clean it adequately. Water pollution is a major problem in some locations.

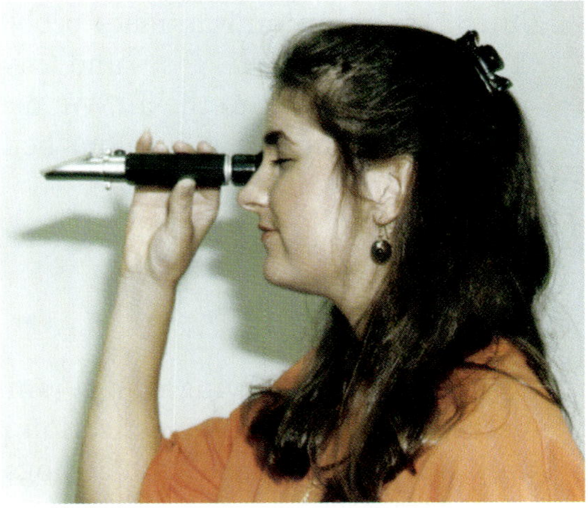

10-5. A salinity refractometer can be used to learn if water contains salt.

AIR

Air is the mixture of gases that covers the earth. It is found everywhere on the earth. The gas mixture contains nitrogen, oxygen, carbon dioxide, hydrogen, argon, and others. More nitrogen is found in the air than any other gas. Most air is 78 percent nitrogen. The air has 21 percent oxygen. Other gases are only 1 percent combined. Dust, water vapor, and pollen are in most air.

Some gases in the air are required for life. Animals need oxygen. Plants need carbon dioxide and oxygen. If these are not present, life cannot go on. Using air changes it. As we know, humans use oxygen from the air and release carbon dioxide. Other organisms are similar. Plants help keep nature in balance. They use carbon dioxide and release oxygen.

10-6. Pollution from a factory is going into the air.

Air has no taste, color, or odor. If it has either or all these, it is polluted. Air pollution is a major problem. Engines release exhaust into the air. The exhaust contains gases and tiny solid particles. Many other activities release substances into the air. We need to be careful about what gets into the air. We may breathe it into our bodies! Some things will make us sick and cause bad diseases.

WILDLIFE

Wildlife is all living things (animals, plants, and others) that have not been domesticated. Domestication is bringing organisms under human control. All farm animals, such as hogs and chickens, were once wild. They are now domesticated. The same is true with our pets and work animals.

Wildlife is important in many ways. Some people think of wildlife only as game animals and the sport of hunting. It is much more than this. Wildlife adds to the balance of life. It helps with natural processes and provides benefits to humans.

Animal Wildlife

Animal wildlife is all animals that have not been domesticated or tamed. Occasionally, individual domesticated animals return to the wild. These are known as feral animals. A ***feral animal*** has gone back to being wild. Feral dogs and swine are common.

10-7. The wild horses on Chincoteague Island, Virginia, are feral—once domesticated. (Courtesy, U.S. Fish and Wildlife Service)

Career Profile

GAME MANAGEMENT TECHNICIAN

Game management technicians work to protect wildlife and natural resources. They set up habitat, monitor wildlife activity, and conduct educational programs. They may also help enforce laws to protect wildlife.

Most game management technicians need college degrees in wildlife management or related area. Some get degrees in environmental science or biology. Advanced masters or doctors degrees are helpful for promotion. Begin planning in high school by taking science classes. Agriculture classes that focus on the environment or wildlife are helpful.

Most jobs are with government agencies, including parks and wildlife conservation districts. Private businesses that use forests or natural resources may employ game management technicians.

Animal wildlife may be terrestrial or aquatic. Terrestrial animals live on the land. Examples are rabbits, deer, birds, and bears. Aquatic animals live in water. Some need saltwater; others need freshwater. Trout, shrimp, oysters, and whales are examples of aquatic animals.

10-8. Fish, such as this yellow perch, are popular aquatic wildlife. (Courtesy, U.S. Fish and Wildlife Service)

Insects have unique roles as wildlife. Many do useful roles, such as the honey bee. Others are pests, such as the armyworm and wasps. Some are known for their beauty, such as butterflies and moths.

Enjoying Animal Wildlife. People enjoy animal wildlife. In doing so, we need to respect what they need to live. Keeping forests and streams clean helps wildlife. In some cases, we can provide feed. Many people put up bird or squirrel feeders. These attract wildlife and provide food for the winter. Nesting boxes help birds find a place to raise young.

Some wildlife animals are hunted. Species hunted are known as ***game***. Game animals may be hunted for food or merely as a sport. Laws exist to protect animals. The laws tell what species can be hunted and when hunting can occur. The laws may limit the number that can be killed. Laws also tell how animals may be taken, such as with guns, traps, and primitive weapons. Every sport hunter should obey the laws. Common game animals include squirrels, rabbits, ducks, and pheasants.

People enjoy animal wildlife in many ways without destroying them. Some people enjoy watching animals. Others like to hear the sounds or songs they make. A few people enjoy photographing animals.

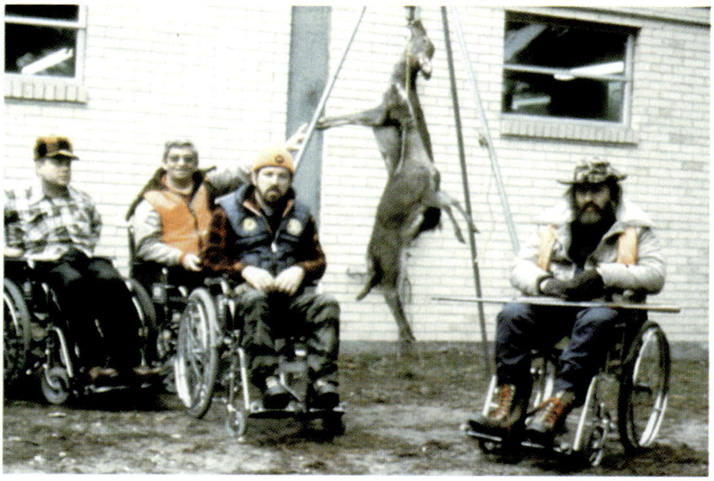

10-9. These hunters are proud of the game they have taken. (Courtesy, U.S. Fish and Wildlife Service)

Endangering Wildlife. People sometimes take advantage of animal wildlife. They kill too many and destroy their homes. Fear develops about the future of a wildlife species. Sometimes, animals become extinct. ***Extinction*** is the complete disappearance of organisms from the earth. Once extinct, an animal cannot be brought back. It is gone for ever. Examples of extinct animals include the passenger pigeon and the dinosaurs.

10-10. A helicopter is being used to lift a gray wolf into the protected area of Yellowstone National Park. (Courtesy, U.S. Fish and Wildlife Service, Luray Parker)

A few animals are endangered. An ***endangered species*** is one threatened with becoming extinct. Their numbers are very low. If not protected the animal species would become extinct. Nearly 400 species have been listed as endangered in the United States. Examples of endangered species are the whooping crane and gray wolf. It has been estimated that 90 percent of all species that have lived on the earth are now extinct!

Plant Wildlife

Plant wildlife includes wild flowers, vines, and trees that are growing naturally. Many people enjoy wildflowers. They go hiking into meadows and woods to see them.

10-11. Mountain Laurel flowers are a sign of spring in the Blue Ridge Mountains.

10-12. An old growth forest in Mount Hood National Forest in Oregon has many kinds of plant wildlife.

Plants are threatened much as wildlife animals. Some have become extinct; others have been endangered. For example, the pitcher plant and Texas snowbell plant are endangered species. Wildlife plants can be protected by steps to assure their growth.

Where wildlife plants grow is their habitat. ***Habitat*** is the physical area where an organism lives. It contains the essentials for life. Because of the close relationship between plants and animals, maintaining plant habitat also provides habitat for animals. Leaving strips of land unplowed or not developed into a shopping center provides habitat for some species.

Other Organisms

Many other organisms are wildlife. This includes fungi, such as mushrooms, and bacteria that help decompose dead plants. Without these organisms, life would be disrupted on the earth. Processes of nature that make life for humans possible would not occur.

10-13. The mushrooms growing in this North Carolina forest are fungi.

MINERALS

A ***mineral*** is a material that has never lived. It is made of inorganic chemical elements, such as calcium and iron. People often think of minerals as rocks.

10-14. Half Dome in Yosemite National Park is an interesting rock structure 8,800 feet above sea level.

Some minerals are valuable, such as copper and gold. These and others are mined as ore. After refining, the elements in minerals are used to manufacture products that people want. All of the things we use that are made of metal began in mineral form. Even the coins in our pockets or purses are made of metals, such as silver, copper, and nickel.

10-15. Copper ore is being removed from a Utah mine. (Two thousand pounds of ore makes only 12 pounds of copper.)

FOSSIL FUELS

A ***fossil fuel*** is a material used to provide energy. These materials were formed over thousands of years by the decay of plant and animal material.

The two major fossil fuels are petroleum (a liquid) and coal (a solid). We all know the importance of gasoline, which is made from petroleum. Fossil fuels are burned to release energy as heat. The engines in tractors, chain saws, and pickup trucks are sometimes known as heat engines.

10-16. A pump is used to lift petroleum from the earth in this California oil field.

PEOPLE

People often do not think of themselves as a natural resource. They are! Humans are the most important natural resource on the earth.

Humans can use the earth's other natural resources to meet their needs. Unfortunately, humans sometimes damage and waste natural resources. All of us can work together to make smart decisions and use these resources wisely.

RENEWABILITY

Renewability is whether or not a natural resource can be renewed. The resource is restored or replenished after it is used. All natural resources are either renewable or nonrenewable.

RENEWABLE NATURAL RESOURCES

A ***renewable natural resource*** is one replaced after it is used. More of the resource is made or used resources are made ready to use again. Renewing may take a long time. Usually, new resources are not made. The amount of

AgriScience Connection

POSTING LAND

Land owners try to protect their property. Posting is the practice of forbidding trespassing. State laws on posting may vary but it is used in most places.

Trespassing is a problem to land owners. It occurs when people go on a person's land without permission. The person who trespasses has violated the private rights of the owner. Trespassers often take things or damage the property. In some cases, they may kill livestock or pick crops.

In general, posting involves putting up a posted sign. The sign should indicate that the property is posted and that trespassing is not allowed. Signs should be placed at regular intervals on property boundaries. They are often placed on fence posts or trees so that the public can see them. In some states, legal notices must be published in the local paper indicating that certain property has been posted.

Get the facts on posting land in your area from a local sheriff or other law enforcement officer. Never go on another person's property without permission!

the resource is constant. Renewing is a matter of restoring the resource for future use. Air, soil, water, and wildlife are renewable natural resources.

Most wildlife will renew itself. Some wildlife animals near extinction have been renewed. A good example is the bald eagle. A few years ago, the number of bald eagles was low. Some people feared they would become extinct. Efforts were made to stop killing bald eagles and make it possible for them to raise young. The efforts worked. Today, the number of bald eagles has increased.

Water and air are different from wildlife. Over time, water and air are renewed by cleaning. The amount of water on the earth never changes. The form it is in and where it is found change. Used water is gradually cleaned

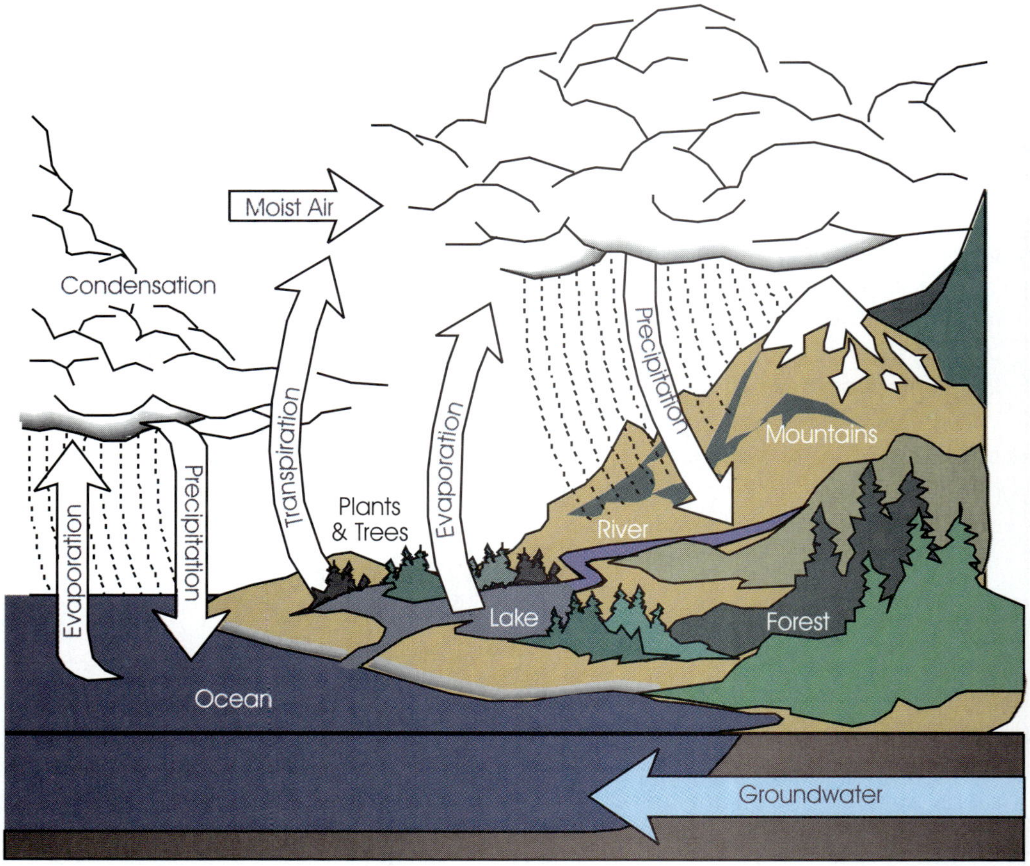

10-17. The water cycle naturally renews water.

and restored as part of the water cycle. We need to be careful and avoid damaging water.

Soil replenishes itself over several years. New soil forms from rock and the decay of leaves and other materials. This is a slow process. Some people use compost bins at their homes to make soil from paper and food scraps.

Soil may be damaged. Harmful substances may get into it. People are sometimes careless. Chemicals or oil may get on the ground. This means that the soil is polluted. It may be too polluted to grow crops.

NONRENEWABLE NATURAL RESOURCES

A ***nonrenewable natural resource*** is not replaced after it is used. Once used, they are gone! Fortunately, some can be recycled after use. Minerals and fossil fuel are nonrenewable natural resources. Of course, some people say that fossil fuel is renewable but an extra long time is required. Further, people are not sure it is happening.

Many minerals have become scarce. Copper, gold, silver, and chromium are examples of scarce metals. Some are still being mined but require much effort to get. A good example is copper. Several copper mines operate in the United States. Large amounts of ore are processed to get a few pounds of copper. At the Bingham Canyon Mine in Utah, a ton of copper ore produces only 12 pounds of copper.

Fortunately, some metals can be recycled. We must save materials made of metal after they have been used. These are reprocessed to get the metal. Iron, aluminum, chromium, copper, and lead are common examples.

10-18. Aluminum cans are often recycled because aluminum is a nonrenewable natural resource.

SUSTAINABLE RESOURCE USE

Sustainable resource use means that natural resources are used so they last a long time. People use no more than they need. Efforts are made to go about life so natural resources are not depleted. For example, why drive a car two blocks when you can easily walk? Driving the car uses fuel and pollutes the air. Besides, walking is good exercise!

People can take steps to sustain resources. Three important steps are renewing, reusing, and recycling.

10-19. A wildlife food plot has been established in this protected area.

RENEW

Renewing is helping natural resources replenish themselves. This is possible only with the renewable natural resources. How this is done varies with the kind of resource. For example, people use wastewater treatment plants to help renew water. Another example is establishing habitat for wildlife.

REUSE

Reusing is using something more than once. Some things can only be used once, but many things can be used again. How is reusing done? For example, grocery bags can be used again—take them back to the store for your groceries the next time you shop. (Reusing also cuts back on trash sent for disposal.)

Caution: Never reuse something improperly. For example, a container that once held a poison should not be used as a waterer for a pet. Some residue might be in the container that would damage the pet. Also, properly dispose of any left over poison, paint, or other hazardous material.

RECYCLE

Recycling is reusing the materials used to make a product. This is very important with nonrenewable natural resources. Glass, plastics, paper, and metals can be recycled.

Collection centers are in most communities. Materials are left for recycling. Occasionally, people are paid to recycle. A good example is aluminum cans, where each empty can is worth 2 to 3 cents.

CONSERVATION

Resources are on the earth for people to use—not abuse. Sometimes, people get confused. Thinking straight about the earth's resources is important. Two important terms are conservation and preservation. Conservation is using resources wisely; preservation is not using resources.

CONSERVATION

Conservation is the wise use of resources. It involves using resources as needed and taking steps to assure they will be available in the future. Conservation applies to all resources—both natural and human made.

Conservation of natural resources is widely accepted. It involves practices that prevent loss and damage. Conservation is much a part of sustainable resource use and sustainable agriculture.

Two important areas in agriculture are soil and water conservation. These are often used together.

Soil Conservation

Soil conservation involves using practices that protect the soil. Steps are taken to prevent erosion. ***Erosion*** is the loss of soil by water or wind. Water that runs off land following a rain may carry tiny particles of soil. These particles have important nutrients for plant growth. Once they are gone, the fertility of the soil has been lowered. Terraces can be placed across hillsides. Land can be plowed on the contour. Sometimes, crops are grown without plowing or with very little plowing.

Growing crops without plowing protects the soil from wind erosion. Strong winds can pick up soil particles and move them about. Sometimes they are blown from fields and lost.

10-20. Small ponds have been built on this Mississippi farm to help prevent soil erosion and collect excess rain water for restoring ground water supplies.

Water Conservation

Water conservation is using practices that prevent the loss of good water. It also includes reducing the use of water. Some practices that prevent soil erosion help conserve water. Terraces slow running water on hillsides. This allows more of it to soak into the land. Farm ponds catch water and allow it to soak into the ground. Some water will eventually reach underground pools known as aquifers.

Water conservation includes the wise use of water. Irrigation uses much water on farms. Knowing when to irrigate is essential. Irrigating without wasting water helps in conservation. Keeping faucets and pipes from leaking conserves water. Even a dripping faucet in your home is wasting water!

RESOURCE PRESERVATION

Preservation is keeping resources without using them. Deciding not to use something is difficult. Get good information. Make informed decisions.

10-21. Yosemite National Park preserves natural areas for visitors to enjoy.

Some resources deteriorate if not used. An example is a forest. The trees in a forest go through a life cycle. If not used, trees will die and the wood will rot—what a waste! On the other hand, some areas are too important to cut. A good example is the giant Sequoia trees in California. People use the Sequoias for beauty and not for wood. Most everyone agrees that a few of these should be protected.

As you can see, not using a forest is a mistake. We will not run out of trees in North America. More trees are now being planted each year than are cut. Tree farms grow improved trees to meet specific needs.

10-22. A forest has been destroyed by fire—what a loss!

Some controversy surrounds wildlife. Some people do not want wildlife animals killed. They want them to live and grow naturally. They do not think that people should use the animals for game. A few wildlife animals prey on pets and farm animals. Some people feel that these animals should be protected though they cause big losses to animal owners.

National parks, national forests, and other areas have been set aside to preserve natural areas. These are for people to visit and enjoy.

Overall, the natural features of the earth are for people to use wisely. Some may need to be preserved for future generations to enjoy. Others need to be used wisely and renewed.

10-23. Harbor seals are interesting to watch at Monterey Bay, California.

REVIEWING

MAIN IDEAS

Natural resources are the things humans use. These resources cannot be manufactured. Wise use of these resources is essential. The natural resources are soil, water, air, wildlife, minerals, fossil fuels, and people.

Some natural resources are renewable; others are not. Renewable resources can be replaced or restored after use. Many years may be required for some to be made ready for use again. The renewable natural resources include soil, air, water, and wildlife.

Nonrenewable natural resources cannot be replaced after their use. Once used, they are gone. Fossil fuels and minerals are the major nonrenewable natural resources.

Fortunately, steps can be taken to sustain resource use. These include renewing, reusing, and recycling. Recycling is particularly important with some metals, such as copper, aluminum, and iron.

People sometimes disagree on the use of resources. Fortunately, getting the facts helps people make good decisions. Most people would agree on the importance of conservation. Some want to go further and practice preservation. Know the resource and its natural cycle in making such decisions. Know how the balance of nature will be affected. Preserving some resources results in a loss of the resource.

QUESTIONS

Answer the following questions. Use correct spelling and complete sentences.

1. What are natural resources?
2. List and briefly define seven groups of natural resources.
3. What is wildlife? What kinds are important?
4. What should hunters know about their sport?
5. What is habitat? What can be done to provide habitat?
6. What is the difference in renewable and nonrenewable natural resources? Give examples of each.
7. What is sustainable resource use?
8. What are three important steps in sustainable resource use? Briefly explain each.
9. What is conservation? Preservation?
10. What are the two major areas of conservation? Briefly describe the areas.

EVALUATING

Match the term with the definition. Write the letter of the term in the blank provided.

a. conservation
b. erosion
c. wildlife
d. feral animal
e. game
f. endangered species
g. habitat
h. mineral
i. fossil fuel
j. recycling

____ 1. Loss of soil by water and wind.

____ 2. Using the materials in a product to make another product.

____ 3. Wise use of resources.

____ 4. All living things that have not been domesticated.

____ 5. A domesticated animal that has reverted to being wild.

____ 6. Wildlife threatened with extinction.

____ 7. Animals hunted for food and sport.

____ 8. The physical area where an organism lives.

____ 9. Product from the remains of decaying plants and animals that is used for fuel.

____10. Material that has never lived.

EXPLORING

1. Have a natural resource conservation officer (game warden) serve as a resource person in class. Have the officer discuss common wildlife in the area and provide a summary of laws about fishing and hunting.

2. Tour a wildlife preserve. Determine the kinds of wildlife in the preserve and the steps taken to preserve it. Prepare a written report on your observations. Take a camera on the tour and include the photographs in your report.

3. Complete a hunter safety education class. Prepare a report on your feelings about the class.

11

AGRICULTURAL MECHANICS

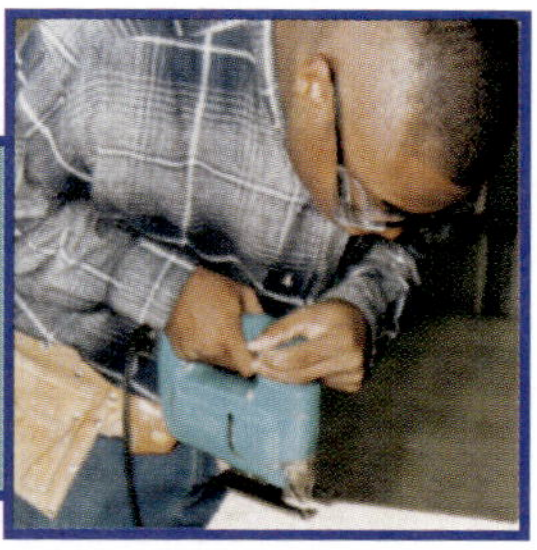

Work has been made easier. Many machines are available to help people do work. In fact, power machines help people do more work in less time. We can use machines better if we know about them.

Most of the work begins with simple hand tools. Once we know how to use common hand tools, we can use power tools. Many principles in using hand and power tools are the same. Power tools just require more attention to safety and operation.

People need simple mechanical skills in their daily living. They need to know the names of hand tools and simple power tools and how to use them. All tools apply science principles.

11-1. Leaf blowers are common power tools in horticulture work. (Courtesy, Husqvarna Forest and Garden Company)

OBJECTIVES

This chapter covers the broad area of agricultural mechanics. This includes hand tools, power tools, and power sources. The objectives of this chapter are:

1. Describe important areas of agricultural mechanics
2. Name and identify important hand tools
3. Name and identify important power tools
4. Explain the basic operation of an internal combustion engine
5. Describe the use of electricity

TERMS

agricultural mechanics
auger
brace
caliper
chisel
circuit
claw hammer
compression stroke
conductor
current electricity
electricity
engine systems
exhaust stroke
hammer
hand tool
insulator
intake stroke
materials
measuring device
pliers
power stroke
power tool
saw
screwdriver
square
tool
wrench

http://www.stanleyworks.com/

(This site has information on hand tools and project plans for using tools and mechanics skills)

MECHANICS

Agricultural mechanics is using mechanical devices to do agricultural jobs. All of the agricultural industry uses mechanical devices. Supplies for crops and animals are produced in automated plants, such as a feed mill. Food processing uses automated systems. Equipment is used in forestry and horticulture. The biggest impact has been on farming. Tilling the soil, planting seed, and harvesting are easier and faster.

11-2. A combine can harvest many acres of wheat in a day. (Courtesy, AGCO Corporation, Duluth, Georgia)

Today, one person using a powered machine can do the work of several people with teams of horses. An individual can do the work of hundreds of people using only simple tools without power. Tractors and equipment have made big changes in agriculture!

The areas of agricultural mechanics are described here.

GENERAL AGRICULTURAL MECHANICS

The general agricultural mechanics area covers the basic skills in selecting and using tools and materials. These are for repairing and building.

A ***tool*** is an implement used to do a mechanical job. Many tools are available. Knowing tools by their names is important. Skill in how they are used is needed. Both hand and power tools are included. Emphasis is on using tools safely.

Many different materials are used. ***Materials*** are the articles or supplies used in constructing. Depending on the structure, materials include lumber, nails, bolts, roofing, hinges, pipe, and cement. Identifying and selecting materials for a job are skills you need.

AGRICULTURAL ELECTRIFICATION

Agricultural electrification is using electricity to do agricultural work. It includes the meaning and safe use of electricity. Electrical controls and motors are also included. Wiring and installing switches are a part of agricultural electrification.

Skills are needed in planning and making wiring connections. Systems are designed to provide electricity as it is needed. How to make splices and attach devices is included. Safety is always important in electricity.

11-3. Electricity is used to operate automatic feeders in a poultry house.

AGRICULTURAL STRUCTURES

The agricultural structures area deals with the facilities used in agriculture: horse barns, greenhouses, fish hatcheries, farm shops, and others. It includes the basics in all areas of building construction. Fences, gates, and similar structures are a part of this area.

Using plans and materials is a part of agricultural structures. Skills are needed in measuring, cutting, and fastening materials. Installing plumbing, heating, and lighting systems is included. Agricultural structures, such as grain bins, milking barns, and poultry houses, are designed for specific uses and must meet specific needs.

11-4. Modern farm buildings often have a steel frame.

11-5. Fish farming in tanks requires mechanical systems to provide water and feed.

AGRICULTURAL POWER

Agricultural power deals with tractors, equipment, and all forms of power. It includes engine operation, maintenance, and repair. Chain saws, lawn mowers, leaf blowers, water pumps, and other kinds of powered equipment are included. Skills are needed to use most of the equipment.

Many kinds of implements are used. Some till the land. Others plant, cultivate, and harvest crops. All must be in good condition to operate properly. The people who operate them must know how to adjust and use the

11-6. Tractors provide the power to pull equipment, such as a manure spreader. (Courtesy, New Holland North America, Inc.)

equipment. Repair skills are needed. Welding may be needed to fasten broken parts. In-depth engine skills are needed by some people.

RESOURCE CONSERVATION PRACTICES

Agricultural mechanics includes practices in resource conservation. Tasks would include designing and building levees, ditches, terraces, drainage tiles, and irrigation systems.

Aquaculture pond construction is a part of this area. Similar skills are used for structures in soil and water conservation.

11-7. Laser-guided systems are used to level land for water management. (Courtesy, Spectra-Physics, Dayton, Ohio)

HAND TOOLS

A ***hand tool*** is a small powerless tool. Many kinds of hand tools are used. They do work that cannot be done with a bare hand. Just think how hard it would be to tighten a screw without a screwdriver!

Hand tools are selected based on the job. Each tool should be cared for properly. Always use the right tool. Keep tools clean and dry. Tools with cutting edges should be protected. Store tools properly after use.

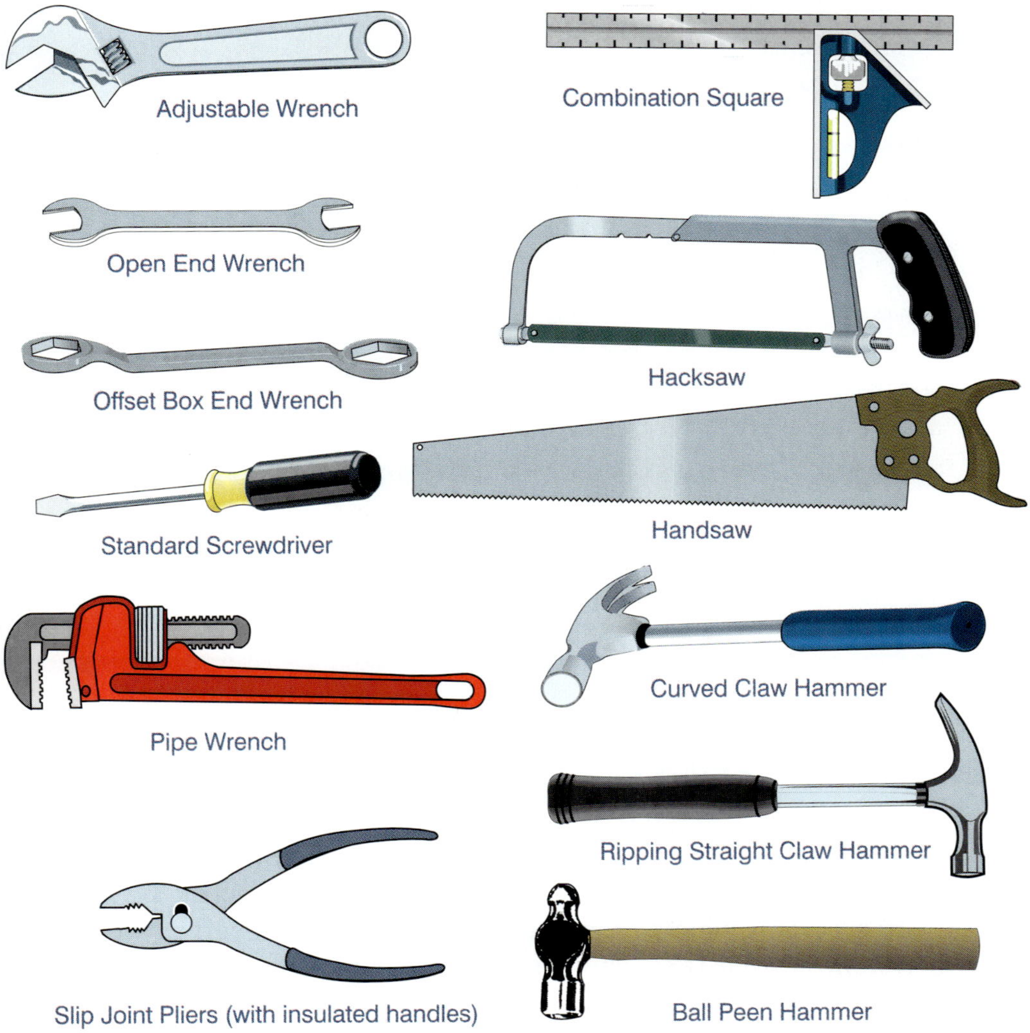

11-8. Examples of common hand tools.

Safety is essential with hand tools. Always know how safely to use a tool. Never use a tool in a way that was not intended, such as to pry with a screwdriver.

Use personal protection equipment. Wear eye protection. Use ear protection. Masks and respirators may be needed with some jobs. Special clothing is needed in welding and similar work.

Hand tools are made for working with all kinds of materials. Table 11-1 lists common areas and some hand tools that may be used.

Table 11-1. Examples of Hand Tools Used in Areas of Agricultural Mechanics

Carpentry	Electricity	Engines and Equipment
squares	long-nose pliers	open-end wrench set
claw hammers	combination pliers	Allen wrench set
ripsaw	measuring tape	screwdrivers
crosscut saw	screwdrivers	ball peen hammer
measuring tape	claw hammer	long-nose pliers
wood chisel	wire stripper	adjustable wrench
brace	keyhole saw	combination pliers
auger bit set	electrician's pliers	hacksaw
wood rasp	electrician's knife	gear puller
nail set	volt and ohm meter	tension (torque) wrench

MEASURING DEVICES

A ***measuring device*** is a tool used in making measurements. Most are marked in standard units in the customary system or metric system. The customary system is widely used in the United States in everyday work. The metric system is used throughout the world and in some instances in the United States.

Skill in measuring is essential and requires accuracy. Lumber, steel pieces, pipe, and other materials are measured before cutting. Incorrect

Table 11-2. Common Measurements and Equivalents

Customary System		Metric System Equivalent
1 inch (in.)	=	2.54 centimeters (cm)
1 foot (ft.) = 12 in.	=	30.48 cm = .3048 meter (m)
1 yard (yd.) = 36 in. = 3 ft.	=	91.44 cm = .9144 m
39.37 in.	=	1 m
1 ounce (oz.)	=	28.3495 grams (g)
1 pound (lb.) = 16 oz.	=	453.59 g
2.2046 lbs.	=	1 kilogram (kg) = 1000 g
1 acre = 43,560 square ft.	=	.405 hectare
2.47 acres	=	1 hectare

measurements result in pieces being cut the wrong length. The pieces will not work. Materials are wasted. Common measuring devices are:

- Ruler—A ruler is a strip of metal, plastic, or wood marked in units, such as feet and inches. A meter stick is marked in millimeters, centimeters, and other units.
- Tape—A measuring tape is a thin strip of metal or plastic wound in a case. It is pulled out for measuring. After the measurement is made, it is rewound.

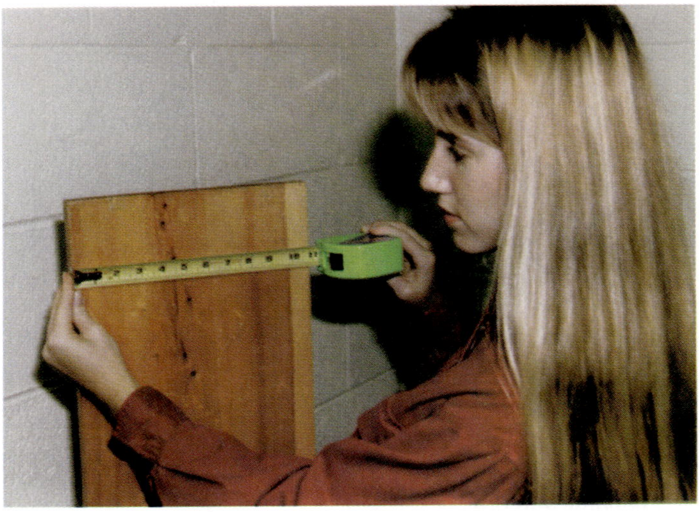

11-9. Using a measuring tape.

11-10. Using an electronic measurer.

- Zigzag ruler—The zigzag ruler is also known as a folder ruler. It is made in sections and unfolded when used.
- Electronic measurer—An electronic measurer is used for straight line distances. For example, the distance from one wall in a room to another wall is a straight line distance. Most use a small battery and are about the size of a handheld calculator.
- Caliper—A ***caliper*** is a tool used for measuring thicknesses or diameters. Some can make very precise measurements. Those used to measure where accuracy is essential are known as micrometers.

Once a measurement has been made, it must be correctly marked. Use a pencil or sharp edge (such as a knife

Career Profile

WELDER

A welder uses welding equipment to join or cut metal. The equipment may use gas or electricity to create heat. Welders need to know different kinds of metal. They must be skilled in measuring, marking, and fitting materials. They must be skilled in adjusting and operating welding equipment.

Welders gain knowledge and skill through welding classes and on-the-job experience. Many begin learning while in high school. Some have apprenticeship programs following high school. A high school diploma is needed. Additional education is beneficial.

Most jobs for welders are in factories or repair shops. Some use equipment that is mobile and moved to a job site. This photo shows steel being cut with oxyacetylene equipment.

blade) on wood. Use a pencil, chalk, or soapstone on steel. In marking wood, a fine line is made at the exact location of the intended cut. Saw across the wood and leave half the line.

SQUARES

Squares are important tools in construction work. A ***square*** is a tool for getting angles and marking materials for straight cuts. Several kinds of squares are used, such as the framing square and try square. Squares should be properly used. If not used properly, angles and cuts may not meet the needs of the work.

11-11. A square is used to mark a board before it is cut.

CUTTING TOOLS

Many different cutting tools are used. Some are used for wood or plastic; others are used for metal, concrete, and glass. Select the right tool for the material to be cut. Always be careful in using cutting tools. Most will readily cut skin and cause injuries.

- *Saws*—A ***saw*** is a cutting tool with sharp teeth on one edge. Many kinds of saws are made. How a saw is made determines its best use. Some saws are for cutting wood and should not be used to cut metal.

The most common is the handsaw, which is used to cut straight edges in wood. Handsaws are made to cut across the grain in wood or in the direction of the grain. Those that cut across the grain are known as crosscut saws. A saw that cuts with the grain is known as a ripsaw. Crosscut saws have 8 to 10 teeth per inch; ripsaws have 4 to 7 teeth per inch. Other kinds of wood saws have special uses. The coping saw cuts curves and shapes. Keyhole and compass saws are made for cutting in small areas where edges are not straight. Hack saws are used to cut metal and plastic.

11-12. Correctly using a crosscut handsaw.

- *Chisels*—A ***chisel*** is a wedge-shaped cutting tool. Most have a sharp blade on one end and a handle on the other. The handle is tapped with an approved hammer. Chisels are specially made for cutting wood and metal. A wood-cutting chisel has a razor-sharp edge. A metal cutting chisel has a somewhat blunt cutting edge. Never use a wood-cutting chisel on metal. The cutting edge will be nicked.

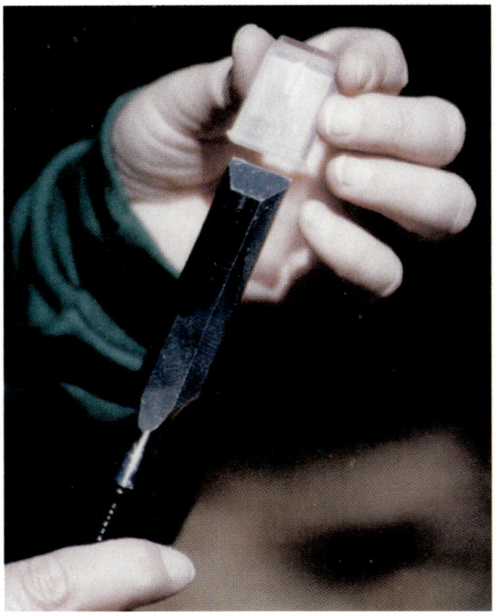

11-13. A wood chisel has a sharp, beveled cutting edge. (The cutting edge should be protected with a tip cover.)

- *Files and Rasps*—Files and rasps are bars of metal with rows of cutting teeth. Files typically have smaller teeth and are used for cutting metal. Rasps have larger teeth and are used for cutting wood. Many shapes and sizes are available. Select the one best

11-14. Using a flat file to sharpen a hoe.

for the job. For example, the flat file used to sharpen a hoe is much different from the round file used to sharpen a chain saw. Wood rasps are often used in removing small amounts of wood to assure a good fit, such as fitting wooden handles in hammers, axes, and other tools.

- *Axes and Hatchets*—Axes and hatchets are used for chopping wood. Axes are larger and have longer handles than hatchets. Most have one cutting edge. A few have two cutting edges and are known as double-bit axes.

- *Planes and Sanders*—Planes and sanders are used to smooth surfaces. Several kinds of planes are available, such as the block plane and jack plane. A plane is made with a metal cutting blade held in position by the body of the plane. Sanders use paper or cloth to which grit has been glued.

11-15. A new handle is being put in a double-bit axe.

BORING TOOLS

Boring tools are used to make holes in wood, metal, and other materials.

11-16. Using a brace with an auger to bore a hole.

11-17. Correctly holding a claw hammer for nailing.

The most common hand-boring tool is the brace with an auger bit. The ***brace*** is the device for holding and turning an auger. An ***auger*** is a steel shaft with screw threads that feed the cutting edge into the material. Threads along the shank remove the cut material from the hole. Augers are selected based on the size hole to be bored. Automatic and hand drills are sometimes used to bore holes.

HAMMERS

A ***hammer*** is a tool made for driving or pounding. It has a head made of metal or another material. A handle fits into the head. Handles are made of wood, metal, or plastic. Many kinds of hammers are available. Only a few are listed here.

- *Claw hammers*—A ***claw hammer*** is a tool used for driving nails into wood. The curved claws are used to pull nails out of the wood. Most claw hammers have metal heads with wooden, metal, or plastic handles. A claw hammer should never be used to strike metal surfaces other than nails.
- *Ball peen hammer*—A ball peen hammer has a flat striking surface on one side of its head and a rounded striking surface on the other side. These hammers are used in shop work for pounding metal and driving punches and metal chisels.
- *Blacksmith's hammer*—A special kind of hammer used for pounding and shaping metal. The metal may be hot.
- *Chipping hammer*—This is a special type of hammer used in metal work. It is used to chip slag away from welds.

WRENCHES

A ***wrench*** is a tool for gripping and turning bolts, screws, and nuts. Wrenches are used to assemble and repair equipment. A wide range of wrenches is available. Selecting the right one for the job is important.

Wrenches have a handle and jaws for holding bolts and nuts. The jaws are either fixed-jaw or adjustable-jaw.

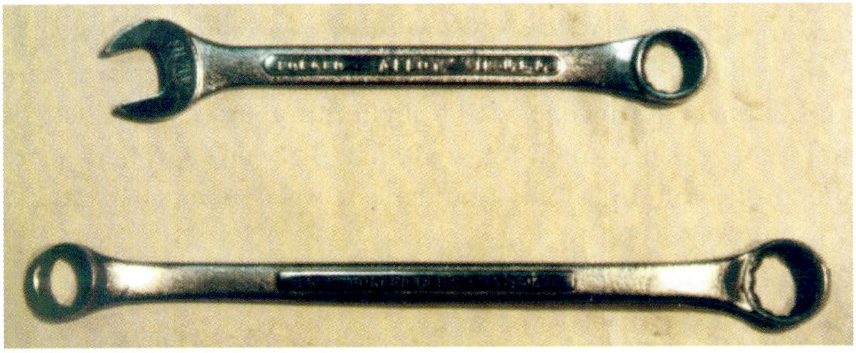

11-18. Two kinds of wrenches are shown here. A combination box-end and open-end wrench is at the top. A box-end wrench is shown at the bottom.

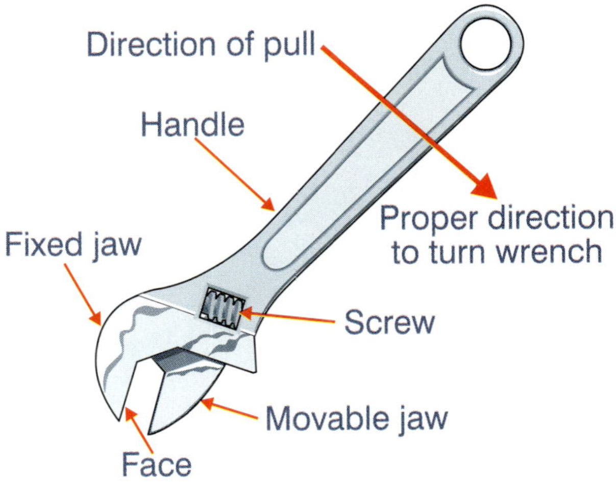

11-19. Parts of an adjustable wrench.

The common fixed-jaw wrenches have either an open-end or a box-end. Socket wrenches are box-end wrenches with changeable handles. Fixed-jaw wrenches are made to fit specific sizes of bolts and nuts. An assortment of wrenches is needed to take apart or assemble bolts and nuts. Adjustable-jaw wrenches have a moveable jaw attached to a screw. This allows the size of the distance between the jaws to be increased or decreased. One adjustable-jaw wrench fits several sizes of bolts and nuts. Most adjustable-jaw wrenches have smooth jaws. Pipe wrenches have jaws with ridges to grip round pipe. All adjustable wrenches should be tightened to a snug fit on the bolt or nut to be turned. If not, the wrench will slip. The bolt or nut will become rounded by slipping. Also, a slipping wrench may result in a personal injury.

PLIERS

Pliers are simple wrenches made of two levers. The levers pivot or rotate on a shaft or bolt. Pliers are designed to hold, bend, and cut materials. Several kinds of pliers are available. The slip joint type is most widely used. Pliers should not be substituted for a wrench. The jaws of pliers will damage bolts and nuts. Sometimes, the shoulders are worn off by pliers making it difficult to unfasten a bolt or nut.

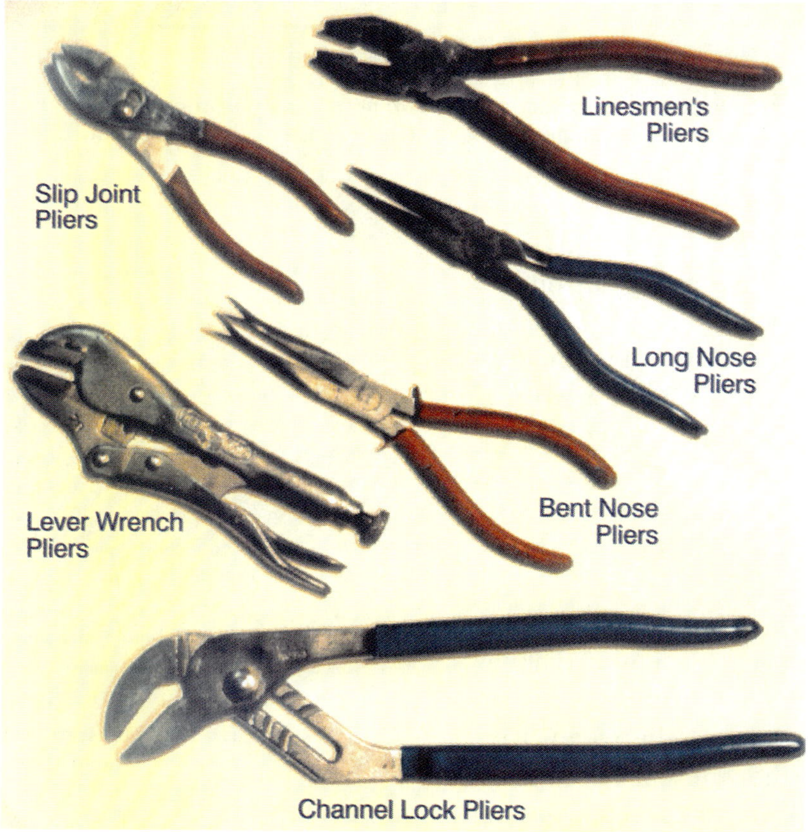

11-20. Kinds of pliers.

SCREWDRIVERS

Screwdrivers are common hand tools. A ***screwdriver*** is a tool with a wood or plastic handle on a long metal shank. One end of the shank is shaped to fit the heads of screws. The most common blade shapes are flat (or standard) and Phillips (cross-slot blade). Always select a screwdriver that fits the head of the screw. A screwdriver that is too large or too small does not work well. Carefully hold a screwdriver in position and turn the handle to loosen or tighten a screw.

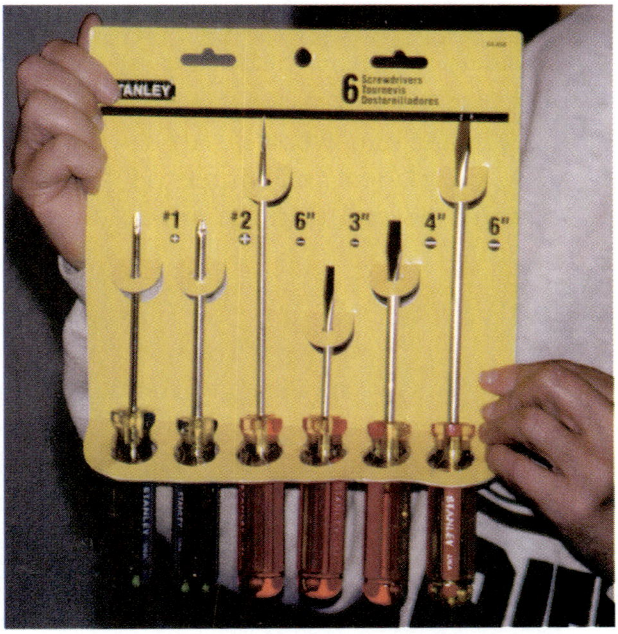

11-21. Screwdriver set with a variety of sizes and tips.

Some screwdrivers have different designs. The offset screwdriver has blades turned to the side.

OTHER HAND TOOLS

Many other hand tools are used. Some have specialized uses, such as concrete work and plumbing. Concrete and masonry work uses levels, trowels, and jointing tools. Plumbing requires pipe cutters, reamers, and threaders.

For more details on all hand tools, refer to *AgriScience Mechanics* (available from Interstate Publishers, Inc.).

POWER TOOLS

A ***power tool*** is any tool with power from a motor or engine. Tools powered by electric motors are most common. Where access to electricity is a problem, power tools may have gasoline engines. A few tools are powered by compressed air or oil.

Safety is extremely important with power tools. Always know and follow all safety rules. Learn the rules by studying operator manuals or from a qualified person.

The major power tools are saws, drills, grinders and sanders, air compressors, and planers and jointers. Horticultural work uses trimmers, edgers, and mowers. Several of these are described in more detail.

POWER SAWS

Power saws are designed to do the same jobs as hand saws but they do so much faster and more accurately. Most power saws have electric motors. A few use gasoline engines, such as the chain saw. Several power saws are briefly described here. Caution: Safety cannot be stressed too much in using power saws. Know what you are doing. Always be careful!

AgriScience Connection

SAFETY WITH POWER TOOLS

Power tools pose many safety hazards. With some knowledge, you can use them safely. Know the dangers of each tool. Follow all safety rules.

- Read the operator's manual for the power tool you are using.
- Observe safety signs on or near equipment (see photo).
- Wear clothing that fits properly, offers protection, and is not loose fitting.
- Wear safety glasses, goggles, or face shields.
- Wear hearing protection, as needed.
- Be sure all safety guards are in place.
- Anticipate any dangers ahead of time.
- Use the tool only for what it was designed to do.
- Be sure all electrical devices are in good condition.

Portable Power Saws

The portable power saws include circular saws, reciprocating saws, and chain saws.

11-22. Using a portable circular saw.

A portable circular saw has a round blade. These saws can be easily moved about. Most portable circular saws are for cutting wood. Some are used for cutting concrete and other materials. Blades are selected based on the material to be cut. Most of these saws used in construction have steel blades.

11-23. Using a sabre saw.

Reciprocating saws have short flat blades that move back and forth. Rapid motion makes quick cuts of wood, light metal, and plastic. The blade is at the end of the saw. A similar saw is the sabre saw. Its blade is underneath the saw body.

Chain saws have an endless chain. Teeth are attached to the chain. Most chain saws cut very quickly but do not offer a high degree of precision. They are widely used in cutting trees, pruning, and rough construction work.

Stationary Power Saws

A stationary power saw is in a fixed location. Several kinds are used, including table saws, radial arm saws, cutoff saws, and band saws.

11-24. Using a table saw to cut plywood. (Note that the safety guard completely covers the blade.)

Table saws and radial arm saws are in many shops. They are used to make straight cuts in lumber and plywood. The round saw blade of a table saw is in a table. The material being cut is pushed into the blade. The round saw blade on a radial arm is suspended over the table and is pulled across the material being cut.

11-25. Using a radial arm saw.

Cutoff saws are also known as chop saws. The round blade is lowered into the material being cut. Cutoff saws make straight cuts in wood, plastic, and metal. Blades are selected for the kind of material cut. Metal blades are used to cut wood. Abrasive discs are used with plastic and metal.

Band saws use long continuous blades. The blade has teeth on one edge. Blades are selected based on the materials to be cut. Band saws are used to make rounded cuts.

11-26. Using a cutoff saw to cut steel.

DRILLS

Power drills are stationary or portable. Stationary power drills are known as drill presses. They are designed for accurate, heavy duty drilling in wood, metal, and similar materials. Most use twist drill bits held in a chuck. The drill bit is lowered into the material being drilled.

Portable drills have many uses. They use drill bits similar to drill presses. Some have specialized bits for making large holes needed in electrical and plumbing work.

11-27. Using a drill press. (Courtesy, Glenn Miller, University of Arizona)

GRINDERS AND SANDERS

Grinders and sanders may be portable or stationary. Grinders are used to remove rough edges and make surfaces smooth. Sanders are used to smooth and shape materials. A special kind of sander is the buffer, which is used to polish surfaces.

Grinders are most commonly used with metals. They have abrasive wheels that turn at high speeds. The grain of the wheels varies. Use a fine-grain wheel for sharpening. A coarse grain is used for shaping and removing rough edges from metals. Stationary grinders are known as bench grinders.

11-28. Using a bench grinder to sharpen a drill bit.

Several kinds of sanders are used. Stationary sanders are used for sanding smaller objects held in position over the sander. Portable sanders are used on larger objects that are not easily moved. Coarse sand paper is used on rough surfaces. Fine sand paper is used to get smoother surfaces.

OTHER POWER TOOLS

Many other power tools are used. Air compressors are needed in most shops. Welding machines are needed in shops where metal work is done. Painting equipment is needed in painting.

INTERNAL COMBUSTION ENGINES

Internal combustion engines are used on tractors and powered equipment, such as lawn mowers, chain saws, and edgers. These engines convert the energy in fuel into power. Most engines are powered by gasoline, diesel, or liquefied petroleum (LP gas).

Air and fuel are mixed and burned. Burning releases gases, which creates heat. When burned, rapid expansion occurs. Engines use tightly sealed cylinders with a piston in one end. As the fuel is burned, the piston is forced backward. A connecting rod attaches the piston to a crankshaft. This converts back-and-forth motion into a rotating motion. Rotating shafts are used to power many tools.

11-29. An internal combustion engine.

ENGINE SYSTEMS

Engines must have several systems to operate. ***Engine systems*** are components that do essential processes for long-term engine operation. The systems are:

- Air—The air system filters and delivers air to the fuel system or cylinder.
- Fuel—The fuel system delivers fuel to the cylinder. The exact amount must be present with air for the engine to operate. Only clean fuel should be used.
- Ignition—The ignition system causes fuel to burn. A spark is used in gasoline and LP gas engines. Diesel engines use high compression (pressure) to ignite fuel. A large electrical system may be a part of the ignition on some engines, such as pickup trucks with many features.
- Exhaust—The exhaust system removes burned gases from an engine.

11-30. A dirty air filter removed from the engine of a lawn mower.

- Lubrication—The lubrication system keeps engine parts oiled. This reduces wear.
- Cooling—The cooling system removes heat from an engine. Larger engines have liquid cooling systems. Smaller engines are air cooled.

ENGINE OPERATION

Most fuel engines operate in a similar way: the straight line motion of a piston is converted into a rotating motion by a crankshaft. Differences are greater in how the fuel is delivered and ignited.

Two types of gasoline engines are common: two-stroke cycle and four-stroke cycle. These are based on the number of strokes a piston makes in a cylinder when the engine is running. A stroke is the back-and-forth motion of a piston. Two-stroke cycles are most common on small engines as used on leaf blowers and trimmers. Engines with four-stroke cycles are larger and are used where more power is needed.

Each stroke of a piston is important in engine operation. With a four-stroke cycle the strokes are:

- Intake stroke—Fuel and air are brought into the cylinder through an open valve during the ***intake stroke***.
- Compression stroke—In the ***compression stroke***, fuel and air are compressed by the cylinder. The valves are closed.

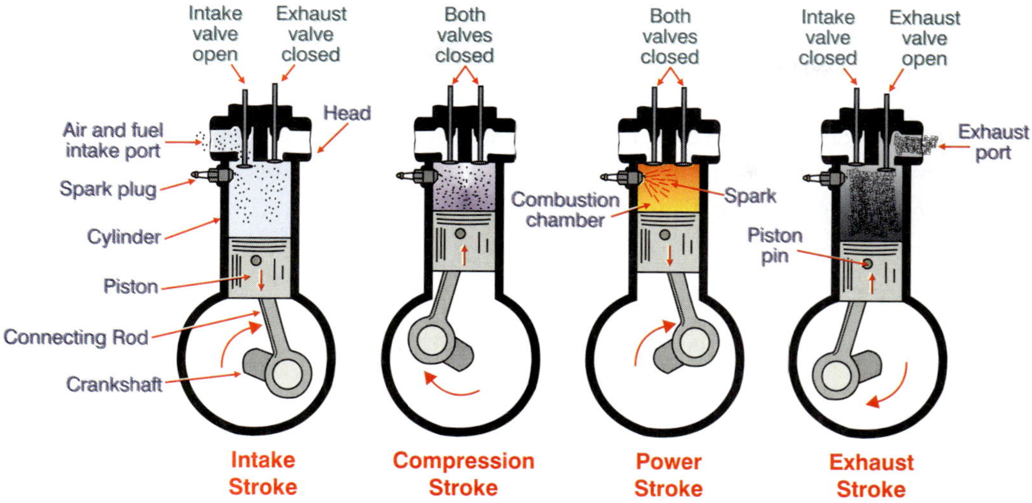

11-31. Strokes of a four-stroke cycle engine.

- Power stroke—The ***power stroke*** is the piston being forced backward. The fuel is burned. Heat is produced.
- Exhaust stroke—The gases from the burned fuel are forced out of the cylinder in the ***exhaust stroke***. The piston pushes the gases out through an open exhaust valve.

MAINTENANCE

Regular maintenance is needed to keep an engine in good condition. Several systems need servicing. None is more important than the air and lubrication systems.

The air system removes dirt from the air before it goes into the engine. A clogged air system will cause the engine to fail. If air cannot get through, the engine will not run. An engine uses more fuel in a partially clogged air system. Air filters need to be replaced or cleaned.

The lubrication system contains engine oil. The oil should be changed at regular intervals. Dirty oil contains grit. This will increase wear on engine parts. Use only the grade of oil recommended for the engine in the operator's manual.

Both air and lubrication systems require more frequent service in dusty conditions.

ELECTRICITY

Electricity is widely used in agriculture. It is used to power motors, cool and warm buildings, and provide light. ***Electricity*** is the flow of electrons in a conductor. The path of flow makes a circuit.

CURRENT ELECTRICITY

Current electricity is flowing electrons in a circuit. A ***circuit*** is the complete path for the flow of the electrons. Two kinds of current electricity are used: direct current (DC) and alternating current (AC).

With DC electricity, the electrons flow in one direction. This type of electricity is produced by batteries and used in flashlights and pickup trucks.

With AC electricity, the electrons flow in alternating directions. The electricity reverses direction of flow. This is the kind of electricity from power lines.

11-32. An electric motor is used to power equipment.

The rate of electrical flow is amperage. This is similar to the rate of flow of water through a pipe. The force of electrical flow is measured in voltage. This is compared with the pressure in a water system. Amperage and voltage are combined to have a useful measure of electricity known as the watt. A watt is a measure of the energy used by an electrical device. We select light bulbs based on wattage. Other electrical devices have a wattage rating. Al-

11-33. Small electric hedge trimmers are easy to use.

ways check the wattage to be sure we do not have appliances that require more wattage than a circuit can provide.

CIRCUITS

Electricity must be safe to use. Circuits must operate properly. To do so, conductors and insulators are needed. A ***conductor*** is a material that transmits electricity. Copper and aluminum wires are frequently used. An ***insulator*** is not a good conductor of electricity. Rubber and plastic are frequently

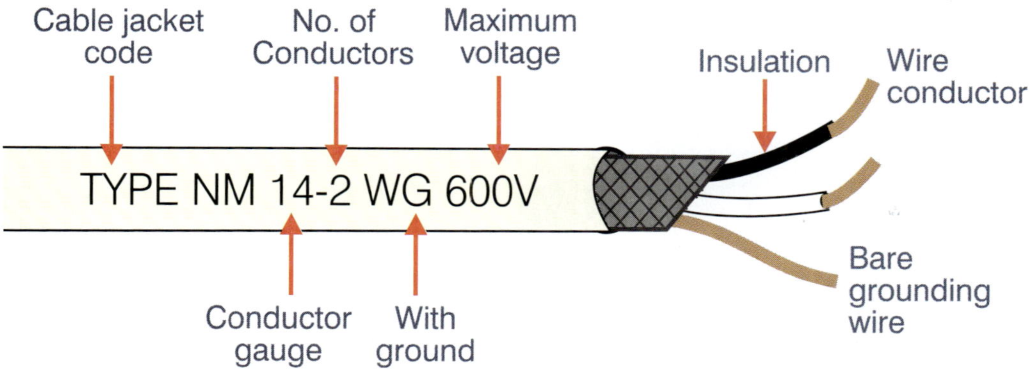

11-34. Electrical cable is used in wiring. (This shows a two-wire cable with grounding wire.)

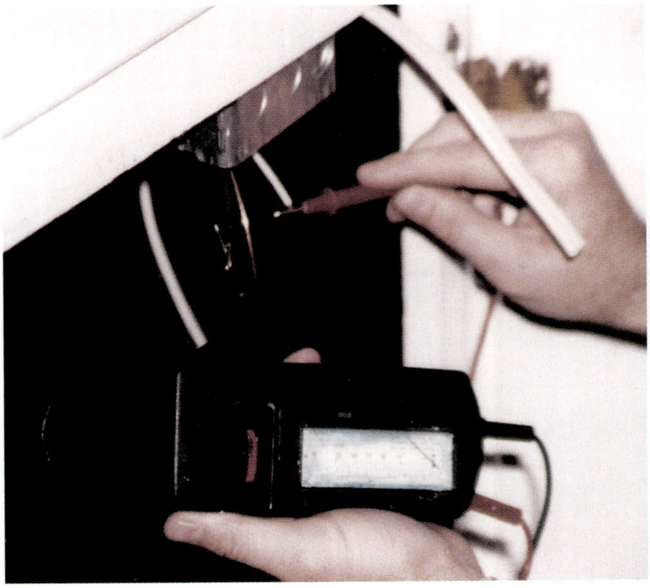

11-35. Testing a wire to see if it is "hot." (Caution: Always be sure a circuit is off before working on it.)

used as insulators. Insulators are placed around conductors so they work properly. The material on a conductor is insulation.

For circuits to work, the path of electrical flow must be complete. At least two wires are needed. One wire carries the electricity from the source. It is known as the hot wire. The other wire carries it back and is known as the neutral wire. Many circuits have ground wires for safety. Carefully designed wiring systems are used to get electricity where it is needed. Various switches and controls are used. These turn electricity on and off. Sometimes, the rate of flow is varied. Lighting may be regulated with a dimmer switch.

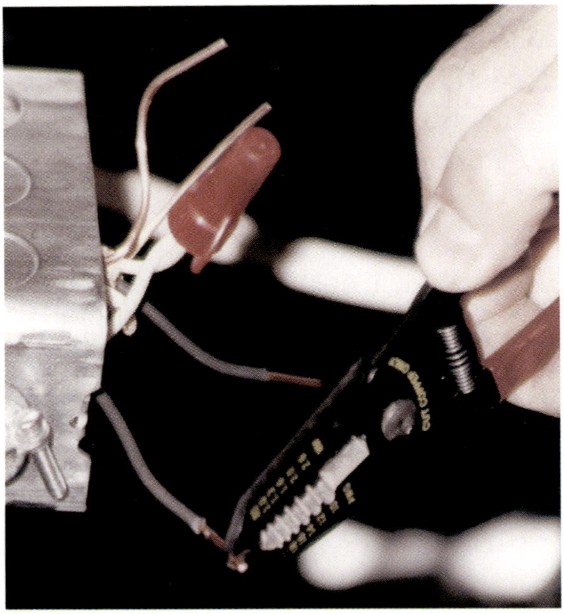

11-36. Splicing electrical wires.

Circuits include appliances that do useful work. Some appliances require 120 volts; others 240 volts. A common light bulb requires 120 volts. An arc welding machine or clothes dryer requires 240 volts.

REVIEWING

MAIN IDEAS

Mechanical devices are used to do many agricultural jobs. These have made it possible for one person to get far more done. Besides general agricultural mechanics, agricultural electrification, structures, power, and resource conservation practices are included in agricultural mechanics.

Hand tools are used for many jobs. Skill in using hand tools helps in using power tools designed for the same work. The most important hand tools are measuring devices, squares, cutting tools, boring tools, hammers, wrenches, pliers, and screwdrivers. Many hand tools are used.

Power tools have a motor or engine. They do work faster than hand tools. The major power tools are power saws, drills, and grinders and sanders. A wide range of power tools are used, such as nailers and painting equipment.

Internal combustion engines and electric motors provide most of the power in agriculture. Engines use a fuel that burns to create heat. The heat expands causing a piston to move in a cylinder. The piston is connected to a crankshaft. The result is a rotating shaft that provides power for many uses.

Electricity is important in agriculture. Current electricity is flowing electrons in a circuit. A conductor must be present. For safety and to prevent unwanted grounding, insulation is used on conductors.

QUESTIONS

Answer the following questions. Use correct spelling and complete sentences.

1. What is agricultural mechanics? Why is it important?
2. What are the major areas of agricultural mechanics? Briefly explain each.
3. What are the major kinds of hand tools?
4. What measurement devices are used?
5. What kinds of cutting tools are used?
6. What kinds of hammers are used?
7. What are the major kinds of power tools?
8. What is the difference in stationary and portable power tools?
9. What are the major systems on an internal combustion engine?
10. What are the four strokes in a four-stroke cycle engine?
11. What is current electricity? What are the two kinds?
12. What is a circuit?

EVALUATING

Match the term with the correct definition. Write the letter of the term in the blank provided.

a. circuit	e. saw	h. pliers
b. caliper	f. chisel	i. insulator
c. tool	g. claw hammer	j. intake stroke
d. materials		

____ 1. Material that is not a good conductor of electricity.

____ 2. Implement used to do a mechanical job.

____ 3. Articles or supplies used in construction.

____ 4. Complete path for the flow of electricity.

____ 5. A cutting tool with sharp teeth on one edge.

____ 6. A tool for measuring thicknesses or diameters.

____ 7. A wedge-shaped cutting tool.

____ 8. Part of engine cycle when fuel and air are brought into a cylinder.

____ 9. Tool made of two levers that pivot.

____10. Tool used for driving nails into wood.

EXPLORING

1. Visit a local hardware store. Study the tools for sale. Get prices and descriptive information on the following hand tools: claw hammer, Phillips screwdriver, ripsaw, micrometer, wood chisel, set of box-end wrenches, slip joint pliers, and a pipe wrench. Prepare a report on your observations.

2. Visit a shop where power tools are used. Observe the use of the tools. Prepare a report on what you see. Caution: Do not touch any tools or stand in the way when tools are used. Wear eye and hearing protection. Follow all safety rules.

3. Service the air cleaner on a small engine. Use the operator's manual as your guide. Also, do the work under the supervision of a qualified person. Be sure to follow all safety rules.

12

SOLVING PROBLEMS

Success in agriculture depends on new ways of doing things. These are based on science. Science is an organized way of gathering and explaining information. Agriculture could not succeed without science.

Science uses the fertile fields of agriculture to provide new ideas for research. As scientists seek answers, they use specific steps. These steps have been developed over time to help keep science organized. The same steps may be used to answer many questions you have yourself. In fact, these steps even work on non-science problems.

Go ahead and be a scientist! This chapter will show you how to start.

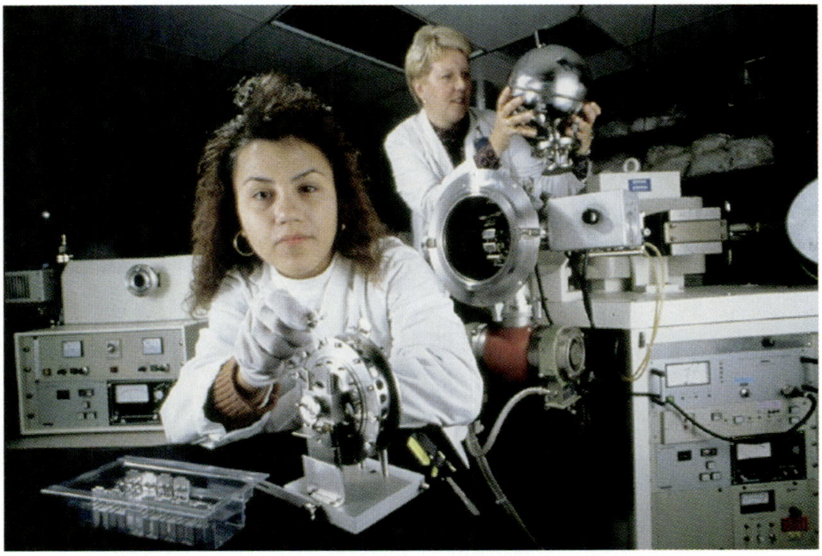

12-1. A food sample is being studied by scientists to learn nutrient content. (Courtesy, Agricultural Research Service, USDA)

OBJECTIVES

This chapter is about the scientific method. It includes how research is used in agriculture. Ways of measuring the quality of research are also covered. The objectives of this chapter are:

1. List and explain the steps in the scientific process (method)
2. Demonstrate the scientific process in a laboratory setting
3. Analyze examples of research to identify steps in the scientific process
4. Explain the role of the scientific process in agriculture

TERMS

applied research
basic research
conclusion
constant
control
data
experiment
experimental variable
hypothesis
popular literature
scientific fact
scientific law
scientific literature
scientific method
theory

http://www.casecorp.com

(This site has information on tractors, power equipment, and agricultural implements.)

THE SCIENTIFIC METHOD

Over time, a step-by-step process of solving scientific questions has developed. This process is commonly called the ***scientific method***. Some people describe this process as linear, which means it starts at a specific point and always moves through the steps the same way. Other people see the process as cyclic, with scientists entering and exiting the cycle at different points. Other people may even arrange the process in a slightly different order. In general, almost everyone agrees the process has seven steps. (Some authorities list five by combining two steps and not including reporting the findings.)

The seven steps in the scientific method are:

- Identify a problem
- Research the problem
- Form a hypothesis
- Design an experiment
- Conduct the experiment and collect data
- Analyze the data and form conclusions
- Report results

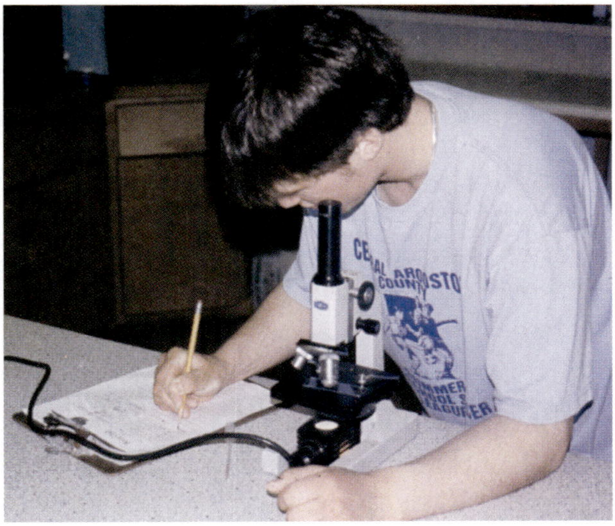

12-2. Problem solving often occurs as an ongoing process.

IDENTIFY A PROBLEM

What needs to be solved? Is it simply a question you are curious about and want an answer? Or do you know someone who has a problem and needs help? Scientists may be involved in ***basic research***, where they attempt to answer a scientific question without immediate use for the answer. Other scientists are involved in ***applied research***, which seeks to resolve a specific question with the results being used immediately to identify solutions.

In the world of agriculture, many problems and opportunities for applied research exist. People raising animals want the animals to be bigger, health-

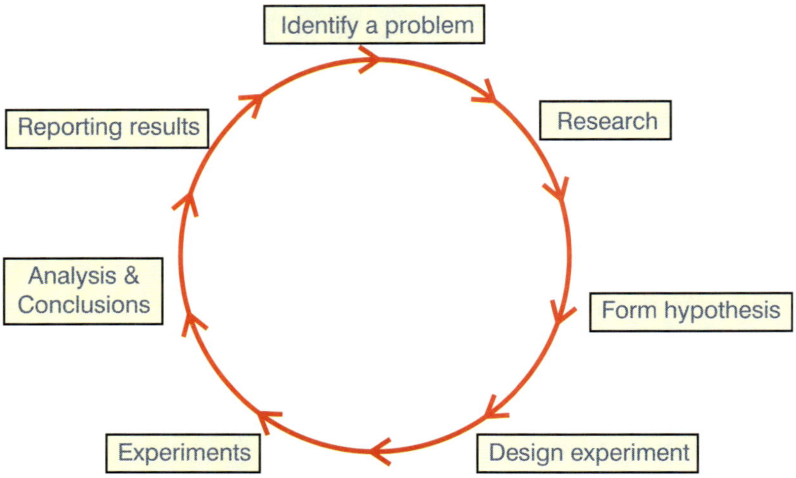

12-3. Questions in agriculture may be answered with a step-by-step process.

ier, and more productive. People who raise plants know the plants could be killed by insects, diseases, or loss of space. Once a product is harvested, safer, faster, and healthier methods of packaging and transporting the food are desirable.

Many scientists work with business people in agriculture to identify problems. They then use the steps outlined to find solutions.

Consider this example. Suppose your teacher kept plants in the classroom. All the students took turns watering and caring for the plants. When your turn came, you noticed that some plants were turning light green and

12-4. Careful observation of the world around you will help as you identify problems.

12-5. Notice the difference in the color of the leaves on these chrysanthemums. What could cause this? Do you think all the plants are healthy?

even yellow. You did not think this was healthy, but you were not sure. You have identified a problem: what is the proper color for plants?

RESEARCH

Scientists around the world are working to solve problems and learn new information. An important step in the scientific method is to learn more about previous research on the same topic and related subjects.

Most scientists do research using a variety of methods. A good scientist is always aware of the work being done by others. Scientists put much effort into seeking information. One method of research is contact with scientists involved with similar work. Most scientists enjoy talking to others with similar interests.

12-6. A range scientist is recording the kinds of plants that sheep eat. (Courtesy, Agricultural Research Service, USDA)

Written information is from many sources. These sources include popular literature, scientific literature, and the electronic media. ***Popular literature*** refers to books, newspapers, and magazines that provide information about what is happening to the public. Popular literature provides an excellent way for identifying problems and for gathering general background about those problems.

Scientific literature, or journals, are published to provide more in-depth information about topics. These journals are often specific to areas of science. Journals use technical language and are much more detailed. An additional source for research that is increasingly important is electronic media. The text of both popular literature and scientific journals may be found on the Internet. Encyclopedias may be purchased on CD-ROM. If your home or school has access to electronic media, much information may be obtained this way.

A fixed rule about the amount of research needed does not exist. Sometimes, research is the last step because it reveals a definite answer to the question or problem. Generally, though, research helps the scientist define

Career Profile

RESEARCH SPECIALIST

A research specialist uses problem-solving methods to improve agriculture. First, the specialist identifies the problem. Then he or she researches the problem, creates a hypothesis, and conducts experiments. The data are analyzed and the information is shared with other people.

Research specialists need college degrees in the areas to be studied. Most have graduate degrees at the masters or doctors level. Practical experience related to the research is very helpful. Begin preparing to be a research specialist by taking courses in agriculture and science in high school.

Research specialists find jobs with agribusinesses, universities, and agriculture agencies. Some work for associations. This photo shows two research specialists working together to collect data on a crop planted in soil with a heavy mulch. (Courtesy, Agricultural Research Service, USDA)

12-7. It is important that you plan your research and search all available information sources.

the question or problem. To succeed, a scientist must become well informed about a topic.

In the example, your next step would be to research plant growth. You might start by reading some basic information from a plant science book. You could take the plant to a florist or a greenhouse producer. You might ask your teacher what his or her ideas on the cause are. You could even go on-line (the Internet) and "talk" to amateur and professional plant care-givers from around the world.

In so doing, you would discover that plants require many nutrients for successful growth. You would probably find that lack of nitrogen causes yel-

12-8. A steer has an opening into its stomach for studying food digestion. (The steer is kept in the chute only a few minutes at a time. It usually grazes in a pasture.)

lowing of the plant's leaves. If you were satisfied with this explanation, you could return to the classroom, add nitrogen in the water, and not worry about the plant anymore.

FORM A HYPOTHESIS

Not satisfied? Want to prove to yourself exactly what is making the plant turn yellow? Good! The next step is to form a hypothesis.

Once the scientist has enough background knowledge about the subject, he or she creates a hypothesis. A ***hypothesis*** is simply an educated guess, which the scientist believes will answer the question or identify a solution to the problem. The hypothesis is traditionally given as a statement of fact.

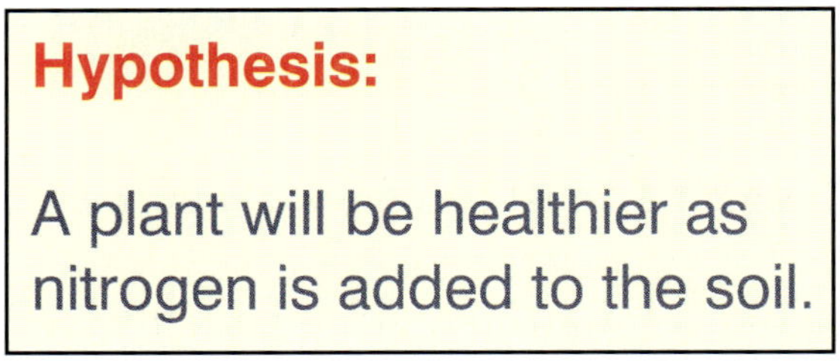

12-9. A hypothesis directs the remainder of the project.

In our example, your hypothesis might be "The regular addition of nitrogen through watering has a measurable impact on plant growth." Or, you might expand your search to consider more plant nutrients and say "The regular addition of plant nutrients through a complete commercial fertilizer has a measurable impact on plant growth."

The hypothesis decides the next steps in the process. It is important that a hypothesis be clear and realistic. A poor hypothesis leads to poor results.

DESIGN AN EXPERIMENT

Once the scientist has developed a hypothesis, he or she decides what steps may be used to test that hypothesis. The steps in testing the hypothesis become an experiment. An ***experiment*** is any method of testing the hypothesis through which the researcher may gather information. Experiments may be very simple and take only seconds to complete, or they may be quite complex and require years.

Either way, the key to successful experiments is careful planning. Before beginning work, the scientist decides exactly what to do. This is usually done in a written form. The scientist states the problem and describes the research that led to the hypothesis. The scientist then provides a detailed outline of

12-10. A Clapper Rail is being color banded for tracking. (The bird will be released and later found. This provides information to help understand Clapper Rails.) (Courtesy, U.S. Fish and Wildlife Service, Roy Lowe)

the experiment. The outline lists all materials and equipment, a time outline, how information will be collected, and how it will be analyzed.

Once the plan is complete, the scientist reviews it and often shares it with colleagues. This step helps the scientist clearly decide what will be done. It also helps identify potential roadblocks to successfully completing the experiment.

It is the scientist's responsibility to decide how precise an experiment to conduct. If the experiment is just to prove something to the scientist, it may not be as organized as one in which the scientist wants to share the results.

You think you know what is causing the yellowing of your teacher's plant. Just telling your teacher the solution might be an adequate experiment. Together you could add plant nutrients and watch the changes.

12-11. Careful planning is a key to successful experimentation. Working with others helps in the planning process.

You might prefer, however, to conduct a more organized experiment. In this case, you would create a detailed written plan. You would identify the materials you would need, including plants of the same species and size and different types of plant nutrients. You would decide exactly what you will measure, such as plant color, height, and stem width. You would also decide how long this experiment will continue.

You would then share your plan with others to get their ideas. Have you considered all the roadblocks? Have you thought of all the equipment? Have you overlooked an issue?

Once you are prepared, it is on to the next step!

EXPERIMENTATION AND DATA COLLECTION

Finally, it is time for the actual work! To this point, the scientific method has involved observing, reading, thinking, and planning. The experiment allows the scientist to put thoughts into action.

In science, an experiment is simply a test through which the scientist collects information, or ***data***. Attention to details is very important. The scientist must be alert to both expected results and to the unexpected. Often, a scientist learns as much from experiments that disprove the hypothesis as from experiments proving the hypothesis.

12-12. Grain sorghum head formation is being studied. (Courtesy, Agricultural Research Service, USDA)

"Serendipity" is an often-used phrase in scientific research. This refers to discovery of new information by accident. In reality, however, serendipitous discoveries are made by careful scientists who simply find unexpected results.

Data collection is a very important step in the experiment. As the experiment is conducted, the scientist records conditions and results. The more details the scientist records, the better. For some scientists, data are recorded first on paper and later transferred to a computer. Other scientists work directly with computers. Either way, careful and complete records are essential. A study is no better than the collected data!

With our example, begin by setting up the experiment. You have decided to test the effectiveness of a complete fertilizer by varying the rate and frequency of application. This is the only difference you will allow between the plants, and is called the ***experimental variable***. The experimental variable is the only change allowed between the plants in the experiment. All other conditions should be the same for every plant throughout the experiment. These conditions, which remain the same, are called ***constants***.

In addition, you have one plant that will not receive any of the fertilizer. This plant will be the control. A ***control*** is any object or group within an experiment that does not receive any contact with the experimental variable. The control allows the researcher to see changes that would have occurred without the presence of the experimental variable.

12-13. Thorough observation is a key to successful experimentation.

As you work, you record the height, color, and stem width of each plant. At the beginning, each characteristic should be the same as the other plants. This helps ensure that the changes you observe are in fact caused by your actions. You are varying the rate and frequency of use, so some plants get small amounts of fertilizer frequently. Some get larger amounts of fertilizer frequently. Others get small and large amounts less often. In your design, you will have specified the number of plants and the frequency of application.

You regularly measure the changes in the plant and record them in a data table. After six weeks, you end the experiment. An example of part of your completed data table is included as Table 12-1.

Table 12-1. Sample Data Table

Date		Plant A	Plant B	Control
February 1	Height	15 cm.	15 cm.	15 cm.
	Color	Dark green	Dark green	Dark green
	Stem width	6 mm.	6 mm.	6 mm.
	Notes	——All plants uniform——		
February 8	Height	19 cm.	18 cm.	17 cm.
	Color	Dark green	Dark green	Light green
	Stem width	8 mm.	8 mm.	7 mm.
	Notes			Yellowing on lower leaves
March 14	Height	35 cm.	25 cm.	20 cm.
	Color	Dark green	Dark green	Yellowing
	Stem width	20 mm.	16 mm.	14 mm.
	Notes	——Strong and healthy——		Spindly and weak in appearance

DATA ANALYSIS AND CONCLUSIONS

Once the experiment has been completed, the scientist must decide what the results mean. Sometimes, this involves complex evaluation or analysis of the data. Computers are often used to do this. Relationships or patterns within the data numbers are observed.

12-14. The result of a successful experiment in agriculture is typically an improved product.

What do you see within the data for our example? Do fertilizers affect plant growth? In what way? How does this relate to our hypothesis? Can we accept the hypothesis, or do the data prove us wrong?

Ultimately, the scientist must decide whether the experiment supported or canceled the hypothesis. Often, experiments will do neither. With either, the scientist must design and conduct new experiments. In any case, the researcher forms a ***conclusion***, which is a statement of the relative value of the hypothesis based upon the experimentation.

12-15. A rice variety with more nutrition is being studied. (This rice was grown using tissue culture. It has high lysine content.) (Courtesy, Agricultural Research Service, USDA)

Your conclusion might be as simple as, "I accept the initial hypothesis regarding the measurable impact of plant nutrients upon plant growth." Your conclusions might be specific, "These plants survive best when given x amount of nutrients on a specific time basis." You might also conclude that providing too many nutrients to a plant is possible. Damage to the plant may be a result.

Over time, information about potential solutions to a problem accumulates. With this information, scientists may create theories. A ***theory*** is a statement that explains the collected data and can fit new data as it is gathered. Theories are never definite and are always being revised.

Scientific information that can be shown any time is called a ***scientific fact***. For example, we know that if we let go of an object, it will fall to the ground. That is a scientific fact. When related scientific facts are combined to make a broad sweeping statement, it is called a ***scientific law***. The law that relates to falling objects is the law of gravity. The law of gravity exists only because of long-term, careful observation by scientists.

AgriScience Connection

AGRISCIENCE FAIRS

The best way to understand the scientific process is by using it. Students are often required to complete science projects during school.

The final product of these projects may be a science display. These displays may be presented through fairs. The scientific content is judged. Students are rewarded for their hard work and learning. Fairs are conducted on many levels, from a single classroom to international competitions.

The National FFA Organization has a National Agriscience Fair. Students compete against others of similar ages and with similar topics. As the Agriscience Fair grows, the awards and recognition a student may earn will also grow.

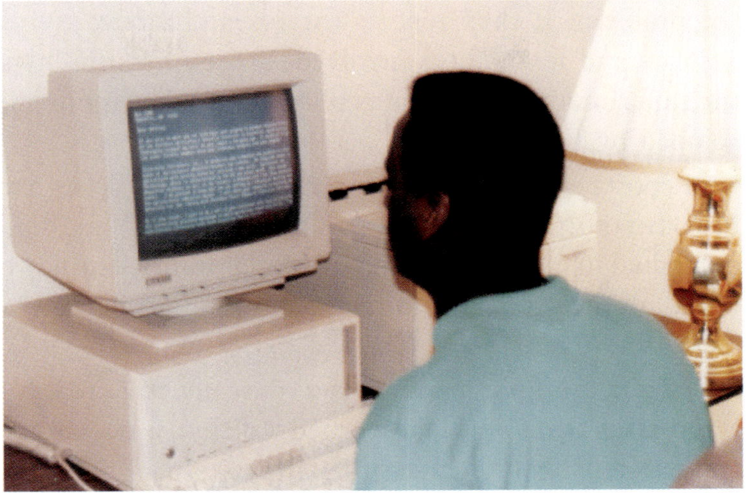

12-16. A computer may be used to prepare a report.

REPORTING RESULTS

The final step in the scientific process is informing others about the research conducted. Research that is not reported has very little value. The report typically describes the work done within each step. It begins by detailing

12-17. One measure of the success of the scientific method is a productive team that is learning and working together.

the problem. It then reports the research done, giving proper credit to the authors and researchers involved. It then lists the hypothesis, the design of the experiment and the data collected. Finally, it provides the conclusions developed by the scientist.

Reports are provided in many different ways. As a student, you may turn in your report to your teacher. Other scientists may submit their work to be published in a journal. Research of major interest may be published as a book or in another separate form. Today, some scientists are even reporting directly to sites on the Internet.

In our example, perhaps the most effective report would be to tell your classmates what you had learned. In that way, you could inform them and your teacher about changes in plant care that will help keep the plants in the classroom healthy.

Remember, the scientific process is often considered cyclic and never-ending. In conducting your research, you may have discovered other plant topics you want to explore further. Or, you may have noticed something odd during your experiment that you would like to research. Whatever the case, if you are curious, use the scientific method to help develop your learning and problem-solving skills!

THE SCIENTIFIC METHOD IN ACTION: GREGOR MENDEL

This next section will describe someone who used the scientific method in solving a problem of interest to him. As you read this section, try to pick out the individual steps within the method.

Gregor Mendel successfully used science to explain a mystery in agriculture. Mendel was a monk in 19th-century Austria. One of his responsibilities at the monastery was to care for the garden. As he worked in the garden, Mendel looked carefully at the plants. One thing Mendel noticed was the difference in the pea plants. The plants had many differences. For example, some plants were very long and tall; others were very short.

Mendel wondered why the plants were different. Although he had access to books, none explained the differences. Mendel chose to discover an explanation for himself.

12-18. Mendel studied the common pea. (Courtesy, U.S. Department of Agriculture)

Over an eight-year period, Mendel conducted experiments with pea plants. He examined seven different contrasting traits, such as height, shape, and flower and seed color. Mendel would take plants with the opposites of the same trait (such as short and tall) and cross them. He would then examine the trait shown by the young.

Mendel was very methodical in his approach. He studied only one trait at a time. As he worked, he carefully recorded each step. He counted and classified each new group of peas carefully.

Eventually, Mendel created mathematical models to explain the outcomes he had identified. He decided that peas had two different sets of instructions for each trait and that some sets were more easily obeyed than others. He also decided that, during reproduction, plants had a way of dividing the two sets into single sets.

Mendel wrote a paper that summarized his ideas, and presented it to the Brunn Natural History Society in 1865. The paper was also published in the society's proceedings in 1866.

Could you identify each step? Mendel did, in fact, follow the same scientific method we use today. He began by identifying a problem. He then sought information to explain that problem. When he could find no explanation, he conducted his own experiments. Upon completing his experiments, Mendel drew conclusions and reported his work.

Ironically, no one realized the importance of Mendel's work until about 15 years after his death. In 1900, three men discovered Mendel's paper and began duplicating his work. They began the science of genetics, which is important today.

THE SCIENTIFIC PROCESS IN AGRICULTURE

Agriculture benefits from both basic and applied research conducted using the scientific method. In fact, much of the knowledge gained by agriculturists is a result of the scientific process. Without careful, methodical research and record-keeping, information is lost and learning stops. The understanding of plants and animals, which is the foundation of agriculture, could only have been achieved through use of the scientific process.

Many scientists are employed in agriculture to conduct applied research and find solutions to agricultural problems. Most of these scientists use the scientific method as a tool to help them in problem solving. Without the scientific process, agriculture would have little knowledge and few avenues for finding solutions.

12-19. The effects of increased carbon dioxide on plant growth are being studied in these growth chambers. The facilities are used by scientists at the National Soil Dynamics Lab at Auburn University.

REVIEWING

MAIN IDEAS

The scientific method is a very useful tool to both scientists and agriculturists. The method is a step-by-step process used for solving problems. The steps include identifying a problem, research, developing a hypothesis, designing an experiment, collecting data, analyzing the data and forming conclusions, and reporting the work to others.

Each of these steps is important. Most work done by scientists follows these principles. To be successful researchers, students should be aware of these steps. The benefits of the scientific method may even be shown in non-scientific settings.

The scientific process is a theoretical foundation of agriculture. Without the process, agriculture could not exist in its present state.

QUESTIONS

Answer the following questions. Use correct spelling and complete sentences.

1. What are the steps in the scientific process?
2. List ten problems you think you could solve through the scientific method.
3. What is the difference between popular and scientific literature? Which is likely to be more accurate?
4. What is the difference between a hypothesis, a theory, and a scientific fact?
5. What factors should a scientist consider when designing an experiment?
6. What roles do experimental variables, constants, and the control have in an experiment?
7. Why is careful data collection important?
8. Why should researchers report their findings?
9. Do you think the steps in the process should always go in the same order? Why or why not?

EVALUATING

Match the term with the correct definition. Write the letter by the term in the blank provided.

a. applied research	d. experiment	g. conclusion
b. hypothesis	e. control	h. theory
c. scientific method	f. data	

____ 1. An educated guess.

____ 2. Seeks to resolve a specific question with immediate results.

____ 3. Statement of the relative value of the hypothesis.

____ 4. Any method of testing a hypothesis.

____ 5. A step-by-step process of solving scientific questions.

____ 6. Statement that explains a large quantity of collected data.

____ 7. Information or results collected through experimentation.

____ 8. Object or group that has no contact with the experimental variable.

EXPLORING

1. Use reference materials to research a scientist whose work has had a profound impact on society. Examples include Linus Pauling, Charles Darwin, Louis Pasteur, Marie Curie, or Alfred Noble. You and your teacher may choose other scientists. As you research, consider the steps in the scientific process. How did this scientist demonstrate those steps? Did they follow a precise order? What impact did their work have upon the agriculture of today?

2. Visit a local scientific facility. Ask how the scientists identify problems and if their work is basic or applied research. Observe their use of the scientific method. Learn what types of research and experimentation they use. Write a short report of your findings.

3. Identify a problem you have been concerned about. Use the scientific method to identify potential experiments, and, when possible, carry out the experimentation. Provide a written or oral report to your class.

13

ANIMAL SCIENCE

You probably have not thought much about how many roles animals play in your life. Think about it now. How do animals affect you?

If you have a pet at home, you have been involved with animal science. If you had ham, eggs, or milk for breakfast, you have been involved with animal science. If you have worn leather shoes or a leather jacket, you have been involved. If you slept on a pillow with feathers last night, you have been involved.

In short, animals contribute to our lives in many ways. We depend on them, and many animals depend on humans. Animals help meet some of our basic needs. This chapter will tell you more about the interdependence between humans and animals.

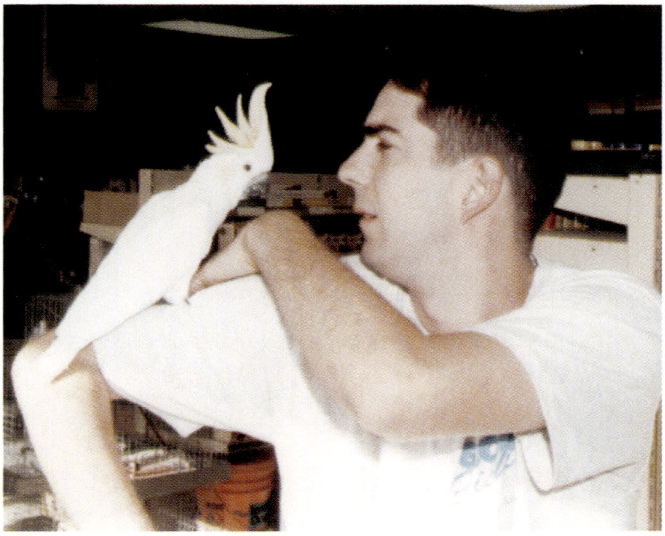

13-1. Animals are fun! (Here an owner is enjoying training a Lesser Sulfur Cockatiel.)

OBJECTIVES

This chapter provides basic information on animal science. It includes the ways animals are used and the types of animals. The basic biology of animals is also included. The objectives of this chapter are:

1. Explain how animals benefit people
2. List and explain the major types of animals
3. Describe animal structures and functions
4. Explain nutrition and feeding
5. Define reproduction and describe the processes involved
6. Explain animal well being

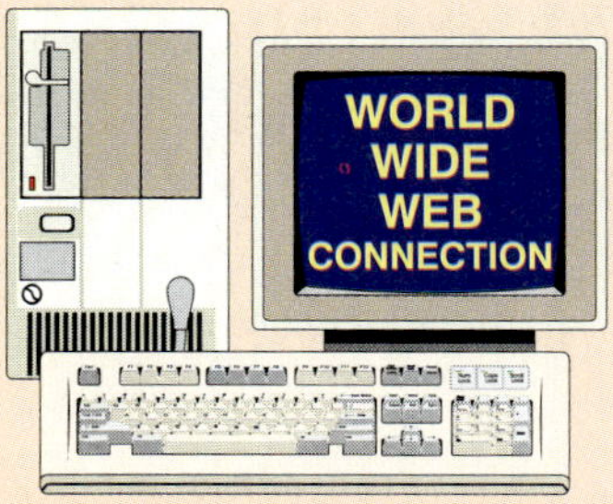

http://www.ansi.okstate.edu/breeds/

(This site has information and photographs of common animals and breeds.)

TERMS

animal by-products
animal well being
artificial insemination
balanced ration
beef
breeds
carcass
castration
cheval
concentrate
debeaking
dehorning
docking
farrowing
feeder animal
gestation
hybrid
lactation
lamb
livestock
meat
monogastric
mutton
nutrient
parturition
pleasure animal
polled animal
pork
poultry
purebred
ration
reproduction
roughage
ruminant
service animal
udder

HOW ANIMALS BENEFIT PEOPLE

Animals and humans interact in many ways. Few people know all of the products from animals. Besides meat, animals are sources of milk, egg and poultry products, wool and mohair, many by-products, and service and pleasure.

13-2. Swarming is an interesting behavior of bees.

MEAT

Cattle, swine, and sheep are the most common animals raised for meat. The meat is called ***beef***, ***pork***, and ***lamb***, respectively. (Meat from sheep more than one year old is called ***mutton*** and is less frequently eaten.) Meat from young milk-fed cattle is veal. Horses are a source of meat, called ***cheval***, in some European countries.

Meat is the muscle tissue of an animal. It is a source of nutrients, including protein, fat, and carbohydrates. Meat is the most valuable part of a carcass.

Some animals are raised from birth for meat. A ***feeder animal*** is a young animal that needs additional feeding before it is processed. Meat animals are raised to a specific size, or finished.

Meat animals are made into products at a processing plant. First, the animals are made unconscious. They are killed painlessly and their internal or-

13-3. A meat case well stocked with wholesome animal products.

gans removed. Every part of an animal is used. The parts other than meat are called by-products. The ***carcass*** is the muscles and bones of an animal. Carcasses are cut into large pieces called wholesale cuts. These cuts are then made into smaller pieces and sold as retail cuts.

MILK AND DAIRY PRODUCTS

Milk is produced by female dairy cattle, goats, and sheep to feed their young. This process is called ***lactation***. In parts of the world, horses, water buffalo, reindeer, and other mammals are used for milk.

Milk is a whitish liquid that contains protein, carbohydrates, vitamins, and minerals. Milk may be modified to reduce or remove the fat. It may also be evaporated or condensed. Milk is the base for dairy products, such as cheese, butter, yogurt, and ice cream.

EGGS AND POULTRY PRODUCTS

Domesticated birds are commonly called ***poultry***. Eggs and meat are most often produced by chickens and turkeys. Geese are kept for feathers and down.

Eggs contain protein and fat. They are prepared in many ways. Not only are they eaten as is, but they are a common ingredient in other foods, such as cakes and pastries.

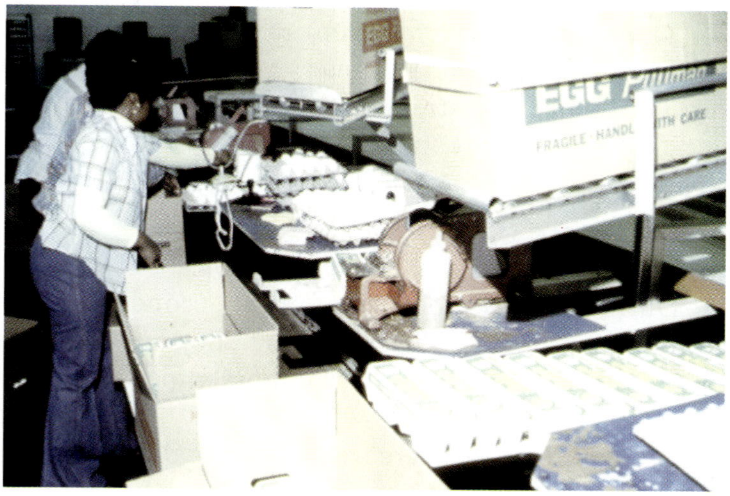

13-4. Eggs are being placed in protective cartons. (Courtesy, Mississippi State University)

Poultry meat, often called white meat, is lower in fat than the meat of cattle, sheep, or swine. It has become more popular in recent years.

WOOL AND MOHAIR

Sheep and goats produce hair fibers that may be used for fabric. These fabrics are extremely warm and durable. Sheep produce wool; special goats (Angora) produce mohair. A less common goat, the Cashmere, produces cashmere.

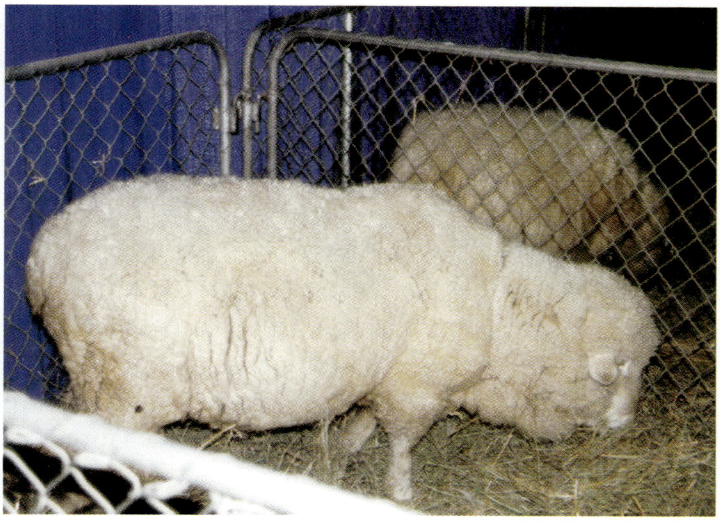

13-5. The wool of sheep is used to produce warm, durable fabric.

The hair fibers from the entire body are fleece. The fleece is removed by shearing the fibers off. The fibers are woven to form thread, yarn, and/or fabric. The length and strength of the fibers are keys to the quality of the product.

ANIMAL BY-PRODUCTS

During animal production, many secondary products are created. These are known as ***animal by-products***. They are both edible and nonedible. Although they are not initially the most important part of the animal, they are valuable.

Some animal parts are used for animal feed. Other parts are used for lard and tallow, which go into shortening, pastries, and candies. Other parts, such as hides, are used for clothing (leather). Hair may be used in paint brushes. More than 20 medicines are made from animal organs. Animals provide many products for our use.

13-6. A three-pound Yorkshire Terrier is fun to watch and hold.

SERVICE AND PLEASURE

Animals may be kept for service or pleasure. A ***service animal*** is one that helps people in some way. Some service animals are used to guard property. Others are used to guide people, such as a dog for a blind person. Animals are also used in laboratories to help people learning about and developing new products. A few animals are used for power, and are known as draft animals. An example is a horse that pulls a wagon.

A ***pleasure animal*** is an animal kept for fun. They are sometimes known as pets or companion animals. Exotic animals in zoos are also kept for pleasure. Pleasure animals include dogs, cats, ornamental fish, horses, and many others.

TYPES OF ANIMALS

Animal scientists group animals based on similarities of growth and use. Three common groups are livestock or large animals, small and companion animals, and poultry. Each group has specific terms for males, females and young animals. Examples of common terms are in Table 13-1.

Table 13-1. Examples of Terms for Various Animal Species

	Male	Female	Young	Castrated	Young Female	Young Male
Cattle	bull	cow	calf	steer	heifer calf	bull calf
Horse	stud	mare	foal	gelding	filly	
Swine	boar	sow	piglet	barrow	gilt	
Sheep	ram	ewe	lamb	wether	ewe lamb	ram lamb
Goat	billy	nanny	kid			
Duck	drake	hen	duckling			
Goose	gander	hen	gosling			
Chicken	rooster	hen	chick	capon	pullet	
Rabbit	buck	doe	kit			
Turkey	tom	hen	poult			

LIVESTOCK

Livestock are large animals raised on farms or ranches for food or other uses. Common livestock are beef cattle, dairy cattle, swine, sheep, goats, and horses. Bison, elk, and llamas have become more important.

Beef Cattle

Beef cattle are raised for meat. They can convert grass and other roughage into meat. Beef animals are selected for their potential to produce meat. Con-

13-7. Three common beef breeds: Angus (top), Polled Hereford (center), and Brahman (bottom). (Courtesy, American Angus Association and American Brahman Breeders Association)

formation is the physical form of the animal. Some parts of the animal are more valuable than others. For example, the middle back or the loin is the source of club, T-bone, and porterhouse steaks. Animals with heavy muscle in their back are more desirable.

Many breeds of beef have been created over time. Some examples are Angus, Hereford, and Brahman. At least 20 breeds of cattle have registries in the United States. An additional 60 to 80 breeds are recognized.

Angus cattle are black polled cattle. A ***polled animal*** is naturally hornless. Angus cattle originated in Scotland. The Angus have a high quality of meat.

Hereford cattle originated in England. They have red bodies with white faces. They have horns and are of a moderate size. Herefords are very hardy and good at finding food. A separate breed, which lack horns, is the Polled Hereford.

Brahman cattle originated in the United States. They descended from Indian cattle and are characterized by a distinctive hump. Brahmans are especially well adapted to heat and humidity. They grow well in the southeastern United States.

Beef cattle are raised in one or more of three environments: cow-calf operations, stocker-yearling systems, and finishing operations.

13-8. These beef animals are groomed and ready for showing at a fair.

The cow-calf operation raises the calf from birth to seven to nine months of age. The cow-calf system focuses on production of strong and healthy animals and relies upon quality cows.

The stocker-yearling system raises the calf to about three-quarters of its final weight. The cattle are generally fed in a pasture-like environment. The animals receive primarily forages. Finishing raises animals to a desired slaughter weight. A much higher quantity of concentrates is fed.

13-9. Two dairy breeds are Holstein (top) and Jersey (bottom). (Courtesy, Pete's Photo, Minnesota)

Dairy Cattle

Dairy cattle are selected for milk production. This ability is shown by a sleek, angular body, a strong mammary system, and strong feet and legs.

The ***udder*** is the milk-producing organ of a cow. At peak lactation, a dairy cow can produce 150 pounds of milk each day. Lactation occurs following the birth of a calf. The cow is milked for up to 305

Career Profile

DAIRY SPECIALIST

A dairy specialist studies ways of improving milk production. The work involves going into milking facilities and other places where cattle are kept. They may work with dairy producers to improve their herds.

Dairy specialists need a college degree in animal or dairy science. Most have masters or doctors degrees. Practical experience on a dairy farm is essential. Most jobs are with research centers, dairy products manufacturers, and large dairy farms.

This shows a dairy specialist observing milking in a dairy barn. (Courtesy, Mississippi State University)

days each year. A dry cow is one that is not producing milk. Cows are dry for two months before they calve again.

Cows are typically milked with machines. The cow udder is cleaned before and after milking. Humans never touch the milk. Milk is about 87 percent water, and a little over 4 percent each of fat, protein, and lactose sugar.

Six major breeds of dairy cows are recognized in the United States. The two most common breeds are Holstein and Jersey. The other four breeds are Guernsey, Brown Swiss, Ayrshire, and Shorthorn. Each breed is different in terms of size, appearance, milk production, and percentage of milk fat.

Holsteins are black and white. The average Holstein produces the highest amount of milk by volume. Jerseys are gray, tan, or fawn colored with dark eyes. Jerseys produce milk with the highest percentage of milk fat.

Dairy cattle production focuses on cows only. Male calves and older, nonproductive cows are used for beef.

Swine

Swine are raised primarily for their meat—pork. The meat may be processed into ham or bacon. Swine are also raised for breeding and continuing the species. One key to swine production is the availability of their primary feed, corn. Swine are monogastrics. They cannot eat large amounts of roughage.

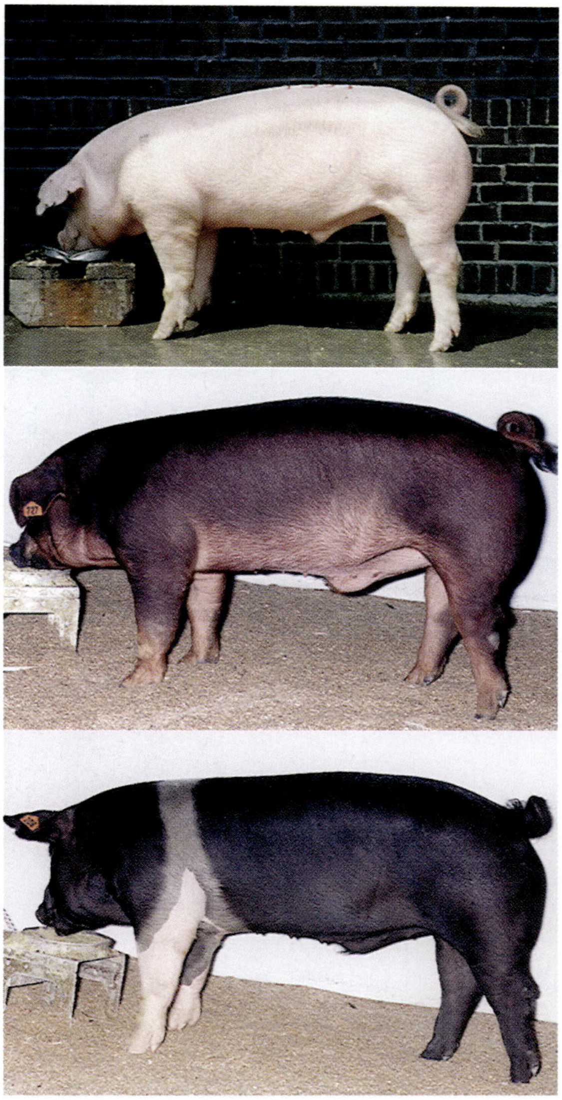

13-10. Three popular hog breeds are Chester White (top), Duroc (middle), and Hampshire (bottom). (Courtesy, *Chester White Record*, National Duroc Swine Registry, and Seedstock Edge)

The birth of piglets is known as ***farrowing***. Sows usually produce seven to twelve piglets twice yearly. The ears are often notched at birth to identify individuals.

Eight major breeds are recognized in the United States. Breeds are distinguished by color, ears, and litter size. Four of the most common breeds are Yorkshire, Duroc, Chester White, and Hampshire. Hybrids are common.

Yorkshire swine have white skin and erect ears. They are well known for their mothering ability.

Durocs are red animals with drooping ears. They have large litters and are also excellent meat animals.

Hampshires are black with a white belt around the front legs. The size of the belt varies tremendously. Hampshires also have erect ears.

Growers raise swine to different sizes. Some raise them from farrowing to finish, while others raise them from farrowing to a feeder weight of 40 pounds. Still others are involved only with finishing operations. A few growers raise purebred animals for breeding stock.

Swine have many organ systems similar to humans. Two examples are the heart and circulatory system and the digestive system. For this reason, swine are used in medical experiments to find treatments for human illness.

13-11. Some students raise and show swine.

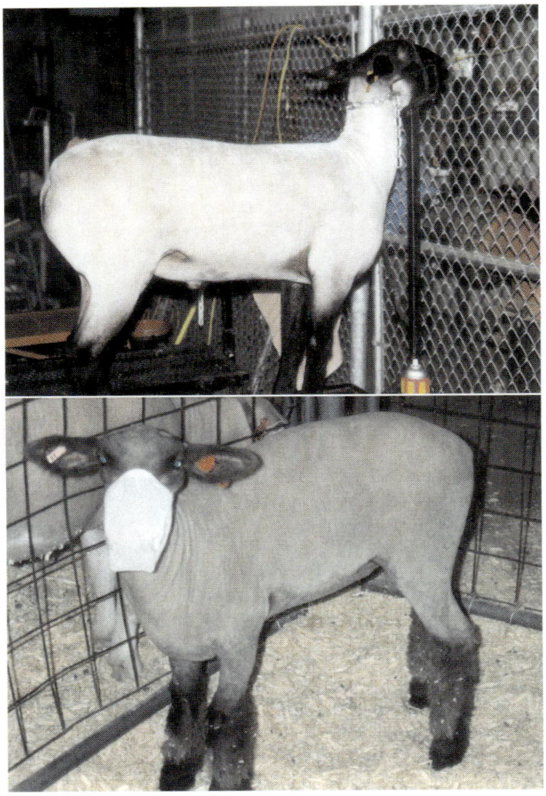

13-12. Two sheep breeds are Suffolk (top) and Hampshire (bottom). (Note: The muzzle on the Hampshire lamb prevents it from eating the bedding made of wood shavings.)

Sheep and Goats

Sheep and goats are grouped together because of similarities in production. They are both ruminants, but graze on different materials.

Sheep and goats are raised for wool, meat, milk, and hides. Their popularity has decreased in the United States. More sheep are raised than goats. The animals are selected depending upon their purpose. Producers may seek fine hair fibers, heavy hair fibers, or meat.

More than 20 breeds of sheep are recognized within the United States. Breeds are characterized by color, size, and relative value of products. Generally, an animal with good meat has poor wool, and vice versa. Therefore, many sheep are crossbred to obtain the best of two breeds. Over time, three groups have been created: wool sheep, meat sheep, and dual use sheep. Sheep may be polled or horned.

Two common breeds are the Suffolk and Hampshire sheep. Suffolks are white with a hairless black face and black legs. Suffolks are excellent meat animals and have short, medium-textured wool.

Hampshires are similar in color, with white wool and black hair on their faces, but with white legs. Hampshires are large and have medium-length, medium-textured wool.

Goats are grouped by use, with distinct breeds within several groups. The groups include Angora, Cashmere, dairy, meat, and pygmy goats. Angora and Cashmere goats are raised for their hair fibers. Dairy and meat goats are raised for those products. Pygmy goats are smaller than the others and are raised as pets.

Sheep and goats are raised in a flock system. The animals live in large groups that may roam freely or may be kept in a pasture. The most common goal is to produce meat. Wool is a secondary product.

Horses

Horses are primarily used for pleasure. Some horses are still used in a working environment, such as managing cattle or pulling loads.

Horses are closely related to donkeys. The two species are sometimes interbred to produce mules, an animal excellent for work. Horses and donkeys are monogastrics with modifications to their systems. They have a large pouch called a cecum between the small and large intestines. This structure continues to be studied, but scientists believe some digestion occurs here.

Horses are selected for their size based on purpose. Horses are measured using hands. One hand is equal to four inches. Horse breeds may be distinguished by purpose or color. The most

13-13. American Quarter Horses are popular light horses. (Courtesy, American Quarter Horse Association)

common groups include light horses, ponies, draft horses, and distinctive color patterns.

Light horses are used for riding and harness. They are used for pleasure, racing, and pulling small vehicles. The American Quarterhorse is a common breed of light horse. Light horses are the intermediate size of horse.

Ponies are the smallest horse breeds. They are commonly ridden by children. Shetland, Welsh, and miniature ponies are the three recognized breeds.

13-14. Miniature donkeys are kept by some people. (Courtesy, American Donkey and Mule Society)

Draft horses are the largest breeds of horses. They are used in working situations, often pulling heavy equipment. Clydesdales are massive horses representative of this group.

Special color patterns distinguish some horse breeds. For example, Palomino horses have a distinctive golden color.

Horses are commonly raised as individuals or small groups. Since horses are not often raised for food in the United States, it is common for a family to keep a single animal as a pet.

13-15. Some people enjoy keeping unusual pets, such as this Rose Hair Tarantula.

SMALL AND COMPANION ANIMALS

Animal science includes small animals. Small animals are kept as companions (pets) and are used in labs and as helpers.

One key with companion animals is a clear understanding of their needs. Their background may influence how they adapt to a new home. Animals removed from the wild to become pets often have trouble adjusting.

Dogs, cats, rabbits, rodents, reptiles, birds, and fish are all examples of small animal species. Some of these are companion animals, others are food sources, and many are used for both.

Dogs

More than 300 breeds of dogs are recognized worldwide. Dogs are monogastric mammals. Usually, dogs are fed a prepared mixture for a balanced ration.

Dogs are most commonly grouped in seven classes based on use. The classes are herding, sporting, tracking, working, terrier, toy, and nonsporting.

Besides being pets, some dogs are service animals. They may help an owner by guiding them, by guarding the home, by herding animals such as sheep, or by finding illegal substances. Dogs are used in law enforcement.

13-16. Border Collies are kept as companion animals and trained to herd livestock.

Cats

Cats are also monogastric mammals. They are commonly fed commercial foods.

Cats may be grouped as long-haired or short-haired. Thirty-six breeds of cats are recognized. Typically, cats are not purebred but are a mix of many breeds. Cats may be kept in a small home and not go outside.

13-17. Mixed-breed kittens.

13-18. Rabbits are colorful animals.

Rabbits

Rabbits are raised for food or as pets. Rabbits produce many young in a short time. Since they are used as meat animals, rabbits are sometimes grouped with chickens.

Rabbits may be purebred or crossbred. The breeds differ in size, ear shape, and color.

Rabbits are often used for human medical study because they have similar diseases. For example, rabbits may suffer from spina bifida or cleft palates, two ailments affecting humans.

Rodents

Rodents are a specific order of mammals. They have large front teeth for chewing. Many rodents are used as pets. They are also common animals for medical research. Gerbils, hamsters, guinea pigs, mice, and rats are examples.

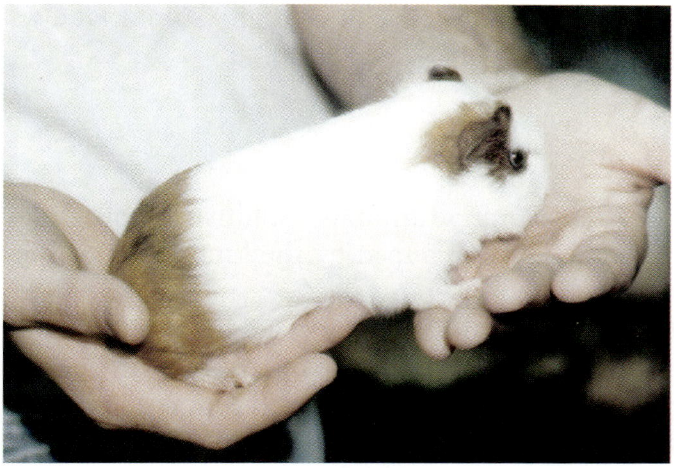

13-19. The Silky Guinea Pig is an appealing animal.

These animals can be kept in cages within the home. As with most animals, they require a clean shelter, good nutrition, water, and attention. Rodents should not be returned to the wild. They can cause much damage to crops and stored grain.

Reptiles

Reptiles are cold-blooded, egg-laying animals. They typically have a dry, scaly skin. Since they cannot adjust their body temperature, most reptiles require heat sources from their environment. A reptile feels cool because its temperature is the same as its surroundings.

Lizards, skinks, snakes, and some turtles are reptiles kept as pets. One species of increasing importance is the iguana—a lizard that may grow as large as 6 feet. Obviously, a pet owner should consider space before acquiring this pet!

13-20. Keeping a Blue-Tongued Skink requires special knowledge. (Courtesy, Susan Osborne, Florida)

Reptiles are one group of animals that may have been removed from the wild. Some reptiles are difficult to raise in captivity. Potential owners should ask whether the animal is wild or farm-raised.

Fish

Fish are used as companion or decorative animals. Goldfish, tropical fish, and saltwater fish are common. Fish are grouped by the kind of water they live in, such as fresh or salt and cold or warm.

One common fish is koi. These are similar to large goldfish and are kept in ponds. Koi-raising is an art form in some areas of Asia, such as Japan.

In recent years, fish and other aquatic animals have become more common as farm animals for food. This is an expanding industry. Catfish, trout, and salmon are now commonly raised in a farm environment.

Fish, like other animals, may cause extreme damage if they are returned to the wild. Many states have laws governing the types of fish that may be raised or sold.

13-21. Fish, like these discus, are popular because they require relatively little space.

Birds

Some birds have been domesticated for food, such as chickens and turkeys. These birds are commonly called poultry.

Other birds may be kept for companionship. One unique feature is that some species can talk. Others sing. Parakeets, cockatiels, and canaries are common bird pets.

The illegal capture of immature or endangered species of birds, including many parrots, continues to be an issue. If you are considering a parrot or other less common species, ask the owner about the origin of the bird. They should be raised in captivity or the owner should have documents proving the legality of the bird.

13-22. The Mexican Redhead Amazon Parrot is a colorful bird.

POULTRY

Domesticated birds may be raised for meat, eggs, and feathers. Poultry may also be called fowl. Birds have beaks, feathers, and lightweight bones. They are egg-laying.

Poultry are monogastrics with a modified digestive system. Since they have no teeth, they swallow only small food particles. Birds have a crop, which stores food. They also have a gizzard, which grinds food into smaller digestible particles. Birds eat small rocks and sand to help break down feed.

Poultry produce eggs rather than live young. The egg is formed entirely in the female. In chickens, the process takes approximately 24 hours. It begins with the yolk and gradually layers of white, membrane, and shell are added. Calcium is often included in a poultry diet to increase the strength of the egg shell.

The common species of poultry are chickens and turkeys. Other species include ducks, geese, guineas, peafowl, emu, and ostrich. The production of game birds, such as partridge, quail, and pheasant, has also increased.

13-23. A colorful rooster provides an early morning wake-up call.

Chickens

Chickens are primarily raised for eggs and meat. Specific breeds of chickens are not used much today. Most chickens are raised in large quantities. Each business has specific varieties they have selected.

Chicken meat comes primarily from young birds raised specifically for that purpose. Those birds are called broilers or fryers. They may be of either sex. Meat may also come from older birds, castrated males, or hens that are no longer laying eggs.

Hens are mature females. Laying hens, or layers, are kept specifically for egg production. The hens usually have no contact with males and the eggs are never fertile.

Chickens are commonly raised in one of four environments: broiler production, egg production, pullet production, and breeding production.

Broiler production raises the birds for meat. The focus is keeping young birds healthy. These birds reach a market size in about six weeks and are feed-efficient.

13-24. Egg production using a cage system. (Courtesy, Mississippi State University)

AgriScience Connection

SHOWING ANIMALS

Some animals are specially managed for showing. This is a form of competition among animal owners to see who has the best animal. A judge carefully studies each animal in terms of an ideal. Each animal is ranked based on how it compares with the others.

Getting an animal ready for a show requires work. The well being of the animal must be considered. Never abuse an animal. Proper feed and water are needed. Good bedding and protection from the weather are essential.

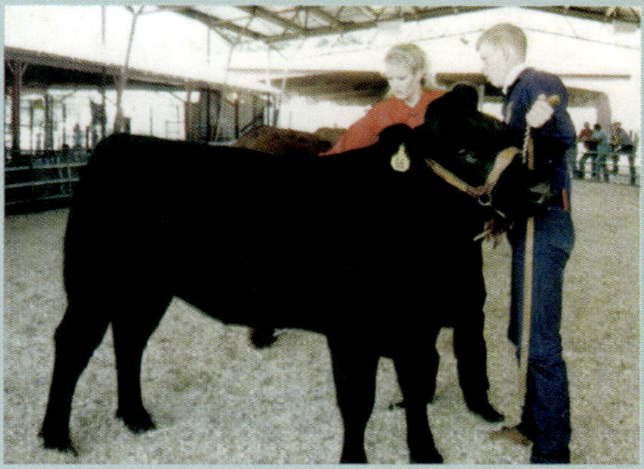

A beef animal is groomed, halter broken, and taught how to lead. The person with the animal must know how to hold the halter and control the animal.

Egg production is to provide high quality eggs to consumers. The goal is to have each hen produce one egg each day. Proper nutrition is a key. The eggs must also be kept clean and undamaged.

Pullet production is raising young female birds and preparing them for laying. This operation raises only females, as males cannot provide eggs.

Breeder production raises young males and females to continue reproduction. The emphasis is on high quality and healthy birds.

Turkeys

Turkeys are raised for white meat. They are usually raised in large flocks. Most are kept in pens. Some turkeys are kept in barns or turkey houses.

Turkeys are marketed in 14 to 18 weeks after hatching. Females are sold earlier than males. Turkeys have been bred so exclusively for the white breast

13-25. An ostrich kept in a small pasture area.

meat that many cannot function normally. Although first considered a holiday food, turkeys are now produced and eaten year-round.

Other Poultry

Because poultry is known for low-fat meat, many other species are raised for meat and eggs in small quantities. Ducks and geese are commonly raised in European countries.

Peafowl, often called peacocks (which is more appropriately the term for males), are raised as companions and for their distinctive feathers. Peafowl are also known for a unique cry, which some people compare to a human scream.

Ostrich and emu are large, nonflying birds. Their meat is especially low in fat, but has a flavor similar to beef. Their feathers may be used decoratively, and their bodies produce oils with medicinal value.

ANIMAL STRUCTURE AND FUNCTIONS

Knowing body structures and functions helps in raising animals. Whether the animal is large or small, similar conditions are needed. All need a good environment for living.

BODY STRUCTURES

All animals have similarities. The structure begins with cells—the smallest building blocks in an animal. Groups of similar cells form tissues. Tissues form organs. Organs with similar functions form organ systems. The organ systems form the organism.

Cells

The cell is the basic building block of life. All living things are made up of cells. Plant cells have walls; animal cells have only membranes. The cell provides information and uses energy.

Probably the most important part of the cell is the nucleus. Within the nucleus are genes that contain complete instructions for the organism. These genes are important in reproduction and in biotechnology uses.

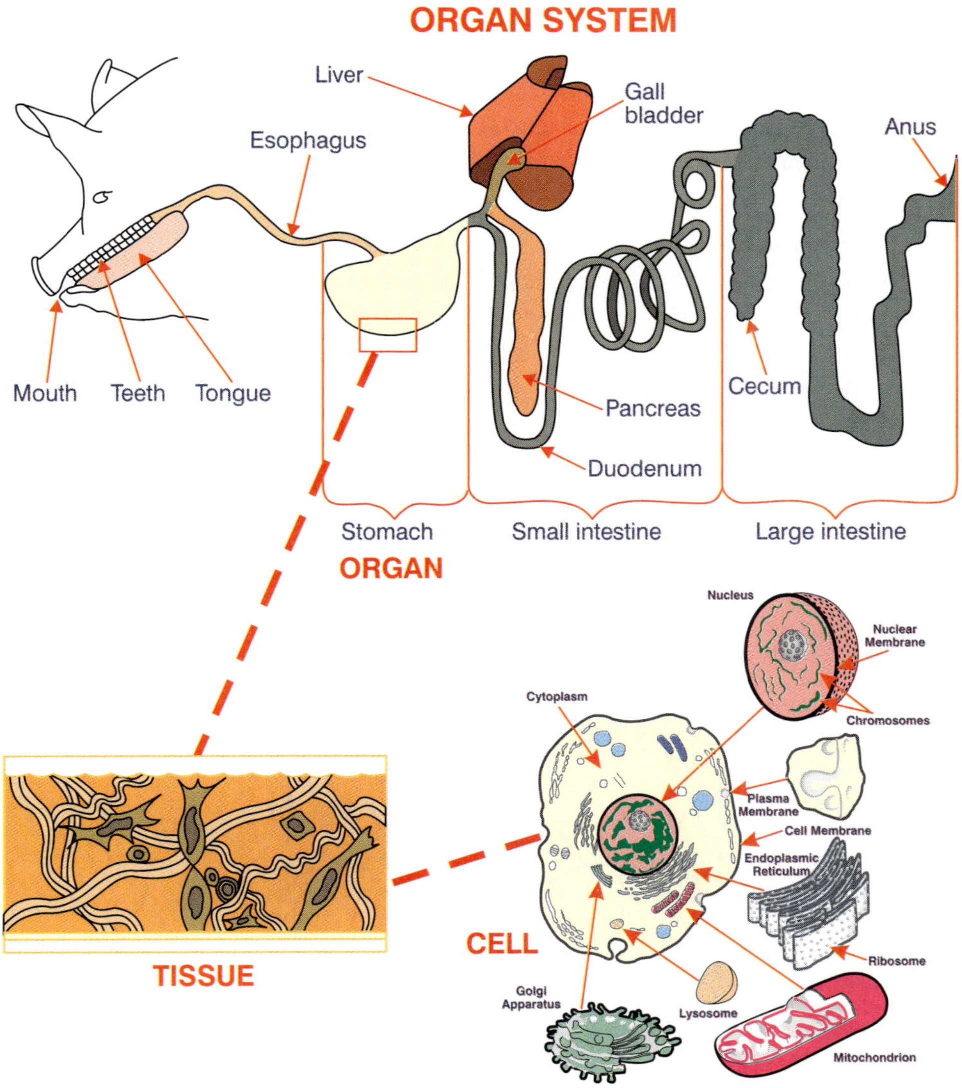

13-26. Cells form tissues. Tissues form organs. Organs working together form an organ system.

Tissues

Four kinds of tissue are in animals. The four tissues are protective, connective, muscular, and nervous. Tissue is a group of cells that do a specific job. Tissues may be protective, like skin. They may be connective, joining various body parts. Another tissue is muscular and aids in movement. The nervous tissue responds to outside factors and transmits information.

Organs

Organs are groups of similar tissues that work together to form a specific function. Animals have many organs, which typically do not work alone. Examples of organs are the heart, liver, and kidney.

Organ Systems. Organs form organ systems. The organs work together as a system to do certain activities. Most animals have ten organ systems. Without the systems, the animal could not survive. For example, the circulatory system consists of a heart, veins, arteries, and capillaries. Another example is the digestive system. It consists of the mouth, stomach, intestines, and other parts.

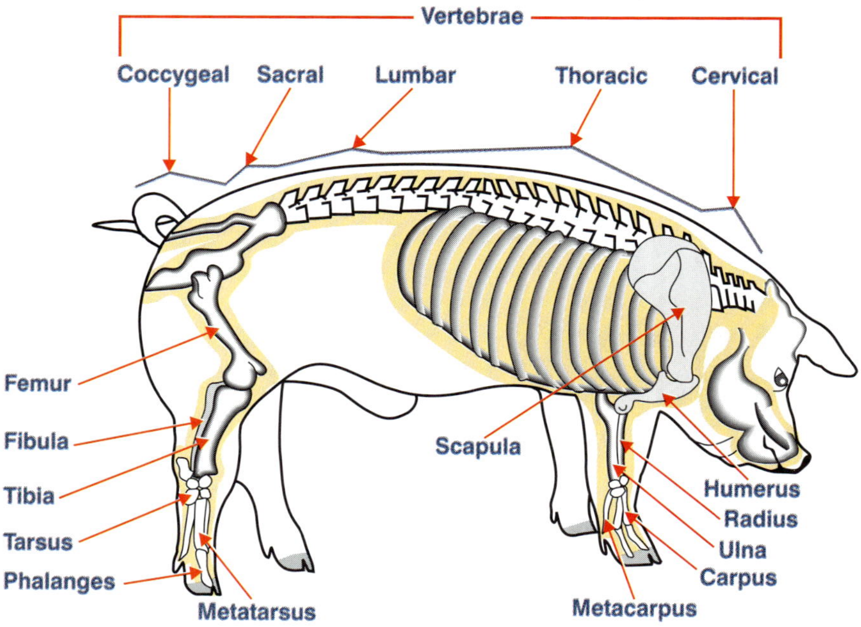

13-27. The skeletal system of a pig supports the muscles and other organs.

Table 13-2. Organ Systems and Functions

System	Function
Skeletal system	provides a framework for the body
Muscular system	creates body movement
Digestive system	prepares food for use
Respiratory system	exchanges gases, especially oxygen and carbon dioxide
Circulatory system	moves blood throughout the body
Excretory system	rids the body of wastes
Lymphatic system	circulates lymph fluid
Nervous system	coordinates body activities
Integumentary system	covers and protects body (skin)
Reproductive system	produces offspring

The organ systems are similar from one animal to another. The systems and major functions are presented in Table 13-2.

NUTRITION AND FEEDING

One organ system of particular importance in raising animals is the digestive system. This system prepares food for use by the body.

Food is used for life processes, including growth, development, reproduction, and work. Three keys to feeding are understanding digestive systems, nutrient requirements, and feeding strategies.

DIGESTIVE SYSTEMS

The digestive system breaks food into usable parts. Two different types of digestive systems are common in animals: the monogastric system and the ruminant system. These may also be called the non-ruminant system and the polygastric system.

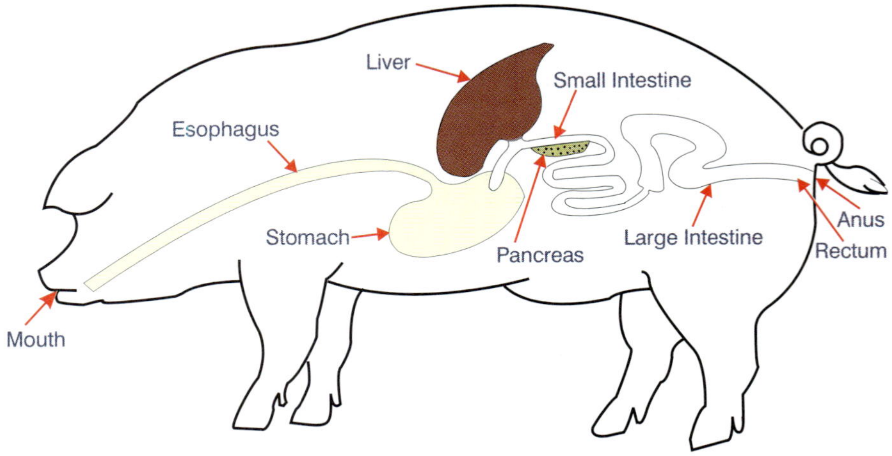

13-28. A monogastric digestive system includes the mouth, stomach, and intestines.

Humans and pigs both have a simple monogastric system. The basic ***monogastric*** (non-ruminant) system consists of the mouth, stomach, small intestine, and large intestine. Until the small intestine, the food simply is broken into smaller and smaller particles. Nutrients are absorbed in the small intestine. The large intestine absorbs water and forms feces—solid waste.

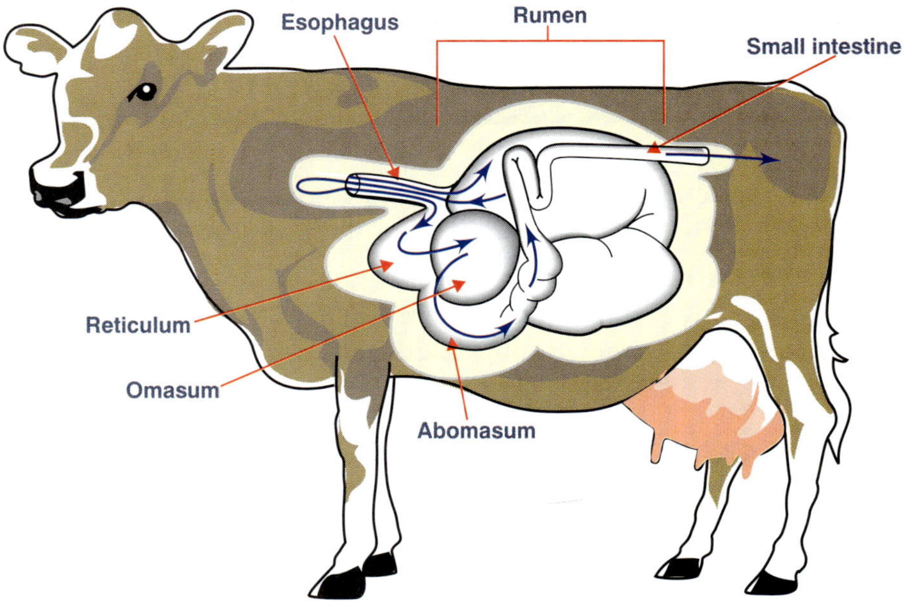

13-29. The ruminant has four separate compartments in its stomach.

Poultry and horses are monogastrics with modified systems. Those modifications covered earlier in the chapter.

Cattle, sheep, and goats are ruminants. A ***ruminant*** has a mouth, small intestine, and large intestine. Their stomachs, however, are divided into four compartments. Those compartments are the rumen, reticulum, omasum, and abomasum. Each compartment performs a slightly different function. The system is good at digesting grass and other roughages.

Ruminants digest their food gradually. The process is assisted by bacteria in the rumen. The bacteria help break down fiber particles. Ruminants also chew their food many times. They regurgitate pieces of food, called cud. The pieces are chewed and reswallowed. The process continues until the food is thoroughly chewed.

Nutrient Requirements

A ***nutrient*** is any substance required for life. Animals need six types of nutrients: water, protein, carbohydrates, fats, vitamins, and minerals. Some scientists include a seventh—air. Nutrient needs vary based on the type and size of animal, as well as the life stage they are in. For example, a lactating cow (producing milk to feed her young) requires more feed.

Water may be obtained from drinking and from the feed given animals. Water is the major component in cells and helps transport other nutrients.

Protein is the building material. It is needed for the growth of muscles, tissues, and bones. Protein also helps repair damage or injury to cells. Protein is important for weight gain, growth, and reproduction.

Carbohydrates may be simple or complex. They provide energy for all animals. Carbohydrates should make up about 75 percent of an animal's diet. Carbohydrates also provide fiber, which helps the digestive system run more smoothly. Ruminants can digest this fiber; non-ruminants cannot.

Fats, also called lipids, are part of animal cells. They provide energy and carry some vitamins. They also help form certain chemicals used in body functions.

Vitamins are nutrients required in very small amounts for specific functions. Sixteen known vitamins are required. These include vitamins A, D, E, K and the B-complex. Ruminants can produce some vitamins within the rumen.

Minerals are chemical elements required by the skeletal system. Other systems also require minerals. The minerals needed in the largest amounts are calcium and phosphorus.

13-30. Animals need feed to grow.

Feeding

Animals are fed differently. A ***ration*** is the feed given an animal in a 24-hour period. A ***balanced ration*** contains all the nutrients in correct proportions. Animals grow best with a balanced ration.

A key for an animal producer is feed efficiency. Feed efficiency is the amount of feed in comparison to production either as weight gain or milk production. For example, 5 pounds of feed yeild 1 pound of pork.

Specific types of rations or diets include maintenance, growth, reproduction, lactation, and work. The requirements vary with each type.

Animals are fed three types of food or feed: concentrates, forages, and supplements. The quantity of each varies based on the type of diet.

A ***concentrate*** is high in energy and low in fiber. Concentrates are feeds such as corn, wheat and other grains. They are easily digestible. They typically are the seed parts of plants. Concentrates are rarely fed in a maintenance ration.

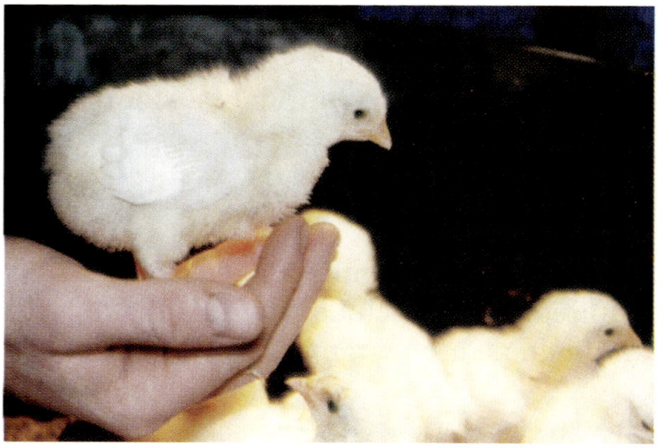

13-31. This day-old chick will need a balanced ration for proper growth.

Forages or ***roughages*** are mostly leaves and tender stems. Forages may be dried or green. Common roughages are grasses, legumes, hay, and silage (chopped plant parts.) Forages are high in fiber and low in protein and energy. Forages cost less than other feeds. They are fed in large quantities to ruminants.

Supplements provide nutrients lacking in the concentrates or forages. They typically are high in a specific nutrient. Many supplements are meals made from crops as a secondary product. Examples include cotton seed and soybean meals, which are high in protein.

Animals may be fed free access, which means they can eat any time. When the animal wants to eat, it can. Water is provided free access. Another way to feed is with scheduled feeding. Feed is available at certain times of the day. Cattle must be fed concentrates on a schedule or they will overeat.

REPRODUCTION

Reproduction is the process by which animals are produced. With many animals, reproduction is the production of sex cells, mating, pregnancy, and birth. Birds and fish differ in that the young usually develop outside the mother's body. Reproduction is not essential for an animal to live.

PRODUCTION OF SEX CELLS

The male sex cell is the sperm. They are produced in the testicles. The female sex cell is an ovum, which is produced in the ovaries. The two sex cells join to form a zygote. As the young develops, it becomes first an embryo and then a fetus.

MATING AND BREEDING

Mating is the physical union of two animals. It is a natural process. Successful mating depends on readiness of the female. Females of most species have reproductive cycles and are only fertile (capable of reproducing) at specific times. A female animal that is fertile is said to be in heat.

Breeding is helping animals reproduce. Animal scientists do this by controlling male and female animals. They may be kept in separate pens or pastures. Hormones are sometimes used to cause fertility.

Animal producers may use ***artificial insemination.*** This is using implements to place sperms in the female mechanically. In this case, the sperms

13-32. This cow-calf pair are purebred registered Simmentals. (Courtesy, Victor A. Winslow and Sons.)

are collected from the male beforehand as semen. The semen may be frozen to be kept for an extended period. One advantage to artificial insemination is that an animal can be bred to a high-quality mate from anywhere in the world without moving the animals. In addition, more offspring may be obtained from each male.

Over time, animal species have been bred for specific qualities. This has led to ***breeds***, which are groups of animals with consistent and distinctive traits. Breed names often come from the region of the world where the breed was developed.

A ***purebred*** animal has two parents with the same set of distinctive characteristics. The animal has a documented pedigree, which is a certificate proving its parentage. The pedigree is obtained from a registering agency for a fee.

A ***hybrid*** animal has parents with different characteristics. In some cases, this results in hybrid vigor. This means that offspring have the best qualities of both parents.

In fish, the union of ovum and sperm may occur within the female's body or outside the body. In some cases, the eggs are laid by the female and the male swims over the eggs and releases sperms.

In chickens and other birds, a single mating may result in the fertilization of many eggs. Eggs mature in the female body in about 24 hours.

PREGNANCY

In mammals, the young develop in the mother's body. The period of development is known as ***gestation***. The time varies from 31 days for rabbits to 336 days for horses.

Once the sperm and the ovum unite, the newly formed organism attaches to the uterus of the mother. During development, the embryo and fetus get nutrients from the mother.

The number of young carried differs by species. Swine may have as many as 14 or more piglets. Cattle and horses typically have only one offspring at a time.

Since some fish and birds do not carry their young within their bodies, they are never pregnant. The young develop within the egg to the hatching stage.

13-33. Some baby animals, such as this bulldog puppy, have sealed eye lids for a few days after birth.

Birth

Birth is ***parturition***. At this time, the young animal is pushed from the mother's body. Other tissues associated with the fetus are also pushed out. Often, the birth naturally occurs and the animal producer does not participate in any way.

One potential problem with birth is the position of the young. In most species, the babies are born head and front feet first. If the baby does not appear this way, the producer may have to help the mother by turning the baby.

Immediately following birth, hormones in the mother's body stimulate the production of milk. This is known as lactation. Lactation is particularly important in dairy cattle where milk is the primary product.

Fish and birds hatch from their eggs. The length of development varies widely. Chickens hatch 21 days after the egg is laid. The animals receive nutrition from the yolk for several days after they hatch.

ANIMAL WELL BEING

Once an animal is born or hatched, the producer is responsible for its well being. ***Animal well being*** is caring for an animal so all its needs are met and it does not suffer. Healthy animals are alert, eat, and have normal body functions.

HOUSING

Animals may be kept in the controlled environment of a barn or in an open pasture. The decision about which type of shelter to provide depends upon the animal and type of operation.

13-34. Chickens grown in houses require proper management. (Courtesy, Michael Stevens, McCarty Farms)

Sometimes, many animals are in a relatively small area. Cattle feedlots, hog farms, and chicken houses are examples. Feed, water, and health are all carefully monitored. Other factors, such as temperature and humidity, may also be controlled. For poultry, light may be controlled as well.

On the other end of the scale, many cattle and other ruminants are allowed to roam large areas of land. They are exposed to the elements and must find their own food.

SANITATION

Sanitation is having a clean environment. Proper sanitation prevents physical damage to the animal. Sanitation also reduces disease.

Most facilities have a way to remove animal waste—a primary cause of uncleanliness. They may also restrict visitors, thereby preventing the introduction of diseases.

13-35. Keeping cages clean is a part of sanitation.

Disinfectants may be used to destroy harmful organisms. Bleach and alcohol are common disinfectants. These and others are part of the sanitation program for animals susceptible to diseases.

DISEASE PREVENTION

A disease is a disorder that prevents normal function. Animal diseases may be caused by poor nutrition, bacteria, viruses, or parasites.

Some diseases impact only one species of animal while others attack several species. Some animal diseases may even be passed to humans.

Diseases are prevented by proper nutrition, monitoring of the animals, vaccination, and prompt attention to problems.

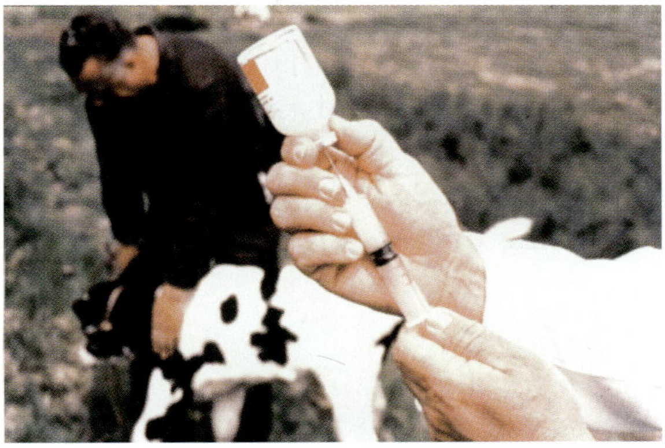

13-36. Preventing disease may involve giving vaccinations. (Courtesy, Mississippi State University)

BEHAVIOR MANAGEMENT

Producers may make changes to animals to protect them or to produce the desired animal product. Some of these changes are castration, docking, branding, ear notching, dehorning, and debeaking.

Castration

Castration is removing the male testicles. It is typically done early in an animal's life. Castration removes the organs that produce sperm and hormones, which cause aggressive behavior. This prevents unwanted breeding, leads to better meat, and results in less fighting among males.

Docking

Docking is removing the tail. It is common with sheep. Sheep cannot move their tail during bowel movements. Without docking, the area around the anus would become dirty and disease could result.

Identification

Branding and ear notching are used to mark animals. Branding identifies an animal in some way. It is especially common for beef cattle. Branding is placing a distinct mark on an animal. The appearance and location of the mark help in identifying the animal.

Ear notching distinguishes individual piglets in a litter. Ear notching is cutting small pieces of the ear away in a numbering system. It is similar to ear piercing in humans. Animals may also be marked by a variety of tags, including those with bar codes.

Dehorning

Dehorning is removing the horns on a young animal. It is most common in cattle and is done with chemicals or by mechanical means. Dehorning protects humans and other animals from being injured by a horn.

13-37. A calf that has recently been dehorned by destroying tissues that will form horns.

Debeaking

Debeaking is removing the tip of a bird's beak. It is common among chickens to prevent them from attacking each other. Debeaking is sometimes used with companion birds to prevent them from attacking their owner.

REVIEWING

MAIN IDEAS

Animals help humans enjoy life. Besides providing meat, they provide dairy products, eggs, wool and mohair, and many by-products.

Animals carry out many functions. Animal scientists study cells, tissues, and organ systems to understand how animals function. Animals must receive proper nutrition to support these functions. Animals use food differently depending upon their digestive system and their life stage. Reproduction is an important, but not essential, life function and includes production of sex cells, mating, pregnancy, and birth. The animal producer controls housing, sanitation, disease, and behavior to secure animal health and well being.

Animals commonly used in animal science may be grouped as livestock, small and companion animals, and poultry. Livestock are large animals raised in a farm or ranch-like setting. They include cattle, swine, sheep, goats, and horses. Small animals are commonly kept in a home or experimental setting. These include dogs, cats, rabbits, rodents, reptiles, fish, and birds. Poultry are domesticated birds raised for meat or eggs in a farm setting.

QUESTIONS

Answer the following questions. Use correct spelling and complete sentences.

1. What are some common products from animals? Which of these products do you use?
2. What is a by-product? What roles do by-products fill?
3. What is the relationship between cells, tissues, organs, and organ systems?
4. Why do animals need food?
5. What nutrients do animals require? What purpose does each serve?
6. Compare and contrast monogastric and ruminant digestive systems.
7. Why do animals require different quantities and amounts of feed at different times of their lives?
8. Why are balanced rations and feed efficiency important to an animal producer?
9. What are the steps in the reproductive process?
10. Why are animal producers concerned with animal well being?
11. What roles do housing and sanitation play in animal well being?
12. Why would a producer take steps such as castration or debeaking?

EVALUATING

Match the term with the correct definitions. Write the letter by the term in the blank provided.

a. livestock	e. anatomy	i. castration
b. meat	f. ruminant	j. poultry
c. by-product	g. forage	k. ration
d. reproduction	h. breed	l. polled

____ 1. Feed of leaves and tender stems.
____ 2. Secondary product of animal production.
____ 3. Domesticated birds, such as chickens.
____ 4. Naturally hornless.
____ 5. Removal of the testicles.
____ 6. Large animals produced in a farm or ranch setting for food.
____ 7. Muscular tissue of the animal.
____ 8. Study of animal structures.
____ 9. Has a stomach with four compartments.
____10. Process by which more animals are produced.
____11. Group of animals with consistent and distinctive characteristics.
____12. Feed given an animal in a 24-hour period.

EXPLORING

1. Research the production of an animal species. Consider all the factors highlighted in this chapter, including terminology, feeding and nutrition, reproduction, and care. Prepare a report for your class that includes this information. Include a visual aid, such as a poster, with your report.

2. Visit one or more animal producers in your area. Have prepared questions about the strategies they use for animal production. Compare their techniques with the information included here. If you visit several producers, compare and contrast their operations. Prepare a report for your class based upon your findings.

3. Volunteer at a veterinarian's office or animal shelter for an afternoon. Learn what strategies they use in providing for the animals' well being. What steps does the vet take to keep animals healthy? What skills and knowledge does it take to become a vet? Prepare a poster or bulletin board illustrating a veterinary career.

14

PLANT SCIENCE

Life depends on plants. They are a vital part of our world. Plants clean the air. They make our surroundings more attractive. Plants provide food for other creatures, including human beings.

Think about the ways you depend on plants. Have you eaten any foods from plants recently? Have you taken medicines made from plants? Do you think the oxygen you just breathed in came from a plant? These questions help us realize plants are an important part of our world.

Food, clothing, and housing products are from plants. Without these products, our lives would be much different.

14-1. Bark is being removed from a Pacific Yew log to make taxol—a drug used to fight cancer. (Courtesy, U.S. Fish and Wildlife Service)

OBJECTIVES

This chapter is about the important role of plants. It includes basics of plant science and the products we get from plants. The objectives of this chapter are:

1. Describe how plants are classified
2. Name and explain the major parts of plants
3. List and describe ways plants are important
4. Explain the conditions needed for plants to grow
5. Define the major plant science industries

TERMS

annual
biennial
binomial nomenclature
dicot
flower
fruit
germination
life cycle
monocot
perennial
phloem
photosynthesis
root
seed
stem
vascular system
winter annual
xylem

http://www.monsanto.com/MonPub/index.html

(This site has information about plants, the environment, and biotechnology.)

PLANT CLASSIFICATION

As plant scientists work, they must talk to each other. To do so, all scientists must use terms others will understand. In naming and grouping plants, scientists use ***binomial nomenclature.*** The binomial nomenclature system was developed by Carolus Linnaeus. Organisms are placed in smaller and smaller groups with more and more similarities.

14-2. Wheat plants provide important cereal grains. (Courtesy, Mississippi State University)

The groups (beginning with the largest) are kingdom, phylum, class, order, family, genus, and species. Many people remember these groupings with phrases, such as "King Philip came over for ginger snaps" (KPCOFGS). The scientific name in this system is the genus and species.

14-3. Poinsettias in this Washington school greenhouse are being grown for the holiday season. (Their scientific name is *Euphorbia pulcherrima.*)

As scientists work, they may group plants in other ways. These are equally important when talking about plants.

PLANT LIFE CYCLES

One way of grouping plants is by life cycle. ***Life cycle*** is the length of time it takes for a plant to complete life from germination to death. The time is usually measured as growing seasons. A growing season refers to the periods of the year in which temperatures are favorable for plant growth. Growing seasons vary with geography.

Annuals

Some plants have fast life cycles. An ***annual*** completes its entire life cycle in a single growing season. In the contiguous 48 states, an annual would be planted between March and June, and would grow, reproduce, and die before the first fall frost.

14-4. Squash are annuals.

One variant of the annual is the winter annual. The ***winter annual*** completes most of its life cycle in a year, but the winter is included. These plants tolerate some cold weather and frost. Winter annuals are planted in the fall, are relatively dormant in the winter, and complete their life cycles early in the next growing season. Many grains, such as wheat and rye, are planted as winter annuals. Winter annuals cannot be planted in areas with long harsh winters.

14-5. Cabbage is a biennial, completing its life cycle in two years. In most situations, however, the cabbage is harvested during the first summer.

Biennials

A ***biennial*** completes its life cycle in two growing seasons. Most biennials grow during the first season and reproduce during the second growing season. One result is that most biennials store food in their roots to maintain the plant during the winter and the start of the second season. Beets and carrots are biennials.

Perennials

A ***perennial*** requires three or more growing seasons to complete the life cycle. Perennials may reproduce often during their life span. Trees, such as ap-

14-6. The flame azalea is a perennial.

AgriScience Connection

WINDOW GARDENING

It is common to have plants in the home. These plants are often an important part of the home decor. In other cases, the home owner may use the plants, such as herbs, in cooking.

Window gardening is simple. Plants should be placed in appropriate containers, given adequate water and nutrients, and kept in a window that provides the right amount of light. Some plants are in hanging baskets. Plants requiring a great deal of light should be placed in eastern or southern exposure windows. Plants typically do not grow well in northern exposure windows.

Plants can contribute to the quality of your life. Consider having a few plants to make your room more attractive.

ples and pines, are perennials. Flowering plants and shrubs, such as roses, are also perennials.

In some climates, plants that would typically be perennials in a more moderate environment may be considered annuals. An example is cotton. Cotton is a perennial but is grown like an annual when planted in climates with frost. This kills the plant and new plants must be started the following year.

SEED LEAVES

A second way of grouping plants is based on how the embryo is formed. Generally, this depends on the number of leaves present when the plant first emerges from the soil. Other characteristics appear as the plant develops. These same characteristics are used in binomial nomenclature, but are not apparent in the name of the plant.

14-7. Cucumber leaves show the branching veins of a dicot.

Dicots

A ***dicot*** is a plant that has two seed leaves when it emerges from the soil. Dicots are also characterized by the grouping of the veins within the stems, branching veins in the leaves, and very attractive or showy flowers. Roses, carnations, and potatoes are all examples of dicots.

14-8. Pearl millet has the parallel veins of a monocot. (Note that the heads are beginning to form for seed production.)

Monocots

A ***monocot*** is a plant that has only one seed leaf when it emerges from the soil. Monocots have scattered veins in the stem, parallel veins in the leaf, and non-showy or plain flowers. Grassy plants, such as corn, timothy, and wheat, are examples of monocots.

WORKING TOGETHER: PLANT PARTS

Plants fill many roles. They are food for humans and animals. They produce fibers used for paper and clothing. They are used to provide shelter, furniture, and packaging. Plants beautify our homes, classrooms, offices, and parks. They are used for medicine, cosmetics, and energy. Plants are essential to the world's well-being!

A plant is a multicellular, highly developed organism that produces its own food. As such, plants have different parts. Each part has a different function. Some parts are vegetative, which means they are only involved with the growth of the plant and not with reproduction. Other parts are reproductive. Their purpose is to create new plants and continue the species.

In a typical plant, the three vegetative parts of the plant are roots, stems, and leaves. These parts are present even in an embryonic plant and are the first to develop. The three reproductive parts are flower, fruit, and seed. As with most organisms, some plants are not typical and lack one or more of the six parts.

THE ROOTS

The ***root*** is a vegetative part of the plant that grows primarily underground. Roots have three functions: anchor plants in the soil media, absorb water and other nutrients, and store food for the plant.

14-9. Healthy roots on most plants are creamy white. (Courtesy, Agricultural Research Service, USDA)

Roots are usually cylindrical and white, tan, or grayish. A root cap is at the end of the root. This is where most new growth occurs. Root hairs are along the length of roots. They absorb water and nutrients.

Roots may be categorized by shape. A fibrous root system has many branches of similar size. A tap root system has one major root, called the primary root, which is much larger than all others. Often, a tap root stores food. The tap roots of some plants are eaten by people.

A root grows downward into the soil. It is one of the first parts of the plant to develop. The root grows because the cells inside are rapidly dividing. As the roots move into the soil, the plant becomes anchored. At the same time, the root hairs begin removing water and other nutrients from the soil. Those items are transported through the roots to the stem and eventually to the leaves, where they are made into food.

14-10. Carrots (on the left) have a tap root system. Corn (on the right) has a fibrous root system.

THE STEM

The ***stem*** is a vegetative part of the plant that supports the leaves, buds, and other organs. Stems transport food, water, and nutrients in a plant. Stems vary in size from the smallest hairlike structures to the trunks of enormous trees.

A first stem is present in a plant embryo. It begins to function early in a plant's development. Stems have important internal and external features.

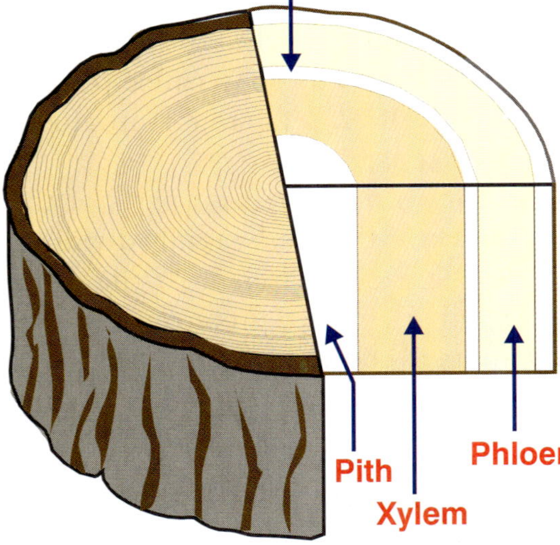

14-11. Cross section of a woody stem.

These properties differ slightly based on the type of plant. The stem has a number of tissues including xylem, phloem, and possibly cambium.

Internal Structure

Xylem and phloem together make up the ***vascular system*** of plants. Just as humans have veins, arteries, and capillaries that carry blood through their bodies, plants have tissues that carry their fluids. Veins and arteries have slightly different jobs. Xylem and phloem do too.

Xylem is a tissue that carries water and minerals from the root hairs throughout the plant. Most fluid movement in the xylem is upward. Many times, xylem is made of cells that are not truly living any more.

Phloem is a tissue that carries plant products, such as glucose, from production sites to other parts of the plant. Usually, this means the fluids are moving from the leaves to the roots. One way to remember xylem and phloem is to remember that fluids in the phloem flow downward just as a river flows downstream.

The third tissue in the stem is the cambium. The cambium is the site of all new cell production in the stem. It creates both xylem and phloem cells. A plant lacking cambium makes all the cells it will ever have during its initial growth. As the plant grows further, these cells just enlarge.

External Structure

Outside, the stem has sections called nodes and internodes. The node is any place where one or more leaves or buds are attached. Sections between nodes are called internodes. The stem also has breathing pores called lenticels.

14-12. This cross section of a corn stem has clearly defined vascular bundles.

Not all stems look the same. Over time, many stems have been modified, or changed through adaptation, to work more efficiently for the needs of a particular plant. For example, the stem of the cactus has become thick and fleshy. It holds water and makes food. The leaves of the cactus have become protective spines, or needles.

Other stems form underground or run across the surface of the ground. One example is the Irish potato, which is a modified underground stem called a tuber. The potato stores food during the winter and produces a new plant in the spring. Another example is the strawberry, which produces stems called stolons or runners. These stems eventually form new roots and another plant. Other stem modifications include rhizomes and bulbs.

THE LEAF

The third vegetative part of the plant is the leaf. Leaves are the "factories" of a plant. Raw materials are changed into usable plant foods in the leaves. Leaves come in many sizes, shapes, and colors.

On the outside, the leaf has two major parts: the blade and the petiole. The blade is the flattened expanded part of the plant. The petiole is the cylindrical part that attaches to the stem.

Leaves are known by their form or shape, the margins or edges, and the arrangement of the leaves on a stem. These features create unique qualities between plant species.

Leaves contain specialized cells and tissues. At the top and bottom is the epidermis—a single layer of cells. This layer, like human skin, protects the internal tissues from water loss and infection. Guard cells are in the bottom

14-13. Leaf shapes of selected trees.

14-14. The needles on this evergreen are the leaves that provide this tree's food.

epidermis. Guard cells control water loss from the plant. Although water loss is important by helping the plant rid itself of wastes, too much water loss causes plant death. The guard cells open and close around a stoma or pore.

The center of a leaf has many cells with chloroplasts. Chloroplasts are responsible for photosynthesis. ***Photosynthesis*** is a chemical process that converts water and carbon dioxide to glucose sugar and oxygen. Some scientists consider photosynthesis the most important chemical process in the world. The food resulting from photosynthesis is used not only by plants, but by animals including humans.

When a plant has grown enough and becomes mature, it then forms the three additional reproductive parts.

REPRODUCTIVE PARTS

The three reproductive parts of plants are flowers, fruit, and seeds.

The Flower

The ***flower*** is the first reproductive part of the plant. It will become the fruit and the seed. The flower is the sign that the plant is mature and ready to produce young. Flowers are formed from highly modified stems.

The flower has four major parts: sepals, petals, the pistil(s), and stamens. Each part has a role in reproduction. Sepals are the outermost section of the flower. They are typically found at the bottom of an open flower and are often green in color. The sepals protect the inner structures of the flower as they develop.

The petals are often brightly colored and useful in attracting insects for pollination. A flower requires pollination for reproduction. Pollination is the

Career Profile

RETAIL FLORIST

Flowers and floral arrangements are used for decoration and gifts. A retail florist is responsible for preparing flowers for special occasions.

A florist cares for flowers on arrival, waits on customers, takes telephone orders, creates arrangements, handles money, and makes deliveries. The florist must have a strong sense of color and be creative.

Florists get on-the-job training. They may also go to specialized seminars at colleges or technical schools. Florists have technical or community college education or a college degree in horticulture. This photo shows a florist finishing an arrangement of roses.

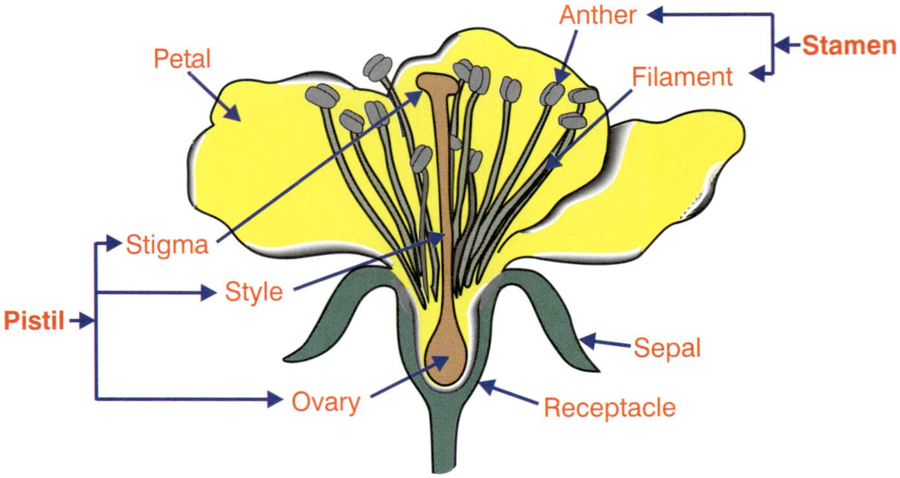

14-15. Major parts of a flower.

14-16. Though small, the bright yellow flowers of the tomato attract insects for pollination.

transfer of the male sex cells to the female organ of the flower. The transfer occurs in many ways. The two most common ways are by insects, such as honeybees, and the action of the wind. Scientists believe petals have markings invisible to the human eye that increase a flower's appeal to insects.

The stamen is the male part of the flower. Most flowers have multiple stamens. An anther is at the top of each stamen. Anthers are tiny sacs that hold pollen—the male sex cells of the plant. The anther is supported by a long, slender tube called the filament. When an anther is mature, pollen is released to be carried to the female part of the flower.

The pistil is the female part of the flower. The number of pistils varies with the type of plant. The pistil has a stigma, style, and ovary. The stigma is a sticky structure at the top of the pistil to which pollen easily adheres. The style is a long slender tube connecting the stigma and the ovary. The ovary contains ovules, the female part of the embryo.

Once pollen is deposited on the stigma (pollination), the pollen cell begins to grow. It produces a tube that grows through the style into the ovary to the ovules. As contact is made, the nucleus of the pollen cell moves through the tube into the ovule and fertilization occurs.

A complete flower has all four parts. Beans, cotton, peanuts, and citrus fruits typically have complete flowers. An incomplete flower is missing one or more of those parts. A perfect flower has both the male and female organs within the same flower. An imperfect flower has either the male or female organs. Corn, squash, and pumpkins are all plants that have imperfect flowers. A sterile flower has neither male nor female parts. It is not involved in reproduction.

Fruit

Following fertilization, the flower parts change. ***Fruit*** forms from the fertilized ovary. The wall of the ovary usually becomes the fleshy part of the fruit. The ovules typically become the seeds.

Fruit is classified by the development of the ovary. A simple fruit, like a peach, forms from a single pistil and is a single mature ovary. An aggregate

14-17. The pineapple is a multiple fruit.

fruit forms from many ovaries clustered as a unit. Aggregate fruits develop from one flower with many pistils. Raspberries and blackberries are common aggregate fruits. Multiple fruits form from many ovaries from many flowers. A pineapple is a multiple fruit.

14-18. Acorn squash are fleshy fruits.

Simple fruits are divided further into fleshy and dry fruits. Fleshy fruits have soft tissue at maturity. Fleshy fruits include apples, tomatoes, citrus fruits, and more. Dry fruits have a hard brittle covering at maturity. Peas, nuts, and many grains are dry fruits.

14-19. Pistachios are dry fruits.

The Seed

Seeds form in fruit. The ***seed*** is the reproductive structure that will become a new plant. Although seeds vary greatly, they all consist of three major parts: embryo, food stores, and protective covering.

14-20. Corn kernels are seeds.

The seed coat is a protective covering surrounding the seed. Within the seed coat is the endosperm, the food stores to be used by the developing plant. The embryo is a complete plant in miniature form. It has one or two leaves, a stem, and a root. At this point, the difference between a dicot and a monocot is the number of leaves within the embryo.

Germination is the process by which a seed starts to grow. It begins with the breaking of the seed coat and the growth of a root into the soil. Germination is complete when the stem and leaves have emerged from the ground and the plant can support itself.

PUTTING PLANTS TO USE

Plants have different uses. Use is based on the nature of the parts of the plants. The six parts of plants are all economically important in different plants. Plants provide food as starch, vegetables, grains, fruits, sugar, oils, nuts, spices, and beverages. Plants provide fiber, timber, fuel, pulp, and an attractive environment. Plants also provide food for the animals that support human life.

HUMAN FOOD CROPS

Human foods are from different plant parts. You probably have a favorite!

The root is an important food source for humans. We eat beets, radishes, turnips, carrots, and sweet potatoes. Roots provide carbohydrates, which are sugars and starches. In fact, one major source of refined sugar is sugar beets.

The stem is eaten as potatoes (a modified stem called a tuber), celery, asparagus, and many others. The stem of sugar cane is pressed to provide the raw materials for sugar. Sap from the stem of maple trees is also used to produce sugar.

The leaves of crops, such as lettuce, spinach, and cabbage, are eaten. Leaves, because of their role for plants, have many nutrients, such as iron. Leaves are very beneficial to people.

It might seem odd, but people eat many flowers. Probably the two most common flowers are the immature flowers of broccoli and cauliflower. Flowers, such as squash blossoms, have also become popular.

14-21. As a flower, broccoli is grown in large fields in the Salinas Valley of California. (This shows a trailer loaded with crates of harvested broccoli next to a field of growing broccoli.)

Fruits are good sources of nutrition. Squash, cucumbers, tomatoes, apples, citrus, and peaches are all fruits. Tropical fruits, which grow only in warmer climates, include dates, pineapples, bananas, and papaya.

Finally, seeds are important in the human diet. Grains, such as wheat and corn, are used in flour, in cereals, and as fresh products. Many seeds are processed into oils and related products. Vegetable oil, margarine, and tofu are seed products. Seeds are also a source of beverages, such as the coffee bean.

14-22. Soybeans are important in many food products. (This shows an open dry pod with beans.) (Courtesy, Agricultural Research Service, USDA)

ANIMAL FOOD PRODUCTS

Animals raised by humans are mostly fed foods of plant origin. Plant products provide protein, fiber, carbohydrates, vitamins, and minerals for animals.

14-23. Cattle graze grasses and other plants for food. (Courtesy, Mississippi State University)

Many animals eat grasses directly from the field or pasture. Here, they consume the leaves and stems of the plants. These plant materials are known as roughage.

At other times, animals are fed grain products, such as corn. These are seeds from the plant. These plant foods are known as concentrates.

FIBER CROPS

Some plants produce fibers that are used to make cloth. This cloth has been the basis of human clothing for many centuries. The two most widely grown fiber crops are cotton and flax.

Cotton fiber is formed in the fruit known as a boll. Each fiber is a long single cell, which originates from the seed coat. Cotton plants require warm temperatures and a 200-day growing season. The fibers are harvested mechanically.

Flax is an annual with very strong fibers in the stem. The fibers are used for linen and paper. Flax grows in cool climates.

14-24. Cotton being harvested. (Courtesy, Mississippi State University)

TIMBER, PULP, AND FUEL CROPS

Many larger plants are sources of products. These include lumber for building and furniture, wood for paper production, and wood as a source of burnable energy. Usually, the usable part of a tree is the big stem known as a trunk. When harvested, the trunk becomes a log or pole.

Plants grown for food may also have nonconsumable parts. These parts may be used as fuel through biomass plants. Crops, such as corn and soybeans, are also being used for fuels, such as ethanol and soy diesel.

14-25. A skidder is pulling logs from the forest.

ENVIRONMENTAL IMPROVEMENTS

Often, plants are used to make an area more attractive. The use of plants outside, such as the lawn around homes and trees in a park, is known as landscaping. The use of plants inside, such as house plants, is interiorscaping.

GROWING CONDITIONS

Plants grow best in a good environment. Growers need to know about the role of water, soil, nutrients, temperature, light, and pest management.

WATER

Water is an important ingredient in plants. It is in every plant cell and tissue. It carries nutrients and chemicals through the plant. It plays a role in

14-26. An aerial view of fields in Colorado shows the effects of irrigation on crops. (Courtesy, Agricultural Research Service, USDA)

photosynthesis and in other chemical processes. Water helps young cells expand. Water even helps the plant cool itself.

Different plants have different water needs. They have adapted to survive on the available water. For example, cacti can survive without water for an extended period. On the other hand, a water lily grows directly in water and cannot survive outside that environment.

Water is available to a plant in two ways: natural sources and irrigation. Natural sources include water in the soil. This water results from precipitation, which varies greatly from one location to another. It must be considered when selecting crops to produce.

Irrigation is providing water to the plant artificially, such as watering a lawn. Many different sources of water and techniques are available. Sometimes irrigation is not practical because of the cost.

SOIL

Soil is the combination of mineral particles, dead plants and animals, water, air, and living organisms required for plant growth. Soil is important in that it is the reservoir for nutrients and water. Soil aids in anchoring plants.

Soil has been successfully replaced with water and nutrient mixes and with inert materials with added nutrients. For example, in hydroponics,

14-27. Inserting a fertilizer spike into the soil of a jade plant provides nutrients over several weeks.

plants are grown directly in water. The greenhouse industry now uses materials like peat moss and bark to make soilless media mixes that, when used with fertilizers, are appropriate for plant growth.

Neither of these replacements, however, is practical on a large scale.

NUTRIENTS

Like humans, plants need specific nutrients. A nutrient is any element that is essential to a plant's growth, development, or reproduction. Nutrients may be grouped as organic, major, and minor. Every needed nutrient must be present for successful plant growth. The difference is in the amount the plant needs.

The organic nutrients are carbon, hydrogen, and oxygen. They are important components in plants. Because they can readily be obtained from the air and soil, they are often overlooked.

The major elements are nitrogen (N), phosphorus (P), and potassium (K). They are often provided to plants through fertilizers.

The list of minor elements varies slightly depending on what is recommended. The nine minor elements most commonly identified as necessary for plant growth are calcium, magnesium, sulfur, iron, manganese, boron, copper, zinc, and chlorine. Other minor elements are vanadium, cobalt, silicon, and sodium.

Plant nutrient supply is one factor in plant production that may be controlled by the producer or scientist. In fact, a successful producer monitors nutrient levels carefully.

TEMPERATURE

Temperature is important in two areas of plant life: the soil and the surrounding air. Temperature requirements vary markedly from plant to plant. Most plant growth, however, occurs in a range from 60° to 90° Fahrenheit. This is true of both soil and air temperatures. Early in the season, the air may be warm, but the soil may still be too cold and plants will not grow. Late in the season, the air may be too cold and even though the soil has retained its warmth the plant will die.

Different plants require different temperatures for growth. Cool temperature plants grow better at the lower end of the scale. Cool temperature plants include beets, lettuce, peas, turnips, and azaleas.

Warm temperature plants grow better at the high end of the scale, above 75 degrees. Warm weather vegetables include corn, melons, peppers, and tomatoes. Warm weather flowers include marigolds, petunias, and most summer annuals.

14-28. Covering the ground with black plastic helps increase the soil temperature.

LIGHT

Light, of course, is necessary for photosynthesis. The amount needed varies significantly. Plants use the red, blue, indigo, violet, and ultraviolet waves of light; they do not use the green light. So, not all artificial sources of light are effective for plant growth. Manufacturers sell plant-grow lights that produce appropriate frequencies.

Light also has an impact on plant blooming and on formation of parts. In addition, light influences the production of hormones in plants.

The light available varies based upon the latitude, season, elevation, and time of day. In an outside setting, agriculturists cannot control the amount of sunlight produced. They do control how much light plants receive, primarily through spacing. In an indoor setting, producers may provide supplemental lighting or they may provide shade to cut down on the light.

PEST MANAGEMENT

Even if all five conditions—water, soil, nutrients, temperature, and light—are met, a plant may still die because of pests. A pest is any unwanted organism. Pests include weeds, destructive insects, mites, fungi, and diseases. Many of these organisms cannot be eliminated from society, but they are controlled in an agricultural setting.

14-29. Boll weevils have damaged this cotton bud (shown greatly enlarged).

Integrated pest management is the most common strategy used by producers. Integrated pest management uses a variety of controls, such as mechanical, cultural, biological, and chemical, to prevent pest damage. The goal is not to eliminate the pest, but to prevent damage that would cause the producer to lose money.

Mechanical Controls

Mechanical controls are those physical steps that may be taken to prevent pests. Examples of mechanical controls include cultivation with equip-

14-30. The white plastic around the tree trunks keeps mice from gnawing away the xylem.

ment, hand hoeing or weeding, removing insects by hand, or burning insects. One mechanical control explored by the potato industry was a vacuum that pulled the insects from the plants.

An extreme example of a mechanical control is plowing a crop under to prevent contamination of surrounding fields.

Cultural Controls

Cultural controls are practices by the producer that prevent pests. Probably the most important control is crop rotation. By planting different crops in a location, insect populations are not allowed to build. Crop rotation also prevents the buildup of disease organisms.

Biological Controls

Biological controls focus on the principle that every organism has predators that will consume it. Those natural predators are released to control the organism. For example, some wasps lay their eggs inside aphids, which suck plant juices. When the eggs hatch, the young destroy the aphid. If the wasps colonize an aphid population, then the aphids no longer threaten the crop.

Chemical Controls

In integrated pest management, chemical controls are the last resort. Pesticides are chemicals that kill pests. Many different kinds of pesticides are available. They work in different ways. The key to chemical controls is to use a chemical only when the damage to the crop surpasses the economic thresh-

14-31. Airplanes are sometimes used to apply pesticides to land and crops.

old. An economic threshold is the level at which damage threatens to make the producer lose money on the crop.

PLANT SCIENCE INDUSTRIES

Many different industries exist within plant science. They may be divided in different ways and often overlap. Three industries are agronomy, forestry, and horticulture.

AGRONOMY

Agronomy focuses on the production of crops in fields, such as grains, forage, and fiber crops. Soil scientists are sometimes also placed in this category.

Agronomists focus on large scale crop production. They manage without crowding the plants or providing a high number of inputs.

Examples of careers in agronomy include crop producers, chemical salespersons, extension specialists, grain elevator operators, and crop researchers.

HORTICULTURE

Horticulture focuses on the production of crops on a small scale with many inputs. Crops produced within horticulture include fruits, vegetables, flowers and decorative plants. Crops in horticulture are typically much more crowded and the environment is much more controlled. All six environmental factors previously mentioned might be controlled in a horticulture setting.

Examples of careers in horticulture are florists, greenhouse inspectors, small fruit producers, and golf course managers.

14-32. These Georgia students are learning about horticulture in the greenhouse at their school.

FORESTRY

Forestry is the management of wooded areas for wood and pulp production and the care of wooded areas for recreation, wildlife, and water. Foresters work with a preexisting resource and try to improve it through practices such as reforestation. Foresters focus more on using what is present rather than on making additional inputs.

Examples of careers in forestry include park rangers, tree farmers, wildlife managers, timber cruisers, and logging equipment operators.

REVIEWING

MAIN IDEAS

Plants have important roles in our environment. They provide food for humans and animals, fiber for clothing, shelter and paper, and make the environment more attractive. Without plants, animals could not survive.

Plants are grouped in many ways. The scientific name comes from binomial nomenclature. Plants may also be grouped as annuals, biennials, or perennials depending on their life span. Dicots and monocots are groups distinguished by embryonic characteristics.

The plant is like a machine, with parts doing different jobs. The roots anchor the plant and absorb nutrients. The stem is the transport system. The leaf is the manufacturing plant, producing glucose through photosynthesis. The flower eventually forms the fruit, which protects the seed. The seed contains an embryonic plant ready to begin the process all over.

Six different factors must be maintained for plant production. Those factors are water, soil, nutrients, temperature, light, and pest management. The degree to which each factor can be manipulated depends upon the locale.

Three industries in plant science are agronomy, forestry, and horticulture. They each vary in the responsibilities and approach to plant production.

QUESTIONS

Answer the following questions. Use correct spelling and complete sentences.

1. What is the reason scientists classify plants?
2. What is the difference between an annual, a biennial, and a perennial?
3. What are the six parts of a plant? Which are vegetative and which are reproductive?
4. Do you think a plant could survive without one or more of its parts? Why or why not?
5. What do you think is the most important function of the plant? Why?
6. What six factors influence plant production?
7. What is integrated pest management?
8. What are the differences between agronomy, horticulture, and forestry?

EVALUATING

Match the term with the correct definition. Write the letter by the term in the blank provided.

a. biennial
b. dicot
c. vegetative
d. photosynthesis
e. phloem
f. agronomy
g. fruit
h. light
i. plant
j. germination

____ 1. Sub-specialty of plant science dealing with field crops.

____ 2. Chemical process through which the plant makes glucose and oxygen.

____ 3. Completes its life cycle in two growing seasons.

____ 4. Forms from the fertilized ovary.

____ 5. Carries nutrients down the stem.

____ 6. Factor important in photosynthesis.

____ 7. Multi-cellular, highly developed organism that produces its own food.

____ 8. Two seed leaves.

____ 9. Process through which the seed starts to grow.

____10. Deals only with growth and not with reproduction.

EXPLORING

1. Research two different crops—one commonly grown in a field and another commonly grown in a greenhouse. Compare and contrast the techniques used with each. What factors are manipulated? What nutrients are added? Prepare a written or oral report describing what you have learned.

2. Purchase an inexpensive flowering plant. Remove the soil and identify major structures throughout the plant. Create a temporary mount or display that illustrates the parts you found. Share the display with your classmates.

3. Visit one or more plant producers in your area. Prepare questions before you visit. Learn about the management practices the producer uses for his or her crops. Prepare a written or oral report on your observations.

15

SOIL SCIENCE

What is soil? Sometimes we get a little confused and call it dirt. Dirt gets on our hands, clothes, and shoes. We wash it off and try to keep it out of our homes. Dirt is something that we think is filthy or unclean. Do we think soil is unclean?

Soil is not dirt. Soil scientists do not call it dirt. They know that it is an important material on the Earth's surface. Soil provides nutrients for plants to grow and make good crops. It needs good care. Soil is fragile and easily degraded. When this happens, the ability of the soil to grow crops is damaged.

We must use practices that sustain and improve the soil. Without good soil, we would likely not have enough food, fiber, and wood products. We do not want that to happen!

15-1. An aerial view of rice harvesting in Arkansas shows a productive field. The pocked places are where the rice has lodged (fallen over). (Courtesy, Agricultural Research Service, USDA)

OBJECTIVES

This chapter describes how soils are grouped, how soil helps plants, and what can be done to prevent soil loss. The objectives of this chapter are:

1. Describe the four major components of soil
2. Explain how soil is classified
3. Describe the chemistry of soil
4. Explain erosion and name ways it can be reduced

TERMS

- bedrock
- clay
- fertilizer
- horizons
- lime
- loam
- major elements
- minor elements
- organic matter
- pH
- sand
- silt
- slope
- soil profile
- soil structure
- soil survey
- soil texture
- soil triangle
- subsoil
- topsoil

http://www.weather.com/twc/homepage.twc

(This site has current weather information.)

SOIL COMPOSITION

Soil is the material that forms the crust or covering of the earth. The materials in soil have nutrients that support plant growth. Plants could not grow without them.

Soil is formed very slowly. It results from the breaking down of rocks and the rotting of dead creatures. Soil not only is the material we walk on, but is important in growing plants.

15-2. Soil is used to produce important crops, such as the corn being cultivated here. (Courtesy, Agricultural Research Service, USDA)

SOIL VALUE

The value of soil depends on what it contains. The amounts of mineral particles, organic matter, air, and water within the soil are important. Each of these is vital to the soil and how it supports plant and animal life.

Soil that has the "right" materials is said to be fertile. Such soil produces good crops. This is because it has the nutrients plants need.

SOIL CONTENTS

Soil contains minerals, organic matter, air and water, and living organisms. The ideal division of these is 45 percent mineral matter, 5 percent organic matter, and 50 percent space for air and water.

Minerals

Minerals in soil are usually very small particles. These are the small pieces that we wrongly call "dirt." The material should never be called dirt—it is soil! Of course, soil on our shoes or under our fingernails is dirt.

Mineral particles are formed by the gradual breakdown of rocks. These are different depending on the parent material. Parent material is the larger

15-3. Action by waves gradually wears rock into mineral particles along the California coast.

Career Profile

SOIL SCIENTIST

Soil scientists study different soils. They collect samples and determine texture and structure. They may conduct chemical tests to determine pH, nutrients, and other conditions. The work may be outside or in laboratories.

A soil scientist needs a baccalaureate degree in agronomy or related area. Many have masters and doctors degrees in the same areas. Practical experience with land, plants, and uses is beneficial. Most soil scientists work with government agencies, research units, or colleges.

This photo shows a soil scientist sampling and mapping soil. (Courtesy, Potash and Phosphate Institute)

rocks that break down. Parent material may result from volcanoes, be deposited by glaciers, be left by an ocean, or be formed at the site of the soil.

Mineral matter provides some nutrients. Its most important function is to provide a place for plant roots to grow. The roots anchor plants and keep them from falling over.

Organic Matter

Organic matter is any product from living creatures. It is formed when a living creature dies and decays. Organic matter may also be waste produced by plants or animals. Humus is organic matter that has decayed to the point it is no longer recognizable. A good place to see organic matter and humus is in the woods. Look under leaves that have fallen. You will see organic matter as it is rotting.

Organic matter provides carbon, hydrogen, and oxygen to the soil, plus other valuable elements. Organic matter helps the soil hold water and nutrients.

15-4. Organic matter is formed by the rotting of stems, leaves, and other materials on this Illinois farm.

Air and Water

The living organisms in soil must have air and water. Too much or too little of either may kill them.

Air and water are in the tiny spaces known as pores between soil particles. The relative amount of water and air varies as conditions change. Shortly after a rain, the pore space may be filled with water. As the soil dries, the amount of water decreases and the amount of air increases.

Water is used by plants in many ways. It is needed in photosynthesis (the process through which a plant makes glucose—food). Water is a part of all living plant tissue. It carries nutrients throughout the plant. It even helps the plant cool itself. Most water used by the plant is absorbed from the soil.

Air is also important to plants, which respire like other living creatures. The roots must be able to gather air from within the soil. It is possible for a plant to "drown" if the soil is saturated for too long.

15-5. Earthworms are shown here in freshly dug soil.

Living Organisms

Soil is home to many organisms. Bacteria, fungi, single-celled protozoa, small worm-like nematodes, and earthworms are in soil. Burrowing animals, like groundhogs and prairie dogs, may also live in the soil. These work together to build soil.

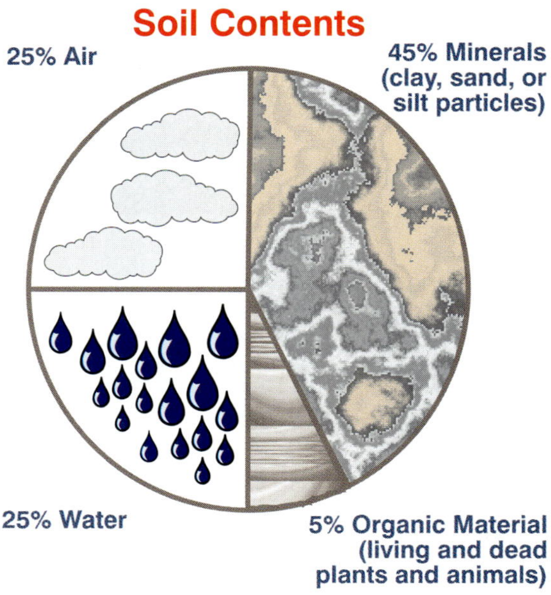

15-6. Soil is made of four major components.

SOIL CLASSIFICATION

Soil varies widely from one site to another. Scientists have tried to group or classify soil. Different ways of doing so have been used. Classification allows soil scientists to identify and discuss soils with their counterparts around the world.

Several different factors are used in the United States. The U.S. Department of Agriculture has developed a system of classification called the Comprehensive Soil Survey System. In this system, soil texture, structure, profile, and color are used to group soils. Soils are grouped into ten orders. The orders are followed by successively smaller groups with more and more similarities.

Over 10,000 different soils have been classified in the United States. Much of this work has been done by the Natural Resources Conservation Service (NRCS), previously the Soil Conservation Service. Once soils have been classified, the NRCS produces soil surveys. Soil surveys are printed in books. A soil survey book has maps, photographs, and information about all the soils in a given area.

15-7. Soil surveys provide technical information about soils.

SOIL TEXTURE

Soil texture is the size of the particles within the soil. Large particles are ***sand***, medium-sized particles are ***silt***, and the smallest particles are ***clay***. The size of the particles tells a lot about soil. The ability to absorb and hold

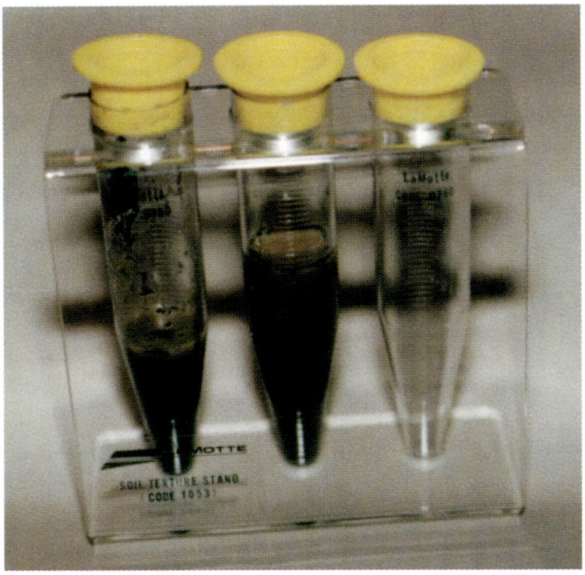

15-8. A soil texture kit helps determine the percentages of sand, silt, and clay in a soil sample.

water is related to texture. Some soils have larger particles, called gravel and stones.

Soil Triangle

Texture is based on the percentages of each particle in a soil sample. The soil triangle shows the divisions. The ***soil triangle*** is a picture graph of soil categories based on the amount of sand, silt, and clay. To use the soil triangle, determine the percentage of each particle present. The three percentages must equal 100.

Two methods are used to learn the proportions of each particle. One, the ribbon method, requires mixing a small soil sample with water and then attempting to create a strip that sticks together. The longer the ribbon, the greater the percentage of clay in the sample. A short ribbon or no ribbon at all indicates a high percentage of sand. The ribbon method requires practice to learn how the strip should feel.

A second method is placing a soil sample in a jar with water. The sample should be shaken thoroughly and then allowed to settle overnight. The particles will divide into layers, with the organic matter (being the lightest) being on top. Often, the organic matter is so light it actually floats on the water. The layers then go in order of weight: clay, silt, and sand. Sand is on the bottom. Determine the percentage of each by measuring each layer and dividing by the total length of the sample.

Once the percentages have been determined, the soil triangle is used to classify the soil.

Best Texture

The best soil based on texture is ***loam.*** A loam has equal parts of sand and silt and a smaller amount of clay. This soil has large enough particles to allow the passage of air and prevent compaction. It can retain water.

Sandy soils tend to lose water very quickly. Clay soils are so tight that roots and air may be unable to penetrate.

SOIL STRUCTURE

Soil structure is based on particle size and the way the particles group themselves. It is how the "pieces" of the soil fit together. Generally, a sandy soil has less structure than a soil with smaller particles.

As soil forms, sand, silt, and clay combine. The combinations are called peds or aggregates. They are secondary soil particles. Aggregates vary in size, shape, and strength. They are beneficial by increasing pore space. This allows soil to retain and/or drain water. The space between peds is a corridor for roots and a pathway for soil animals.

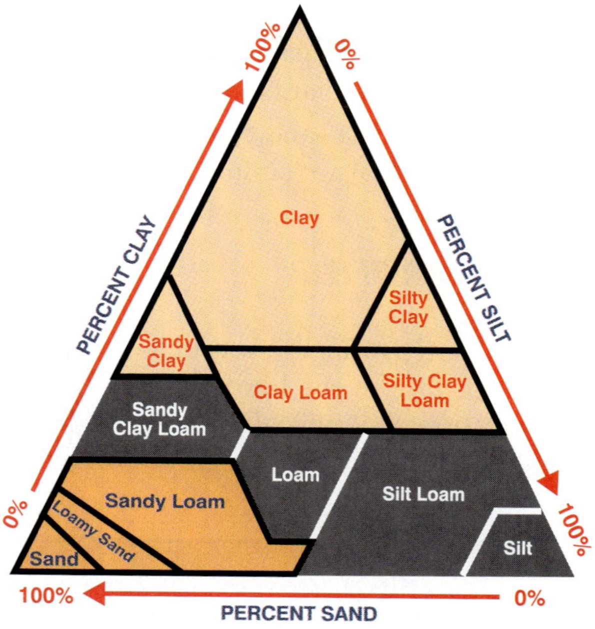

15-9. The soil triangle is used to classify soil textures.

15-10. A corn plant is beginning growth in fertile soil. (Courtesy, U.S. Department of Agriculture)

The grade of ped depends on the degree to which an aggregate will stay together. Four grades are possible: structureless, weak, moderate, or strong. The soil grade varies based on the amount of moisture in soil.

Grade, class, and type are used to create a descriptive name for the structure of soil. An example is moderate medium granular structure.

SOIL PROFILE

A ***soil profile*** is a way of looking at a cross section of the soil. This is from the surface to the lowest levels. A profile is determined by soil texture and structure. Soil typically has visible layers, or ***horizons***, as you dig through the soil. The degree to which one horizon may be distinguished from another

AgriScience Connection

HOME COMPOSTING

Composting is recycling organic materials. Plant and animal materials will rot rather quickly. Metal, glass, and plastic do not. As formerly living things decay, they produce organic materials beneficial to soil.

You can speed the process of decay with a compost bin. A compost bin can be used in a small amount of space. This shows a bin that is five feet square made of posts and wire. Layers of paper, food scraps, leaves, twigs, and similar materials compost in a few weeks when fertilizer and soil layers are used.

Worms can be used in compost pits to speed the process. Earthworms help in the decomposition of food scraps. The worms can later be used as fish bait. The compost can be used in gardens and flower beds to improve the soil.

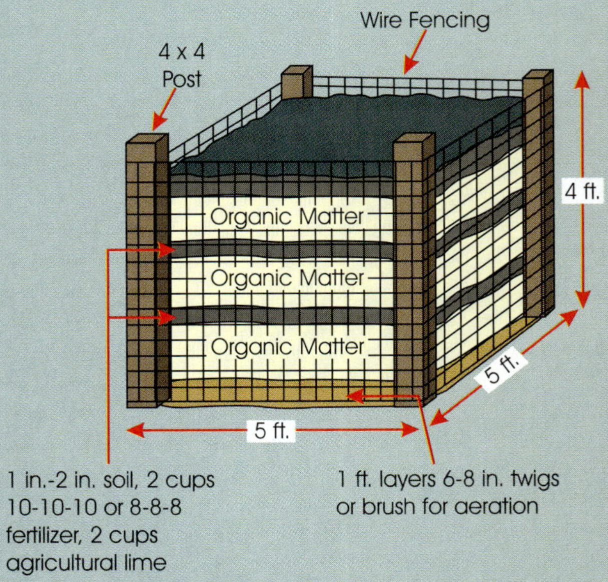

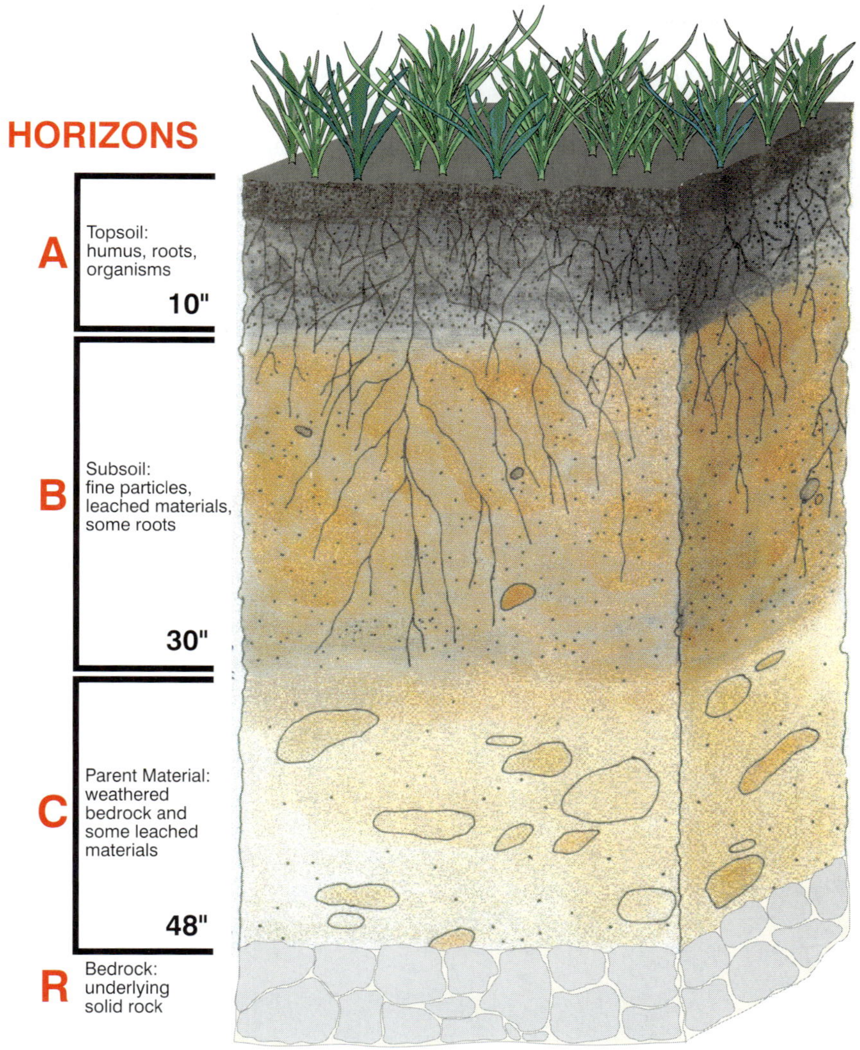

15-11. Soil horizons in a profile.

varies from soil to soil. Soil structure also varies between horizons. Larger particles typically are found deeper in the soil.

Each horizon has specific traits that aid in classification. The uppermost layer is ***topsoil*** or the A layer. It has the highest organic matter. The B layer, or ***subsoil***, is made of smaller particles. A third layer, the C layer, which is parent material, is below the B layer and above the bedrock. Parent material is rock from which mineral particles are forming. ***Bedrock*** is solid rock. The depth of each layer and the components help in grouping and studying soils.

15-12. Identify the parts of this soil profile using figure 15-11.

Agriculture and other activities can influence soil profile and structure. One common result is hardpan. Hardpan is a layer of soil, typically in the lower A or B horizon, that is strongly cemented. Water and plant roots are unable to penetrate the hardpan. Plant growth is stunted. Hardpan is caused and made worse by the weight of heavy equipment and tractors.

SOIL COLOR

Soil color is from mineral and organic matter. Color indicates qualities of the soil that are difficult to measure, such as drainage. The colors result from distinctive causes.

Dark brown or black soil is the typical color for a soil high in organic matter. The darker the soil, the higher the organic matter.

15-13. The reddish color in the soil around the gully is due to iron. (Courtesy, U.S. Department of Agriculture)

Soils that are red or yellow are high in iron compounds. Red soils are similar in color to rust on metal because they are being influenced by the same chemical process. On the other hand, yellow soil could suggest poor internal drainage.

Soils that are gray generally have poor drainage. Since water is regularly in the soil, the nutrients are lost. White soils have low nutrient levels and lack fertility.

A soil that has a spotted pattern, called mottling, has variability in soil moisture. As the moisture levels change, nutrients and minerals change as well.

SOIL CHEMISTRY

In addition to physical characteristics, soil chemical qualities are studied. The most important are pH and nutrients.

SOIL pH

pH is acidity or basicity. pH is on a scale of 1 to 14, with 1 being most acidic and 14 being most basic. A pH of 7 is neutral. Strongly acidic and strongly basic compounds are caustic and can burn tissue.

In soil, pH is rarely so drastic. Most soil is between 4 and 9. pH is important because the nutrients plants need are available differently based on pH. Most plants do well in a slightly acidic soil, with optimum pH being 6.5.

Some plants thrive in acidic soils. Examples are blueberries, azaleas, and rhododendrons. Other plants, such as potatoes, may be less likely to have diseases in acidic soils.

pH is measured several ways. The simplest technique is to mix soil with neutral water and then place litmus paper in the mixture. Litmus paper is small strips of chemically reactive paper. The color change

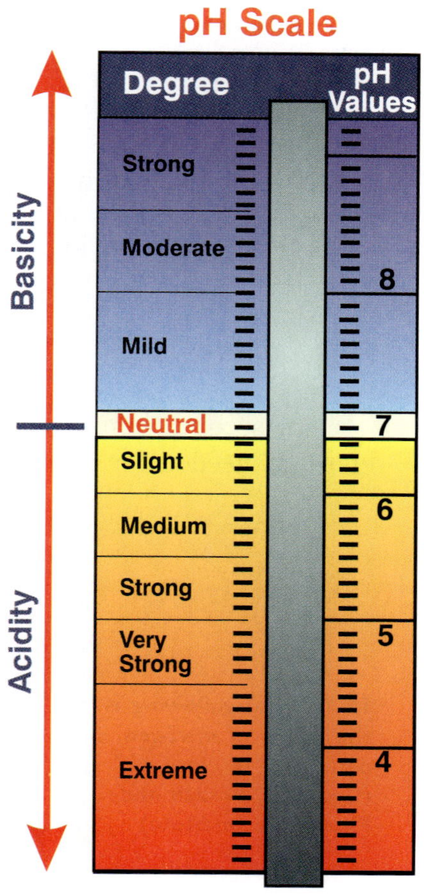

15-14. Few plants can survive in soil with a pH below 4.0 or higher than 10.0.

15-15. Granular dolomitic limestone is used to raise soil pH.

of the paper is compared with standard pH colors. Electronic pH meters give accurate results.

Since soil pH varies, ways are used to raise or lower soil pH. If soil is too acidic, lime can be added. ***Lime*** comes in several forms. One common form is a grayish powder made from ground limestone that is mixed with the soil. The amount of lime depends on the type of soil, the type of lime, and how much the pH needs to be raised.

Sometimes, soil may need to be made more acidic (pH lowered). Sulfur compounds are used. These react with the soil and lower the pH.

Extreme pH readings indicate soil that is not good for plant life. Toxic amounts of elements may prevent plant growth or plants may be unable to absorb needed elements. In either case, the pH must be changed.

PLANT NUTRIENTS

The second chemical aspect of soil is the nutrient level. A nutrient is any element needed for plants to grow.

15-16. Leaf yellowing is the result of nitrogen deficiency in corn. (Courtesy, Potash and Phosphate Institute)

15-17. These cotton leaves show a potassium deficiency. (Courtesy, Potash and Phosphate Institute)

Nutrients are grouped based by the amount plants require. The three groups are the organic elements, major elements, and minor elements. The organic elements are carbon, hydrogen, and oxygen. The ***major elements*** (sometimes called macronutrients) are nitrogen, phosphorus, and potassium. The ***minor*** or trace ***elements*** (also called micronutrients) are other elements required in small amounts for plant growth.

The primary source for organic elements is air and water in the soil. These elements are the building blocks of plant cells and play a role in almost every process within a plant. For this reason, water is the most important nutrient for plants. Since these elements are typically readily available, they are often omitted from studies of plant nutrition.

15-18. A grape leaf with a deficiency of magnesium. (Courtesy, Potash and Phosphate Institute)

15-19. A bag of mixed fertilizer. (The numbers 16-4-8 mean 16 percent nitrogen, 4 percent phosphate, and 8 percent potash.)

The major elements have different roles in plants. Nitrogen promotes the growth of stems and leaves. It gives plants a healthy, dark green color. Phosphorus encourages the growth of roots, flowers, and seeds. It improves the quality of seed crops. Potassium helps plants resist disease and make starches and chlorophyll.

The nine minor elements most needed for plant growth are calcium, magnesium, sulfur, iron, manganese, boron, copper, zinc, and chlorine. Other minor elements are vanadium, cobalt, silicon, and sodium.

Deficiencies and excesses of any elements result in unhealthy plants. Since each element has a different role, the symptoms of problems vary. An excess can present an appearance similar to that of a deficiency. Soil scientists can measure the elements present by testing either the soil or parts of the plant. The tests are used to recommend the fertilizers needed.

The best way to protect soil nutrients is prevention. Soil and plant tissues can be tested. Nutrients should be added to the soil before a problem exists. Nutrients are added as ***fertilizer.*** Fertilizer is mineral or organic material that contains nutrients.

15-20. Special equipment is used to apply liquid fertilizers.

Organic fertilizers are from natural sources. They include animal manure, bone meal, dried blood, composted leaves and green manure crops. Organic fertilizers are slowly available to plants.

Mineral fertilizers are sometimes called chemical fertilizers. These are in many forms, including solids, liquids, and gases. A complete fertilizer is one that contains all three major elements. The choice of fertilizer depends on the needs of the soil and plants. No more should be applied than is needed.

SOIL EROSION

Using the soil can cause wear or loss. Erosion is the wearing away of the topsoil. It may be physical or chemical loss. Erodibility is how susceptible soil is to the forces that cause erosion. One factor is slope of the land. ***Slope*** is the steepness of the land. It is measured as feet of rise to feet of run, such as 1 foot rise per 100 feet of run.

PHYSICAL EROSION

Physical erosion is actual loss of the topsoil. Wind and water are forces that move topsoil from land. Snowmelt and frost also cause erosion. Although topsoil is a renewable resource, the continuous use may accelerate loss to a point it cannot be renewed.

Physical erosion may be reduced in several ways. Having a vegetative cover protects the soil. Using no-till or minimum-tillage farming prevents

15-21. The greater the slope of land, the higher the possibility of erosion. (Courtesy, U.S. Department of Agriculture)

15-22. The topsoil and corn crop were lost due to water erosion. (Courtesy, Kathy Gunderman)

exposure of soil to erosive water and air. Mulching reduces exposure. Mulching is covering bare soil with materials, such as straw and crop residue. Trees can be planted along the edges of fields to break the wind. Land with a steep slope should not be plowed. Strip cropping, contour tillage, and terracing reduce soil loss.

No-till or Minimum Till

Traditionally, tools have been used to plow or break the soil. This is done before planting. It makes a good place for seed. Plowing is also used to reduce weed growth and aerate the soil.

15-23. Soybeans are growing through wheat stubble. (The land was not plowed for planting.) (Courtesy, U.S. Department of Agriculture)

In so doing, the soil is exposed and made susceptible to erosion. Scientists and farmers have now developed strategies that reduce the use of tillage equipment. Crops are planted through the stubble of previous crops. Nonharvested cover crops may also be planted to cover the soil. The stubble or cover crop prevents the soil from exposure to the forces of wind or water.

Mulching

A mulch is any material spread on the soil to protect either the soil or the plant roots. Mulch is used for a variety of purposes, including frost protection and increasing organic matter.

Many different materials are used for mulch. The most common are plant residues, such as straw, sawdust, and leaves. Plastic films have also become popular.

In erosion prevention, mulch is a short-term solution used until other plant material has grown. One site where mulch is used is during construction, such as highways, where a large amount of soil is moved.

15-24. Crop residue is used to mulch the soil between rows of staked tomatoes in black plastic.

Windbreaks

Windbreaks form a barrier between the wind and the soil. Windbreaks are primarily made of fast-growing shrubs or trees, such as willows. The rows of trees are planted across the area at a right angle to the direction of the most common or prevailing winds. Windbreaks may be nonliving items, such as fences or screens. One negative of living windbreaks is the length of time they take to become established. Trees may also have a large root system, which robs crops of nutrients.

Conservation Reserve Program

Sometimes, the best way to prevent erosion is by not plowing land. As an incentive, the federal government has programs that pay for not planting highly erodible land. These programs may provide help in preventing erosion. The details of such programs often change. Information is available from a local office of the Farm Service Agency of the U.S. Department of Agriculture.

Strip Cropping

Strip cropping is planting small plots or strips of land to different crops. The strips alternate crops that require tillage, such as corn, with crops that do not, such as wheat. The width of the strips varies depending on erodibility

15-25. A grassy strip to reduce soil loss has been left between corn fields on this Indiana farm.

of the land. The non-tilled strips prevent fast runoff of water and catch soil that is moving in water.

15-26. The raspberries on this Maine farm were planted across the slope to slow the flow of water.

Contour Tillage

Contour tillage is planting crops across the slope of the land. As the crops are planted, the equipment follows the natural contours of the earth. If crops are planted with (up and down) the slope, water flows much faster and more easily. In effect, contour tillage uses the crops themselves as a water and wind break. Contour tillage is often used with strip cropping.

Terracing

Terracing is creating banks of earth across the slope of the land. The number of terraces and the interval at which they are placed depend on slope and erodibility. Areas between terraces

15-27. A newly-installed terrace on this Mississippi farm will slow water runoff.

15-28. This grass waterway slows the erosion from surrounding cropland.

are used for crops. Terraces are planted with grass or other plants that hold the soil in place.

15-29. Silt fences are used to control erosion around construction sites.

Waterways and Diversion Ditches

Waterways and diversion ditches are used to redirect water flow and minimize erosion. By controlling the water flow, soil loss is reduced. The areas are smoothed and, generally, planted with grasses. These areas are not tilled. They make the surrounding area less likely to erode.

CHEMICAL EROSION

Chemical erosion is the loss of soil nutrients by harvesting crops. Each time a plant is removed, the nutrients it used are taken away. In addition, nutrients may leach or be drained from the soil as water drains.

Chemical erosion is reduced by a good fertilization program. Producers measure the impact a crop has by testing the soil and adding fertilizer. Several other ways of reducing chemical erosion are used. Crop rotation is planting different crops on a plot in succeeding years. Green manure crops are planted and plowed back into the ground. Composting and similar strategies return crop residues (anything left) to the field.

Fertilization Programs

In the past, fertilizers were added to the soil on a regular basis. Now, both testing the soil and plants is used before adding fertilizers to make certain they are needed. If a fertilizer is added and it is not needed, it can leach from the soil. This results in damage to ground or surface water surrounding the site. It is also a waste of money spent on the fertilizer.

Fertilizers are applied in a variety of ways. They may be applied over the surface, which is called broadcasting. Fertilizers may be placed in bands beside the seed at planting or after the plants are growing. Fertilizers may be liquids, gases, or dry materials.

Use care to ensure that strong fertilizers do not come in direct contact with plants. If they do, the plants will be "burned." Some fertilizers are not as concentrated and can be sprayed on plants.

Crop Rotation

Crop rotation is planting different crops on the land in succeeding years. Rotation is the opposite of planting the same crop year after year.

The reason for crop rotation is that crops have different impacts on the land. They remove different elements. Different residues are left.

Rotations differ from site to site and are based on individual soil types. If the crop is changed consistently despite location, the health of the land is much better.

Green Manure Crops

Green manure crops are grown and plowed into the soil. These crops improve the organic matter and return nutrients to the soil. Some green manure crops have the unique capacity to convert nitrogen to a usable form. Common crops used as green manure include alfalfa, clovers, and grains. At maturity, the crops are usually plowed into the soil.

15-30. Vetch is a common green manure crop.

Composting

Composting is taking plant residues and other organic materials (such as egg shells) and encouraging their decomposition. The result is compost. The compost is applied to the land. Compost has the same purpose as organic fertilizers, and, in fact, compost is sometimes called artificial manure. Composting is a method of recycling organic items. It can be done on even a very small scale, such as at your home.

Since land is needed to continue having crops in the future, it is best to take steps to prevent soil erosion.

15-31. Composting is used to dispose of crop residue. (This shows potatoes and sawdust being composted.) (Courtesy, Smith's Farm)

REVIEWING

MAIN IDEAS

Soil is a vital resource for all of us. Its main components are mineral particles (small pieces of rocks), organic matter, water, air, and animals. These parts each do important jobs.

Soil is grouped based upon many of its characteristics. Soil texture refers to the size of the particles in the soil, including sand, silt, and clay. A loam is soil with equal amounts of sand and silt and a smaller amount of clay. Soil structure refers to the way particles join within the soil. A soil profile allows one to look at all levels of the soil. Soil color indicates many qualities of the soil that may not otherwise be measured.

Soil has many chemical features, including pH and nutrient levels. pH refers to the acidity, neutrality, or basicity of the soil. Nutrients are elements necessary for successful plant growth. The three most important elements are nitrogen, phosphorus, and potassium.

One threat to soil is erosion, the physical or chemical loss of the fertile part of the soil. Erosion is caused by many factors, including water, wind, and poor management. Many specialized techniques are available to prevent the loss of this valuable resource.

QUESTIONS

Answer the following questions. Use correct spelling and complete sentences.

1. What is soil?
2. What four components are in soil? What does each do?
3. Do you think animals are part of the soil? Why or why not?
4. Why is soil classification important to soil scientists? What would the field be like without classification?
5. What is the difference between soil texture and soil structure?
6. How does pH affect soil and plants?
7. List the macroelements needed by plants. How do they help plants?
8. What is erosion? What is the difference between physical erosion and chemical erosion?
9. Do you think physical or chemical erosion is more important? Why?

EVALUATING

Match the term with the correct definition. Write the letter by the term in the blank provided.

a. clay	d. physical erosion	g. pH
b. mineral particles	e. major elements	h. horizons
c. soil texture	f. composting	

____ 1. Small pieces in soil we would think of as dirt.

____ 2. The smallest soil particle.

____ 3. A measure of the relative acidity or basicity of a compound.

____ 4. Encouraging the decomposition of plant materials and other organic matter.

____ 5. Visible layers in a cross section of soil.

____ 6. Size of the particles within soil.

____ 7. Actual loss of topsoil.

____ 8. Nitrogen, phosphorus, and potassium.

EXPLORING

1. Fill a clear jar halfway with soil from a site near your home. Add water and shake thoroughly. Measure the relative amounts of sand, silt, clay, and organic matter. Use the soil triangle to determine the texture of your sample.

2. Visit a local farmer to learn about the conservation practices he or she uses. Ask about the soil types, the degree of erodibility, and the physical and chemical modifications they make. Ask if the farmer has a written land management plan that describes the steps they have identified for conservation. Prepare a written or oral report to be shared with your classmates.

3. Visit the local Natural Resource Conservation Service office. Learn about the qualifications and job responsibilities of the staff there. Ask them about the positive and negative aspects of their careers. Prepare a written or oral report to be shared with your classmates.

16

FOOD SCIENCE

Growing plants and raising animals is only part of having food. How the plant and animal products become healthy, tasty food is important to each person at every meal.

We are lucky to live today. Many advances have been made in our food in the last century. The field of food science has come about. We now know so much about food. We have ways to keep food safe to eat and easy to prepare.

Food scientists take raw agricultural products and change them into desired foods. They create products that are tasty, nutritious, and safe. They invent new products, monitor quality, and move agricultural products from the producer to the consumer.

16-1. Food is an important part of our lives. Eating is often the center of our activities.

OBJECTIVES

This chapter introduces food science. It covers the importance of wholesome food, how food spoils, food preservation, packaging, and risks in processing. The objectives of this chapter are:

1. Explain the meaning of wholesome food
2. Describe how food spoils
3. List and explain methods of food preservation
4. Explain food packaging
5. Identify risks in food processing

TERMS

canning
cold processing
dehydration
fabrication
fermentation
food additive
food poisoning
food processing
freezing
grading
irradiation
packaging
pasteurization
pathogen
perishable food
pickling
preserving
spoilage
toxins
wholesome food

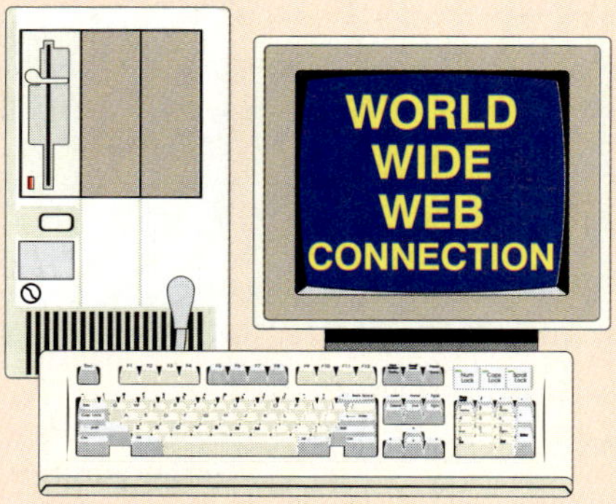

http://www.ag.uiuc.edu/~food-lab/nat/

(This site has step-by-step information to guide personal diet assessment.)

HAVING WHOLESOME FOOD

As with plants and other animals, humans require food. Food provides the nutrients to live and repair cells. Food also serves an important social need. Families gather at meal time. Friends eat together. Meals are special events in the lives of people.

WHOLESOME FOOD

Wholesome food is nutritious and contributes to our health and well-being. Eating wholesome food is safe. It does not make us sick. Wholesome food is free of substances that damage human health. It is prepared in a way to retain needed nutrients.

16-2. Wholesome food is provided in many settings. (Courtesy, ARA Services)

Perishable Food

Wholesome food may become unwholesome. This happens by either the loss of nutrients or spoilage.

Some foods are perishable. This means they do not keep well. A ***perishable food*** is highly susceptible to spoiling and must be used in a short time—often only a few hours or days. Raw milk is highly perishable. Nonper-

16-3. Spoiled oranges are not to be eaten.

ishable foods are less susceptible to spoilage and may be kept for months. Uncooked dry pasta, for example, does not readily spoil.

Spoilage is the invasion of bacteria or other foreign materials. These include rodents, insects, and chemicals. Spoiled food should not be eaten. Spoilage may be due to the presence of a microorganism or to chemicals it produces, known as toxins. A ***toxin*** is a chemical by-product that sickens or kills other organisms.

16-4. A spoiled fluid milk product.

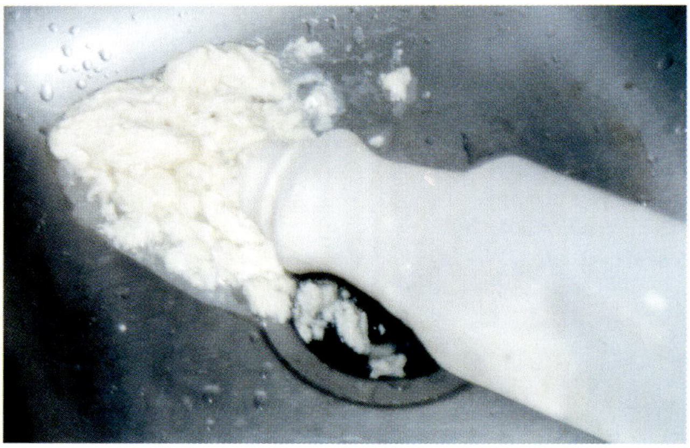

Food Poisoning

Eating unwholesome food may cause ***food poisoning***. Food poisoning is an illness caused by eating bad food. It may be caused either by an organism or by a toxin the organism produces. Symptoms of food poisoning are vomiting, diarrhea, abdominal cramping, fever, and headache.

Common sources of food poisoning are bacteria. Bacteria may grow in meat, milk and dairy products, seafood, poultry, and egg products. Bacteria and other microscopic organisms are often called microbes.

Processing

Food processing is a series of steps through which raw food is made usable. It is taking an animal or plant product and changing it in some way to make it more desirable. Possible steps in processing include grading, fabricating, preserving, and packaging. Food may also be made convenient and easy to prepare. It may be packaged in amounts best used by consumers.

Consumers want products of high quality and similar size. ***Grading*** is sorting products by size, kind, and quality. Items with blemishes or other problems may be used differently or discarded. One advantage to grading is that bad products are removed before a contaminant can spread. For example, a half rotten apple is removed before the rot can spread to other apples.

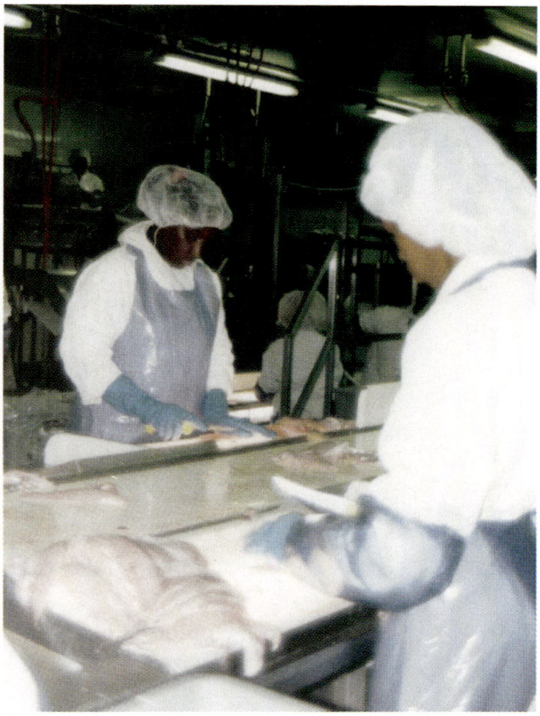

16-5. Food processing plants are clean and automated.

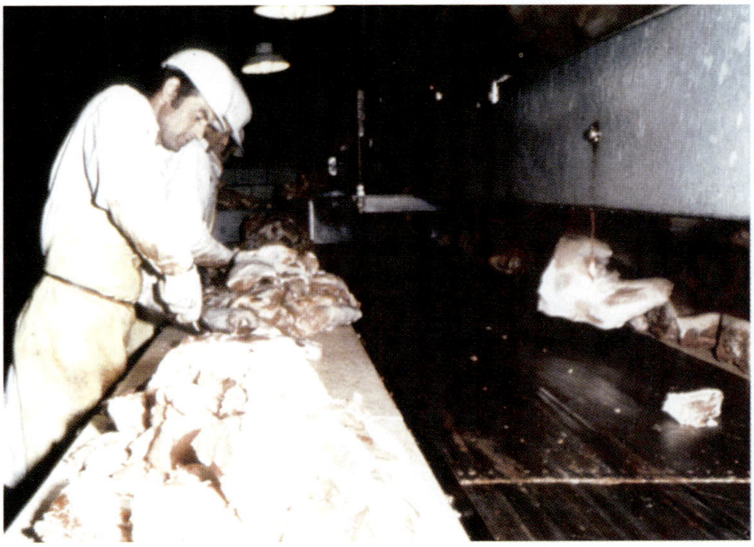

16-6. Fabricating beef into cuts.

Some products must be changed before they are usable by consumers. Can you imagine going to the supermarket and buying a whole pig? That is not very realistic. ***Fabrication*** is used to make the product more convenient. With meat, the animal must be slaughtered and cut up before it is usable.

16-7. Two steps in fabricating potatoes into French fries are shown here. (Courtesy, RDO Frozen, Park Rapids, Minnesota)

Preserving is any technique to prevent food from spoiling. Freezing, canning, and drying are forms of preservation.

Packaging is important for both food safety and consumer sales. ***Packaging*** is placing food in a container appropriate for a food product. Packaging

16-8. An attractive package is used for green peanuts.

protects food by keeping nutrients in and contaminants out. It also divides food into quantities appropriate for consumers.

Processing may be simple or complex. Simple processing may be nothing more than cutting the leaves off and packaging carrots. Complex processing may involve food chemistry to create artificial cheese from soybeans. In either case, processing may reduce the nutritional value of a product. Usually, processing increases the safety and nutritional value of food.

If processing does reduce the nutritional value, the result is a less useful product. If we eat large quantities of food without receiving nutrients, we will ultimately be unhealthy. A food product that does not provide nutrients has essentially the same effect as one that is spoiled—our health is damaged.

16-9. Americans have an abundant, high-quality food supply.

FOOD SPOILAGE

Spoilage makes food unfit to eat. Five factors are involved in food spoilage:

- the source of the contaminant,
- the nature of the food,
- the moisture content of the food,
- time, and
- temperature

Each of the spoilage factors has an impact in the processing procedure—from harvest of the crop or animal through consumption of the final product. In fact, you can have spoilage as easily at your home as a processor can in a plant. Controlling the spoilage factors insures food wholesomeness.

SOURCE

A source is anything that introduces contaminants. A contaminant that can cause spoiling must be present. A ***pathogen*** is something that causes diseases. Common pathogens include types of bacteria, molds, and fungi. These contaminants get into the processing facility in some way. The most common sources are humans, equipment, air currents, dust, and vermin (rats, roaches, etc.). If these sources can be prevented, the possibility of contamination may be eliminated.

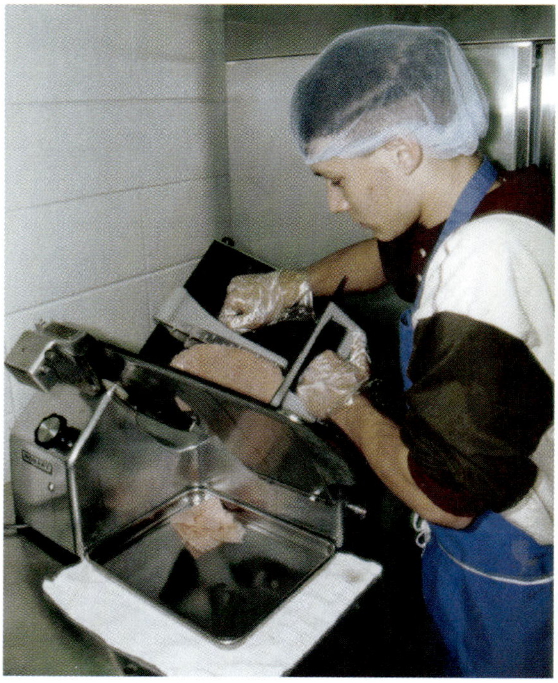

16-10. Apron, hairnet, and plastic gloves are worn to avoid food contamination.

The lack of sanitation leads to contamination. This is why workers in food industries must wash their hands before beginning work, after using the restroom, or anytime when the hands may contact contaminants. The level of human decontamination varies with the industry. Using disinfectants and special clothing helps prevent contamination. A person who wears dirty clothing while working with food will contaminate the food.

FOOD

The pathogen must have a host. A host is a place to grow. Often, this is the food being processed. Sometimes, it is just a small amount of residue left after processing. The pathogen grows in the residue and the residue later comes into contact with the food.

Pathogens prefer non-acidic foods high in protein. Meat, milk, and eggs are prime targets. Many vegetables are less susceptible because they have a high acid content. Cleanliness and quick work are keys. Food products should not be allowed to sit exposed for any time. This is true as much in a home kitchen as in a processing facility. Bacteria or other microorganisms invade quickly.

16-11. Molded cheese is not fit to eat.

MOISTURE

Like animals and plants, most pathogens need water to grow. High moisture foods are more susceptible to spoilage. Foods such as dry cereal, sugar, flour, and biscuits do not have enough moisture for bacteria growth.

Sometimes, water may be removed from the food. This is a processing technique known as dehydration (drying). In addition, the humidity (water vapor in the air) is often controlled in a processing area. Most refrigerators remove moisture to reduce the potential for spoilage.

TIME

Pathogens reproduce over time. When food is first infected, only a few microorganisms may be present. As time passes, however, these first few may create a colony of millions. The pathogens may themselves cause illness or their by-products may cause illness. In either case, the result is inedible food that, if eaten, causes sickness or death.

The solution is to reduce the time available for pathogen reproduction. Speed is essential in processing, preparation, and consumption. It is as important that you eat food soon after it is cooked as it is that the food is not left sitting at the processing facility.

16-12. Poor storage has caused these potatoes to shrivel.

TEMPERATURE

Most pathogens reproduce only at 40° to 140°F. This means that foods should be kept outside that range. Heated foods should be hotter than 140°F and chilled foods should be kept below 40°F. Although this may not completely prevent pathogens, it will decrease the number present.

PRESERVATION METHODS

Many methods have been developed to keep food wholesome. These are continually being improved by food scientists. Each method controls one or more of the factors of food spoilage. The primary strategies include freezing, cold processing, heat processing, and irradiation.

No matter the method, two cautions hold true. Never use a product from a damaged or bulging container. Never eat a food product that smells spoiled. Use good judgment with all foods.

FREEZING

Freezing is a medium-length preservation technique. It requires temperatures of less than 0°F, an airtight environment, and quickness. Freezing slows microbial growth. It kills some pathogens, while others become dormant. Freezing stops the chemical reactions pathogens need. Freezing may be done with a variety of methods, including immersion, indirect contact, sharp freezing, and freeze-drying.

Immersion Freezing

Immersion means to cover completely. In this process, the product is placed directly into a very cold liquid solution. The most common solution is liquid nitrogen. This technique is fast and works well on irregularly shaped objects. It is an expensive method of freezing.

Indirect Contact

With indirect contact, the product is placed into tubes next to the freezing agent. The freezing agent is itself contained within metal shelves. As the food product is pushed past the shelves, the freezing occurs. The most common products using this technique are liquids and purees, such as fruit concentrates.

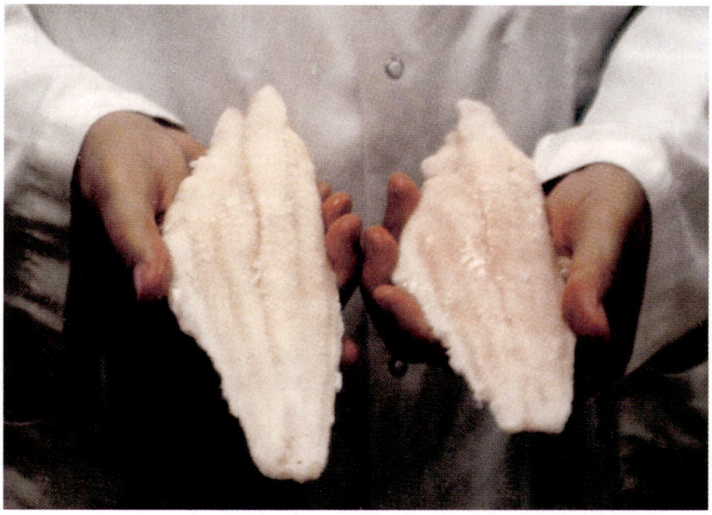

16-13. Individually quick-frozen fish fillets.

Sharp Freezing

Sharp freezing is done directly in the air at temperatures of -23° to -30°F. The product is frozen with blasts of air and then poured into packaging. Peas and beans are commonly preserved this way.

16-14. Many meat products are marketed frozen to allow for an extended shelf life.

Freeze Drying

Freeze drying is a kind of freezing. After freezing, the water is removed from the product. The product is frozen very quickly in a special chamber. Low pressure causes ice crystals in the food to become water vapor. The water vapor is then removed from the room. Freeze-dried food can be kept for extended periods because it weighs very little. (Most of the weight of food is water. Once the water is removed, the remaining food has little weight.)

COLD PROCESSING

Cold processing, also called refrigeration, is common. It is used when transporting and storing perishable foods for a few hours or days. Many products that would deteriorate in a few days are refrigerated. Refrigeration has had a big impact on the transportation of perishable foods, such as fruit to distant markets.

Cold processing typically maintains the product's nutritional value and taste. In addition to keeping a temperature slightly above freezing, refrigerator units also decrease the humidity and oxygen and increase carbon dioxide. This slows deterioration of the product.

HEAT PROCESSING

Increasing the temperature to above 140°F kills most pathogens and insures the safety of the food. This is particularly true when the heating is quickly done. Three common techniques are used in heat processing: canning, dehydration, and aseptic canning.

Canning

Canning is heating both the food product and the container in which the food has been placed. The hot product is placed into the container and sealed. The most common container is the metal can. The resulting can should not be dented or misshaped, should be free from corrosion, and should have a vacuum seal that breaks when opened.

The product inside a can should be true to color and should not have an offensive odor. One drawback to canning is that it can remove nutrients from the food. Canned food can be stored for months or a few years.

16-15. A bulged can is a sign of spoilage. (Do not eat the food!)

AgriScience Connection

YOU ARE WHAT YOU EAT

You are what you eat. It is an old adage, and not completely true. What you eat, however, determines the quality of your body. Just as plants and animals need proper nutrition, your body requires proper nutrition. Energy is needed for being active. Clear skin, healthy hair, and a smiling face result from eating properly.

Some teenagers struggle with obesity. Others face eating disorders, such as anorexia nervosa or bulimia. All are unhealthy. Eat appropriate food each day.

Keep an eating journal for one week. Record everything you eat or drink. Then use the information provided on food labels to evaluate your diet. Are you eating enough? How much junk food did you eat? What strategies could you use to improve your diet?

Dehydration

Dehydration (drying) is the nearly complete removal of water from the food product. It may be used on fruits, vegetables, and meats. Although it may be done at home using the sun, most commercial processors use heat.

Dehydration requires that the product be cut into small pieces or thin layers. The resulting food is lightweight, convenient, and high quality.

Aseptic Canning

Aseptic canning sterilizes food before it is packaged. The food is heated very quickly using special equipment. It is then placed in a container and immediately sealed. Aseptic canning maintains nutrient levels.

Aseptic canning depends on a sterile environment. This means that everything in the surrounding area must be free from any microorganisms. Humans cannot enter the areas where aseptic canning takes place.

IRRADIATION

Irradiation is exposing food to radiant energy to kill microorganisms without contaminating food. Irradiation kills pathogens throughout the product—from the surface to the center. Food that has been irradiated has

16-16. Onions being unloaded at a processing plant.

very low levels of microorganisms but maintains nutrients. In addition, these products have a longer shelf life. Strawberries are sometimes irradiated.

OTHER METHODS

Several other methods of food preservation are used. Some of these methods have been used for many years. These methods include:

- ***fermentation***, which involves the growth of beneficial microorganisms;
- ***pickling***, which involves high salt concentrations; and
- ***pasteurization***, which is used to kill microorganisms in most milk and sometimes in other liquids

FOOD PACKAGING

An additional factor in food preservation is packaging. The container in which food products are stored is a factor in the length and success of preservation.

Packaging must meet a number of criteria. It must protect the product without releasing any toxins of its own. It must prevent contamination from microorganisms, light, moisture, odor, and gases. It must resist impact and

16-17. Consumers want quality products, such as these roasted pistachios. (Courtesy, California Pistachio Commission)

tampering. Packaging should be easy to work with, inexpensive, and attractive to the consumer. Recently, it has become increasingly important that packaging can be recycled.

At least seven different materials may be used for packaging. Primary containers come in direct contact with the food. Secondary containers are outer boxes. Metal, glass, paper, plastic, laminates, plastic pouches, and edible films are all potential packaging materials.

Some food scientists do nothing but evaluate potential packaging. The selection of appropriate packaging depends a great deal on the product inside. Scientists measure all materials for their strength and durability. Obviously, plastic and metal are less likely to break than glass. On the other hand, glass is virtually impenetrable and withstands tampering.

One step in packaging is deciding how much food to put in the container. This may involve portioning, which is dividing the product into the most appropriate amount for one serving. Portioning makes a product easier for a consumer to use.

16-18. These potatoes are in both primary and secondary packages for shipping.

Another key to packaging is labeling. The Food and Drug Administration (FDA) requires that the package contain information regarding the nutritional value of the product. Food scientists measure the calories, vitamins, minerals, and other nutrients in the product and provide that information on the label. A consumer can learn much about the relative value of the product by reading the label.

PROCESSING RISKS

A risk is simply an implied threat. It can be physical, mental, or monetary (related to money). We are constantly exposed to activities with some degree of risk. Our responsibility is to identify the true amount of risk involved and whether we consider that amount to be worth becoming involved. For example, riding in a car involves risk. If you never rode in a car or other automobile, however, you would be severely limited in the places you could visit.

Almost everything in life involves some risk. The issue is what degree of risk an individual is willing to take. Eating food is a risk, but no one can survive without food. Different people perceive the risks differently. What can you think of that are risks with food?

Career Profile

FOOD INSPECTOR

Americans are concerned about the safety of their food. One career in food safety is the food inspector. Many agricultural industries, from meat processing plants to vegetable packaging facilities, employ inspectors.

The responsibilities of food inspectors vary from place to place. Usually, inspectors check the sanitation of a food facility. They determine the wholesomeness of the food product. Inspectors may also provide written reports of their work.

Inspectors must have high expectations. They need strong moral standards and an ability to understand the special requirements of the industry. Inspectors must have good communication skills.

Risks may be summarized many different ways. Risks may be known or unknown. They may also be controllable or uncontrollable. Risks may be placed in four groups: known and controllable, known but uncontrollable, unknown and controllable, and unknown and uncontrollable.

Acceptable risk is risk you are willing to take for the benefits you perceive. Acceptable risk may only be assessed through education. For example, did you know you are at greater risk of being killed while walking than while riding in an airplane?

In the past, some common risks with food came from improper processing and natural toxins. Food scientists have emphasized research on these risks and have sought to decrease the potential for harm.

Currently, threats to food safety include pathogenic contamination, pesticide residues, malnutrition, and food additives.

16-19. Inspection ensures the quality of products and reduces risk to the consumer.

PATHOGENIC CONTAMINATION

Bacteria continue to be a risk in food consumption. *E. coli* and *Salmonella* bacteria can be toxic. These bacteria have been the documented cause of illness in the United States.

Bacteria can be controlled through management of the five factors involved in food spoilage. To enforce this management, the federal government maintains regulations on sanitation in food processing plants.

PESTICIDE RESIDUES

Pesticide residues are small amounts of chemicals left on the product after harvest. Many people identify pesticide residues as the highest risk in food processing. In fact, few people have experienced illness because of residues.

Before pesticides, food was scarce because of crop losses in the field and in storage. With pesticides, the quantity and quality of food have increased significantly.

To protect consumers, the Food and Drug Administration (FDA) maintains an ADI (or acceptable daily intake), which is the acceptable amount of residue. The ADI is one one-hundredth of the amount that causes no impact on laboratory animals. Less than 1 percent of American products exceed the ADI.

The risk of pesticide residues may also be reduced by simple strategies in the home. Washing and peeling fresh produce removes most residues.

16-20. Fresh vegetables, such as radishes, have the potential for pesticide residues.

MALNUTRITION

Malnutrition is poor or inadequate consumption of nutrients. It continues to be an issue in the United States and throughout the world. One factor in malnutrition is availability of food. We are lucky in the United States that food is generally available. Pesticides and management practices have helped increase our food supply. Many other areas of the world are not as lucky. In these areas, the chance that food might not be available is a significant risk.

Often, the risk of malnutrition is controllable. Careful consumption of appropriate foods greatly lowers the risk. If the risk is because of food availability, the use of knowledge and technology to increase production may then be used to reduce the risk.

FOOD ADDITIVES

Food additives are substances added to food to increase taste, improve the consistency, aid in food preservation, or meet another need. Some of these products have been connected to allergic reactions. The risk of food additives varies significantly from one person to another.

16-21. Salt is a common additive used at the dining table.

Control of the risk of food additives depends on education and attention to detail. An individual may have to give up specific products to maintain their health.

Of course, other people have food allergies not related to additives and they also have to avoid specific products.

REVIEWING

MAIN IDEAS

Wholesome food is a key to human life. Spoilage and loss of nutrients can make food unwholesome. Food may be spoiled by contaminants, such as bacteria, insects, rodents, or chemicals. Food processing delays spoilage by grading, fabricating, preserving, and packaging food products.

Five separate factors affect food spoilage. These factors are the source, food, moisture, time, and temperature. Food scientists control one or more of these factors to prevent spoilage.

Many methods exist for successful food preservation. These methods include freezing, cold processing, heat processing, and irradiation.

Packaging is often a key to wholesome food. Packaging must meet many conditions to protect the food product inside. Metal, glass, plastic, and cardboard are all used to package food products.

A risk is an implied threat. Real and perceived risks exist with food processing. These risks include contamination, malnutrition, and food additives. A consumer needs to be well informed. Dealing with the facts and not with incorrect information is essential.

QUESTIONS

Answer the following questions. Use correct spelling and complete sentences.

1. What nutrients do humans need?
2. Why do humans need nutrients?
3. What are some causes of food spoilage?
4. What is the difference between perishable and nonperishable food?
5. What factors contribute to food spoilage? How could these factors be controlled?
6. What is the difference between a pathogen and a toxin?
7. What types of food are most susceptible to spoilage?
8. Why is food preservation important?
9. What are important factors when selecting packaging? Why?
10. What is a risk? What are some risks associated with food?
11. What can a consumer do to learn more about food risks?
12. What are some keys to preventing food spoilage in your home?

EVALUATING

Match the term with the correct definition. Write the letter by the term in the blank provided.

a. freezing
b. malnutrition
c. irradiation
d. packaging
e. cold processing
f. toxins
g. risk
h. processing

___ 1. Chemical by-products.
___ 2. Raw materials are made more usable.
___ 3. Temperature of less than 0°F.
___ 4. Refrigeration.
___ 5. Exposure to radiant energy.
___ 6. Container in which a food product is stored.
___ 7. Implied threat.
___ 8. Poor or inadequate consumption of nutrients.

EXPLORING

1. Tour a local grocery store. Talk with the managers of different departments about the different types of packaging and preservation used for products in their departments. Ask if the employees take special steps to maintain wholesome food. Pay particular attention to perishable products in the dairy, meat, and bakery areas. Prepare a report describing what you learned and share with your classmates.

2. Visit a local slaughter and packing plant. Observe the techniques used to maintain cleanliness. Notice also the uses for non-meat portions of the animal. Compare the appearance of the carcass with the cuts of meat you are accustomed to seeing in the store. Prepare a report with a visual aid that describes what you learned.

3. Contact a food scientist at your state's land-grant university. Since this individual may be a long way from you, use the advances in telecommunication to help. Learn what the scientist is researching. Find out the particular problems the individual faces. Find out how he or she is solving those problems. When possible, arrange to visit the scientist and observe him or her at work.

17

TECHNOLOGY SYSTEMS

What makes agriculture better today than 200 years ago? How can people eat without having to grow the food themselves? Technology! It is everywhere!

Technology is an essential tool in agriculture. It helps producers be competitive in a global economy. Technology helps consumers have what they would be unable to get otherwise. Agricultural scientists create systems that make agriculture always productive.

Without technology systems, American agriculture would be much less productive. Americans would have to work harder to succeed. Most of us like it the way it is!

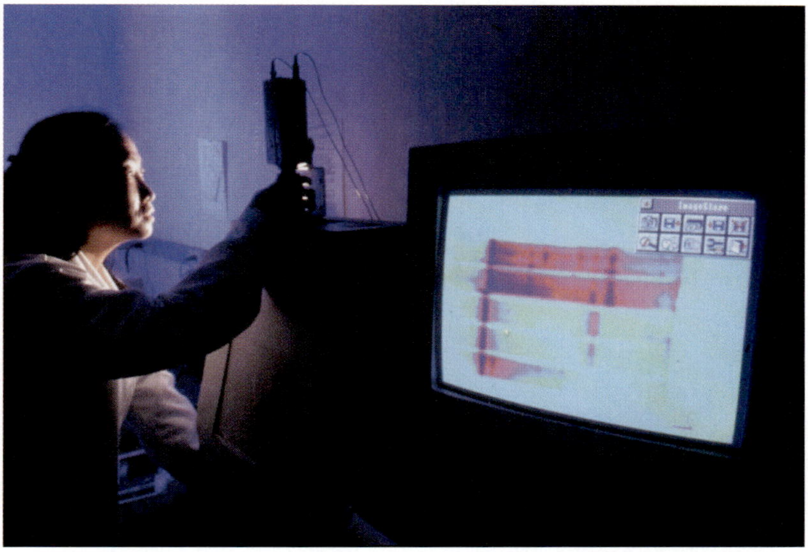

17-1. Video display technology is used to study a tomato disease—infectious chlorosis virus. (Courtesy, Agricultural Research Service, USDA)

OBJECTIVES

This chapter is about important areas of technology. It describes how technology helps people get more done with less effort. The objectives of this chapter are:

1. Explain technology systems in agriculture
2. Describe research and development in technology systems
3. Explain variable rate technology
4. Describe how innovations influence labor requirements
5. Explain how technology is related to sustainable agriculture

TERMS

- global positioning system (GPS)
- precision farming
- proactive
- proprietary R&D
- R&D
- remote sensing
- safety testing
- sustainability
- system
- technology system
- variable rate technology (VRT)

http://www.agriculture.com

(This site has information and links in many areas of agriculture.)

TECHNOLOGY AND SYSTEMS

Technology is the use of inventions to solve problems and help do work. Technology is a direct result of problem-solving research. Technology includes not only the invention of new pieces of equipment, but also the refinement of information.

17-2. Technological systems are the foundation of our food supply. This shows a combine with a corn head emptying into a wagon.

SYSTEMS

Many things work together in agriculture. These form systems. A ***system*** is a thing or process made of several parts. All of the parts or elements work together. All parts must do what they are intended to do. If not, the system fails to operate properly.

17-3. Making ethanol from agricultural products uses a system of technology. (Courtesy, Agricultural Research Service, USDA)

To understand what a system is, think of a water system. It has a source, a pump, storage tank, pipes, faucets, and fixtures. These work together for the water to be available as needed. If any of the parts fail, the system fails.

Systems are used in agriculture. Growing a crop is a system. It involves land preparation, planting, controlling pests, and harvesting, among others. All functions must be carried out. All materials must be in place. If any function or material is missing, the system does not work well or may fail. Technology has a big role in agricultural systems.

AGRICULTURE TECHNOLOGY SYSTEMS

Producers combine tools and processes to have good yields. The combinations relate to a similar problem. They create a system of technology.

A ***technology system*** is combining ideas and machines to achieve a purpose. The purpose may be to raise animals, grow plants, or to achieve other

AgriScience Connection

WEATHER WATCHING

The weather has much to do with work in agriculture. Good information is needed. Weather monitoring systems can be helpful. By getting information ahead of time, crop damage can sometimes be avoided. The photo shows an automated weather station being used to predict cotton yields and harvest dates. (Courtesy, Agricultural Research Service, USDA)

Weather affects the lives of all citizens. Heavy rains, blizzards, hurricanes, and high winds can all damage homes and businesses. Luckily, we have access to early warning systems. In addition to local forecasts, weather information is available through satellite information systems and the Internet.

Think how life would be different without weather information. We would have no way of predicting what the weather would be tomorrow. We could not plan ahead for the field day or school dance.

goals. Farm production uses many technology systems. These systems include knowledge, tools and equipment, and human beings.

In the past, a farmer might take drastic steps to control a minor problem. By using technology systems, the producer or scientist recognizes that each problem or issue is part of a larger environment. In solving problems, he or she considers the larger picture and not just the current problem.

Every person thinks a little differently. Each individual uses a different approach. What a person does is based on what a person thinks and knows. Often, there is neither a right nor wrong way to do something, just a different way. For example, some cultures, such as the Amish, reject some modern technology as too drastic. Another example is a superhuman person. Most people feel that using genetics to create a superhuman is not the best thing to do. The only difference between rejecting agricultural technology and creating a superhuman is the point at which one draws the line. This point is not a real line but is one that a person thinks exists.

SYSTEMS APPROACHES

The systems approach is used in agriculture and nonagricultural business. The principle remains the same—use knowledge, tools, and resources to solve problems. Decisions must be made based on the whole picture rather than a small segment.

The systems approach is used in every facet of agriculture. The number of systems and the variations are almost infinite. While we could present many examples, we have chosen to provide examples related to the previous chapters on soils, plants, animals, and processing.

17-4. Technology systems in harvesting potatoes have changed the way potatoes are harvested. (Courtesy, R. D. Offutt Farms, Minnesota)

Soil Systems

The soil management system is a common system in agriculture. Producers use a series of strategies to protect and maintain soil resources. Tools in that system include testing for nutrients, fertilization, crop rotation, planting practices, and conservation techniques.

With the systems approach, the elements used are selected as appropriate for the situation. Elements change as the environment changes. Soil managers change as they adopt new practices.

17-5. Tillage equipment uses technology in soil management systems. (Courtesy, AGCO, Corporation)

Plant Production Systems

Plants are produced for the benefit of humans, animals, and the environment. Over time, many strategies have been developed to make plant production systems better. Most plant growers use specialized species or varieties that meet their needs.

In greenhouses, producers use intensive systems that control every aspect of the environment. They are proactive. ***Proactive*** means that actions are taken to prevent problems before they appear. Tools in the production system must be used wisely. These include understanding the production needs of individual crops, using automatic irrigation sensors that measure the water in the soil, testing leaf tissue for nutrient measurement, using special lighting, having computers control the temperature, and others. The more producers combine their knowledge and the physical tools, the more effective the system.

17-6. Sensors and controllers are used in having a greenhouse environment.

With crops, producers may use computers, satellites, weather stations, and global positioning to help better manage their crops. It is common practice to combine these tools, highly specialized equipment, and an in-depth understanding of the plant's growth patterns to make management decisions.

Animal Production Systems

Animal technology begins even before an egg and sperm join. The animal manager makes many decisions. He or she considers different species and breeds, breeding strategies, parental characteristics, nutrition, and environment.

Animal production has benefited from biotechnology. Cloning can be used to create identical twins or multiple offspring. Hormones can be used to modify reproduction. Genes can be cut and spliced to produce organisms with new traits. All producers decide what elements will be used in their system. They look at what is acceptable in their situation.

17-7. Milking may use technology systems. (Courtesy, Joseph Farms, Atwater California)

AgriScience Connection

EMBRYO INJECTION

← How the egg is injected

Newly hatched chicks are vaccinated to prevent disease. Most are vaccinated the day of hatching. Handling and vaccinating can be rough on a small chick.

A new technology is now being used in hatcheries to keep chicks healthy. It uses *in-ovo* techniques. Embryos are injected three days before hatching. A needle makes a tiny hole through the egg shell and into the embryo. Vaccine is delivered into the developing chick. New technology allows one machine to do up to 30,000 eggs an hour!

Chicks are no longer subjected to vaccination on their first day. They hatch vaccinated! The photo shows an egg injection system being used. (Courtesy, Embrex, Inc., Research Triangle Park, North Carolina)

Personal likes and dislikes of the manager or owner are important. One person may be okay with a large, highly technical system. Lights, feed, temperature, humidity, and other environmental factors may be controlled by computer. Another person may choose a simpler system in which the animals must seek out their own food and live in a natural environment.

Processing Systems

Years ago, food was a major concern of most people. Each family grew enough for themselves. There was little extra. As late as 1850, one farmer provided only enough food for four other people! The development of technology in processing was the key to changes in production. Today, food may move from the producer to the consumer without ever contacting human hands. Tools in processing are based on knowledge and skill. They begin with an understanding of microbiology and include preservation, refrigeration, grading, storage, and the equipment to deal with large quantities.

17-8. Processing systems are used to prepare potato products. (Courtesy, RDO Frozen, Minnesota)

Supply Systems

Every area of agriculture uses supplies. Producers depend on supply systems. To produce crops or animals efficiently, producers must quickly get inputs at reasonable prices. What is needed depends on what is produced. Inputs include seed, fuel, chemicals, equipment, and replacement parts.

Agricultural supply systems depend on technology to maintain inventory. Good transportation must be available. Inputs must arrive when needed. This applies to all areas of the agricultural industry. Feed must be delivered so that plenty is always available. For example, a grower has a big problem with 100,000 chickens and no feed. Proper inventory allows work to be done on time.

Many systems exist in agriculture. Each producer may create a different system. The difference is to meet the individual needs of the farm or business. In addition, scientists are constantly trying to improve agricultural technology systems through research and development.

17-9. Bulk feed is delivered to poultry farms when needed. (Courtesy, Michael Stevens, McCarty Farms)

RESEARCH AND DEVELOPMENT: NEW TECHNOLOGY

17-10. An elm tree is being injected with an experimental vaccine to prevent Dutch elm disease. (Courtesy, Agricultural Research Service, USDA)

Technology systems do not appear by accident. These systems result from basic and applied research. Much of the research is conducted by research scientists. Every grower can use research to help answer individual problems.

Research and development are used together to create new technology. Known as ***R&D***, research and development has a big role in agriculture. All areas of agriculture are continually being changed by R&D.

Research is using scientific experimentation and investigation. Development uses research findings. Together, research and development create new technology.

WHO DOES R&D?

R&D is carried out at colleges and universities, corporations, government agencies, and independent labs receiving grant support. These sites may work together or independently. They may have different goals, but the outcome is the same: greater efficiency.

Why are these groups involved in research and development? Each has something to gain. One scientist might work simply for the joy of scientific discovery. Another might hope to improve society and create a better future. A third might hope for recognition and financial rewards. In truth, most scientists want all of those things to some degree.

The larger groups have similar goals. The first group to develop new technology may patent or copyright it. This sets ownership. It allows the owner to control where and how the technology is used. Often, the group then profits from the sale of the new technology. In recent years, corporations have become more involved in technology development because of the potential for profit.

Private businesses engage in R&D to come up with new products to make and sell. These products often have quick impact on agriculture. Private

R&D is often known as proprietary R&D. ***Proprietary R&D*** means whatever is developed is owned by the private business. The private business has an opportunity to gain profit from the R&D. Much new technology has resulted from private efforts.

HOW R&D IS DONE

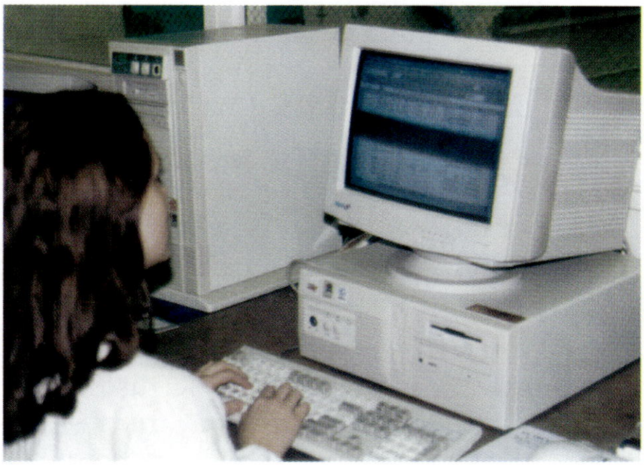

17-11. Computers have a big role in analyzing data and managing technology.

Technology is developed using problem-solving strategies. It is a time-consuming process. A scientist must first gather information or basic research data related to the problem. He or she must then consider how that information may be used to change what is being done in the industry. The scientist will spend much time developing, testing, and eliminating ineffective solutions.

Technology does not develop immediately. The length of time between a new scientific discovery and its practical application may be more than 20 years. Technology is created using trial and error and other experimentation techniques. A scientist must find a solution that is practical and effective.

SAFETY TESTING

An important area is safety. Even if a scientist finds a solution that works, he or she must still show that the product is safe. In particular, food products and technology used on food products go through extensive testing. Testing typically involves computer modeling (another example of technology!), animal research, and human research.

Safety testing is a step the government requires to protect people and the environment. It has come about gradually because technology has sometimes been harmful. For example, the pesticide DDT was effective on insects. It caused problems in other parts of the food chain. Safeguards are now in place to prevent similar incidents with other materials.

R&D is key to continuing success in agriculture. Scientists are seeking solutions that will lead to a larger and safer food supply.

VARIABLE RATE TECHNOLOGY

Variable rate technology (VRT) is varying crop practices based on field conditions. Simply, plant needs are not the same in all the fields of a farm. In fact, plant needs are not the same in all the areas of a field. VRT adjusts the rate of application based on conditions in the field.

Using fertilizer is a good example. Better use of fertilizer results if the rate of use is based on need. Some fields may not need fertilizer; other areas may be quite deficient. Fertilizer is wasted if all areas have the same rate of application. The same general approach applies in the use of pesticides. VRT is also known as site specific farming.

With VRT, cropping practices vary with the needs of the site. It is said to be environmentally sound. VRT uses no more chemicals than needed. Water runoff into streams does not carry residues of unused chemicals. The air is not filled with pesticides that are not needed. Applying the same rates to all areas of a field or farm is wasteful.

Career Profile

AGRICULTURAL ENGINEER

Agricultural engineers develop technology systems. While engineers have different roles, their goal is to make systems better.

Engineers identify problems in a system and use problem-solving strategies to create solutions. They may create new equipment, modify existing technology, or advise farmers. Engineers must be imaginative and willing to take risks.

Engineers must also have strong mathematical and scientific interests, be willing to do research, and enjoy challenging and intense environments. They need college degrees in agricultural engineering or closely related areas. The photo shows experimental equipment being tested to uproot plants after harvest in a way that prevents water loss and insect growth. (Courtesy, Agricultural Research Service, USDA)

KINDS OF VRT

VRT can be used in different ways. Some ways are more technical than others. Further, some are not practical in all situations. Each situation is different. Producers have to study their situations and make decisions about what is best for them.

Two major kinds of VRT are used: precision farming and prescription farming. Of these, precision farming uses the most high-tech methods.

Precision Farming

Precision farming is VRT that combines information and technology to manage crop production. High-tech systems of satellites, computers, and controls are used. VRT can manage the following practices: fertilizing, planting, tilling, controlling pests, and harvesting. With harvesting, a yield map is made that shows the yield of small areas in a field.

Two factors are important in precision farming: variability and timing of practices. Sampling is used to study the soil, pest levels, and maturity of the crop. Soil variability is studied by sampling and testing the soil. Results of testing tell nutrient needs. Soil moisture may be regularly tested. This allows efficient use of irrigation water.

Applications should be at a time that produces the best returns. Irrigation is an example. Irrigating before a crop suffers is best. Waiting until the plants wilt is too late. The plants have been damaged though irrigation water is applied.

17-12. A sensor is being checked on a combine. (Courtesy, Agricultural Research Service, USDA)

17-13. Potash is being applied by a variable rate applicator. (Courtesy, Agricultural Research Service, USDA)

Prescription Farming

Prescription farming is VRT based on applying inputs based on needs in particular areas of the field. Maps and computer-controlled equipment vary rates of application. Prescription farming is not as high-tech as precision farming. Satellites are not used in positioning the equipment. Some producers find prescription farming better for their situations.

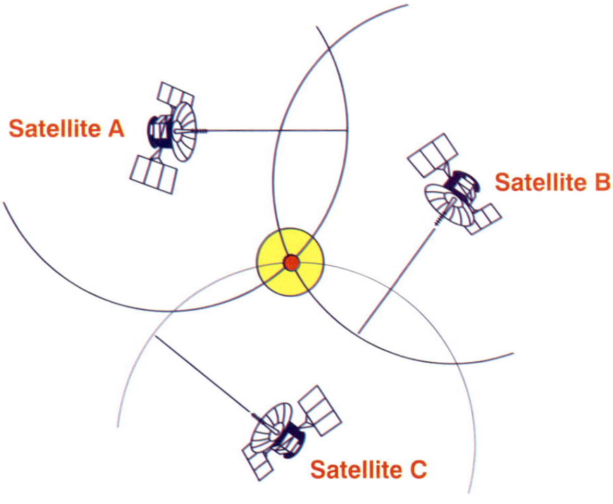

17-14. Satellites are used to provide information with GPS.

SYSTEM NEEDS

Many high-tech practices are used with VRT. This includes equipment plus sampling, testing, scouting, and analysis. Here are the major items:

- Positioning system—A positioning system identifies and records locations. These often use high-tech electronic equipment. The most common kind of positioning system is global positioning. A ***global positioning system*** (GPS) is

17-15. Using GPS in a field. (Courtesy, Agricultural Research Service, USDA)

a satellite-based approach to find exact positions in fields. It works to within three feet or so. GPS was developed by the military to locate specific points. Three satellites are connected by electronic signal with a receiver on the ground. (A fourth satellite is used to verify that the information is accurate.) Distances from satellites to the receiver are quickly measured. Computers calculate the exact location of the receiver. The receiver is typically located on equipment that moves over a field.

- Maps—A map is a representation of the features of the earth. Two kinds of maps are most important with VRT: yield and soil. Yield maps are prepared

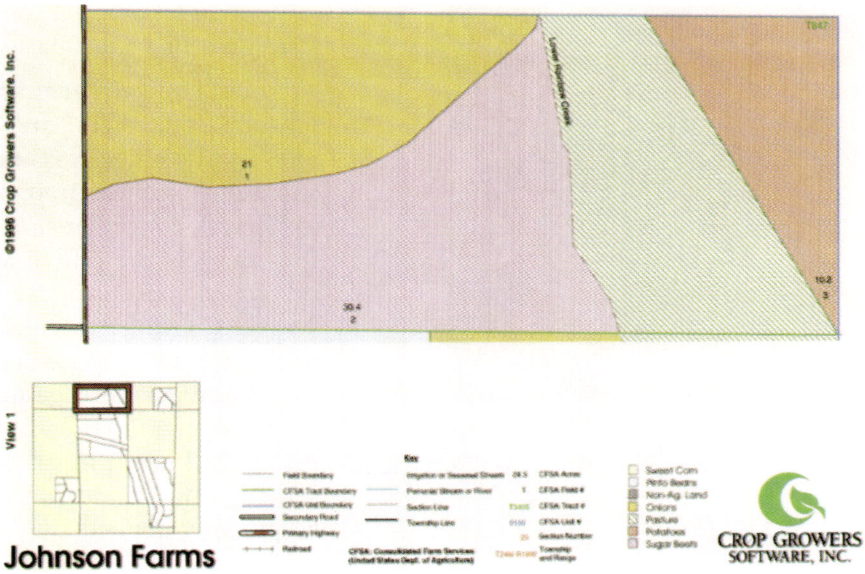

17-16. Map of a farm by crops. (Courtesy, Crop Growers Software, Inc.)

at harvest time. The information is stored and used for the next season's crop. Various sensors are used in harvesting equipment. The sensors record the amount harvested as related to a specific location in a field.

Soil maps have information about the soil in a field. This includes nutrient needs and other characteristics. The information is from sampling and testing the soil. It is plotted on a computer-based map.

Geographic information systems (GIS) are used to process information into color maps. The maps are prepared by computers using data from several sources. These maps tell much about a field. They help producers make decisions. They aid in applying inputs.

- Remote Sensing—***Remote sensing*** is collecting information about something from a distance. No physical contact is needed. For example, soil moisture information can be collected without going into a field. Information can be used to vary rates of fertilizer and other crop inputs.

- Application Technology—High-tech application equipment is used with precision farming. Computer controlled equipment uses all available information. Materials are automatically applied to the land based on the information. Good information gives good results. The reverse is true with bad information.

The equipment used has sensors, controllers, and actuators. A sensor detects conditions and sends information to controllers. A controller is a device that changes the rate of application based on information from a sensor. Controllers send signals to actuators. An actuator is a device that regulates the rate of application of material in a field. All of the system must work together. If not, materials will not be properly applied.

17-17. Information on water collected by remote sensing. (Courtesy, Agricultural Research Service, USDA)

LABOR VERSUS TECHNOLOGY

One concern is how technology affects workers. Technology is a double-edged sword. It has positive and negative effects. Most technology requires fewer people to do work. For example, years ago farmers could grow only enough food for their families. Some farm families even ran out of food. Today, one American farmer produces enough for many more people.

How is this possible? Technology allows for faster work and more efficient use of time. For example, one person using a hand tool to remove weeds can do only a small area at a time. This person can only work as quickly as he or she can walk and hoe. On the other hand, a tractor operator can remove weeds from several rows at once. The speed of the tractor and size of the equipment determines the area a worker can cover. Technology allows one person to get much more done.

Examples of timesaving technology and technological systems are found in almost every aspect of our lives. Cars, washing machines, and computers have all helped change America. Many jobs that were once essential are now unnecessary.

On the other hand, people must be trained to use technology. It is not true that fewer jobs exist. Many jobs that were in agricultural production have now moved to other parts of our work force. For example, more people

17-18. Compare the difference in human effort for hand hoeing and tilling.

are involved in R&D. In addition, some people have more free and leisure time.

Technology is one reason people need education. People used to succeed with little or no formal education. Times have changed. A person must deal with complex technology.

TECHNOLOGY AND SUSTAINABILITY

American agriculture has two needs: to continue to increase production and to reduce and prevent loss of resources. Some people say both needs cannot be met. Others say that technology is a source of solutions.

Land, air, water, and natural environments are resources that need to be used wisely. ***Sustainability*** is using resources in a way that guarantees continued opportunities in that area. For example, using sustainable agriculture systems takes steps to prevent soil loss and water contamination. Conservation is the wise use of resources. By so doing, the land and water are protected for future use.

Technology is used to provide solutions to important issues. Some relate to sustainable practices. Since technology involves both information and tools, researchers are experimenting to learn more about sustainable practices and sustainable techniques.

17-19. A balance in technology will sustain resources for future generations. (Courtesy, Smith's Farm, Maine)

REVIEWING

MAIN IDEAS

Agriculture uses technology systems that include information, tools, and people. These systems help increase efficiency and productivity. Specific strategies and the technology used differ widely. Technology systems are tools used in every phase of agriculture.

Technology is created through research and development. Scientists use information to create new solutions to problems. R&D is a very time-consuming process. One important aspect is safety testing.

VRT helps make wise use of inputs. It protects the environment and sustains resources.

Technology has greatly reduced labor requirements in agriculture. It has allowed people to take jobs in other areas of society. Also, all citizens must be better educated because of technology.

One goal of agriculture is sustainability. Resources should be used wisely and conserved for the future. Technology is a tool in achieving sustainability.

QUESTIONS

Answer the following questions. Use correct spelling and complete sentences.

1. What is a technology system?
2. How do technology systems help agriculture?
3. How is crop growing a system?
4. Select one of the agricultural technology systems and describe why it is beneficial.
5. What is research and development?
6. Name some groups involved in research and development. What do you think they achieve?
7. What are the steps of research and development?
8. Why does the government require safety testing?
9. How does technology reduce labor requirements?
10. Why do you need to get as much education as possible?
11. What is sustainability? How does technology help achieve this goal?
12. What is VRT? Why is it important?

EVALUATING

Match the term with the correct definition. Write the letter by the term in the blank provided.

a. variable rate technology
b. remote sensing
c. global positioning system
d. precision farming
e. technology systems
f. proactive
g. safety testing

____ 1. Collecting information from a distance.
____ 2. Preventing problems before they appear.
____ 3. VRT that uses information and technology to manage crops.
____ 4. Integrated ideas and machines.
____ 5. A step in research required by government that proves a product does no harm.
____ 6. Varying cropping practices based on field conditions.
____ 7. A way to find exact positions in fields using satellites.

EXPLORING

1. Visit an agricultural producer of your choice. Identify the technology systems used. Ask why the systems were selected and how the systems decrease labor. Are any negatives perceived in the system? Prepare a written or oral report regarding your visit and share it with your classmates.

2. Visit a facility involved in research and development in your area. Observe researchers at work. Ask what sort of process they use in their research and development projects. If possible, work with the researchers several hours to learn more about the work they do. Prepare a visual explanation, such as a bulletin board or poster, about what you learned.

3. Research examples of technology that have been accepted or rejected by agriculture. Have your teacher help you come up with examples. Compare and contrast the successes and the failures. What made the difference? Prepare a report describing what you have learned.

4. Use your knowledge of agriculture to identify a problem. Describe steps you would take as a scientist to develop new technology to help solve the problem. If possible, describe a tool you would like to invent. Create a poster and description of the process and the tool. Share your invention with your classmates.

18

MARKETING AND MANAGEMENT

Success in agriculture does not happen by chance. Good products require hard work. Planning, supervision, and effort are required. Being organized is essential.

A successful operation must be able to sell its products. At one time, producers could sell anything they grew. Today, a producer must be sure someone will want the product before it is produced.

Management and marketing have a big impact on success. They are the factors that keep an agricultural enterprise from failing. You can begin developing these important skills now!

18-1. Marketing in a flower shop involves making arrangements and greeting customers.

OBJECTIVES

This chapter introduces the meaning and importance of management and marketing. The objectives of this chapter are:

1. Explain management and list important management areas
2. List and define resources managed
3. Explain areas of management
4. Describe two functions in management
5. Explain the role of consumers in marketing
6. Explain the four Ps in marketing

TERMS

accounting
audit
capital
collateral
competitors
compliance
consumer-driven market
debt
direct marketing
inventory
loan
management
manager
promotion
resource
retail market
staff
value-added product
wholesale market

http://www.kcbt.com

(This site connects with a wealth of information through the Kansas City Board of Trade.)

MANAGEMENT FUNCTIONS

Management is the way an agribusiness (including a farm) is run. It is managing the business. It involves doing things that help an agribusiness achieve its goals. The person responsible for management is a manager.

18-2. Managing a ranch often involves supervising a work crew. (Courtesy, Texas Department of Agriculture)

MANAGERS

A ***manager*** is a person who makes decisions and uses resources to reach goals. The actions of managers are guided by a plan of action known as a business plan. Written business plans are common.

18-3. The manager and a technician discuss the operation of this bioreactor. (Courtesy, Agricultural Research Service, USDA)

18-4. Agricultural cooperatives are important for services, purchasing, and marketing.

How a manager goes about the job varies with the business. Size, structure, and personal philosophy are part of management. In a large business, a person may be responsible for only one part of management. In a small business, an individual may be responsible for all of the management.

The owners of the business determine the general approach to management. Many businesses create visual charts that show the roles of individuals. Managers need to have high energy, be friendly, know how to organize activities, and be honest and fair.

MANAGEMENT FUNCTIONS

18-5. Planning is a continuing step in management.

Managers have different roles. All of the roles are needed for an agribusiness to be successful. These roles are achieved by management functions. The five functions of management are:

- Planning—Planning is deciding what is to be done and how it will be done. Planning is done each day. A daily plan is part of a yearly plan. Yearly plans are part of long-term plans. Planning includes creating goals and objectives. It also includes ways of reaching the goals.

- Organizing—Organizing is grouping resources. A manager who organizes takes a plan and identifies the methods to use to meet the goals. Events are put in sequence. Activities are coordinated.
- Staffing—Staffing is having the people to do the work. It includes hiring people and helping them learn how to do the work. It also includes paying people for what they do.
- Leading—Leading is providing conditions so goals can be achieved. Resources are in place for people to do their work. Leadership is getting people to do what needs to be done.
- Controlling—Controlling is making sure the work is being carried out as intended. It involves evaluation. Controlling includes providing rewards for jobs well done. Being sure that money and other resources have been wisely used is a part of controlling.

AgriScience Connection

ENTREPRENEURSHIP

Did you ever have a lemonade stand? Have you ever conned your friends and family into buying pet rocks? You might be an entrepreneur!

An entrepreneur is a businessperson who creates new opportunities. Typically, an entrepreneur develops a new, innovative product and enters business for himself or herself. Entrepreneurs must understand management and marketing. They must be willing to take a risk.

Many schools encourage their students to start entrepreneurial enterprises. Sometimes, students work with the support of the school to provide goods or services. Consider starting a business of your own—you will be amazed at what you learn!

This photo shows a watermelon producer unloading a truck. (Courtesy, U.S. Department of Agriculture)

RESOURCE MANAGEMENT

Managers are responsible for using resources. In large businesses, management duties may be organized by resource area. A ***resource*** is any material with value. Agribusinesses have many resources. The most common are finances, staff, and inventory. Marketing is sometimes included here. In this book, it is listed separately.

FINANCES

A manager must handle financial resources wisely. Good production may not result in a profit if money is wasted. Financial management is carefully using the money of a business. This includes receiving money, paying debts, and paying taxes and fees.

One area is risk management. Risk is simply an implied threat of loss. Risk management reduces or controls the threat. A manager must not lose money.

Insurance is used in risk management. The business pays another company money to insure their product, facilities, or staff. If something happens, the insurer must pay. If nothing happens, the insurer makes money. Insuring a crop is common in agriculture. If a disaster destroys the crop, the producer may get some money for the lost crop.

18-6. A good relationship between a manager and a financial institution is important.

STAFF

Staff is the people who work on a farm or in a business. They may be known as employees or human resources. The employees of a business are important. A manager who deals with these resources must insure their safety, training, and satisfaction. A staff that lacks any of these is inadequate.

In a large business, the personnel manager hires staff. He or she usually looks for people with specific skills. Often, training is given after hiring.

The personnel manager helps people feel good about their work. Staff members are often concerned about pay, benefits, and work environment. Pay should be fair. The staff should get appropriate benefits. The work environment should be clean and healthy. Employees should be treated fairly.

INVENTORY

Inventory is the goods on hand. It includes both merchandise for sale and materials required for production. A person who manages inventory may deal with inputs, finished products, or both.

Inputs are the raw materials needed to produce products. Examples are animals, feed for animals, fertilizer for plants, and packaging for products. A manager insures that adequate inputs are available at right times. The manager may need to obtain different inputs depending on the season. A manager must see that materials are of high quality and stored properly.

Finished products (merchandise) are often stored for a short time. Warehousing is the physical storage of materials. It includes inputs and products. It is designed to insure quality and prevent loss.

An inventory manager is also responsible for preventing loss of the products. This loss may be caused by natural forces or by theft. The manager must maintain records of the quantity of items owned by the company and check those quantities regularly.

18-7. Microfiche and computers are useful in managing inventory.

AREAS OF MANAGEMENT

Management includes three important areas: capital, operations, and marketing. Without these areas, an agribusiness would not succeed.

CAPITAL

Capital is the resources of an agribusiness. Capital is more than money. It includes land, equipment, animals, and other "things" used in production. The capital available is sometimes less than what is needed to achieve the goals that have been set.

Capital may need to be borrowed. If a farm needs a new tractor, money to buy it may not be on hand. The money will be borrowed. A ***loan*** is money or other capital that has been borrowed. The borrower promises to pay it back. Written agreements are signed in getting a loan. Loans create debts. ***Debt*** is what is borrowed from others and owed to them. Financial management includes getting loans as needed and keeping track of debts.

18-8. Loan officers may visit farms to see the operation for themselves. (Courtesy, Agricultural Research Service, USDA)

Bases for Loans

Getting a loan is not always easy. Loans are made based on the ability of the borrower to repay. Lenders often use the four C's: credit, collateral, capacity, and character.

- Credit—Credit refers to records on previous personal and business loans. It involves the history of an individual or agribusiness. People who repay loans and pay their bills on time are said to have "good credit." People who do not, have "bad credit."
- Collateral—***Collateral*** is what is promised to the lender if the loan is not repaid. For example, a farmer may offer land as collateral for a loan. If the loan is not repaid, the lender can take the land. Of course, laws apply to how this is carried out.
- Capacity—Capacity is the ability of a borrower to repay a loan. A manager should never borrow money that cannot be repaid. Lenders look at the ability of a borrower to pay before making a loan.
- Character—Character is the reputation of the borrower. A person who is honest, hard working, and pays bills on time is more likely to get a loan. The request for a loan by a person who does not have good character may be rejected.

Sources of Loans

Sources of loans are banks, credit unions, the Farm Credit Administration, government agencies, businesses, or individuals. The source depends on the amount borrowed and the length of time for which it is borrowed. Always check on the rate of interest charged. Get the lowest interest you can. Borrow money only when necessary.

18-9. Some financial institutions serve agriculture needs.

RECORD KEEPING

Managers must have good records of their business. This involves accounting. ***Accounting*** is a system of recording financial transactions. The records include all exchanges of money. Careful record keeping allows accurate evaluation of the financial situation. Records are kept on machinery, animals, wages, and other areas.

Good record keeping involves an annual audit. An ***audit*** is an examination of the financial records by someone with special training and without ties to the agribusiness. A report of findings by the person doing the audit is made.

OPERATIONS

Operation is the day-to-day activity of an agribusiness. The manager deals primarily with activities that will lead to reaching the goals set. For example, an operations manager at a swine business would focus on the production of piglets and raising them to an appropriate size.

The operations manager would deal with all five functions of management as they relate to the day-to-day activities of the business. He or she would be involved with planning, organizing, staffing, leading, and controlling. The operations manager would also work with many resources of the business.

One responsibility that typically falls to the operations manager is compliance. ***Compliance*** is fulfilling all the legal regulations associated with a business. Sometimes, compliance is visible in the physical environment. In other cases, the manager may have to complete paperwork to comply with the law.

Three levels of government typically make laws that affect agriculture. Local, state, and federal lawmakers may all establish laws. Usually, these laws are intended to protect the business, employees, and/or the public.

Although state and local laws vary, all agribusinesses in the United States must comply with federal laws. Agencies that regulate agriculture include the Environmental Protection Agency (EPA), the Food and Drug Administration (FDA), offices within the U.S. Department of Agriculture (USDA), and the Occupational Safety and Health Administration (OSHA).

18-10. Managers must see that pesticide regulations are followed.

Regulations in agriculture are both similar and different from regulations for other businesses. For example, agribusinesses pay taxes like other businesses. They must provide a safe working environment. Agribusinesses, however, must comply with specific regulations on chemicals used in producing crops and/or animals. In processing, agribusinesses must take special precautions to maintain a safe food supply. Meat producers, for example, are supervised by federal monitors in packing plants.

As you might guess by the number of agencies involved, the regulations surrounding agriculture can be very complex. Sometimes, a manager must spend substantial time to ensure compliance.

MARKETING

Marketing is moving products from producers to consumers. It has become more important in recent years. Many steps are involved in the marketing process.

18-11. Reputable farm businesses serve consumers better.

Marketing is the final process in getting food, clothing, and housing materials to consumers. A producer must have a product consumers want. If not, the product has no value. We have a global economy that is consumer driven. A consumer-driven market is the foundation of free enterprise. In a ***consumer-driven market***, decisions are made by the buyers of goods or services. If someone is willing to buy a product, a reason exists to produce it. If no one is willing to buy a product, it should not be grown or made.

Marketing plans guide the marketing process. Small producers may have a plan but it may not be written. Large agribusinesses have written plans. A plan covers all areas of the marketing process.

THE MARKETING PROCESS

Marketing a product is often a complex process. The first step is developing a plan. Other steps may include assembling, grading, transporting, storing, processing, packaging, advertising, selling, and distributing the product. An agricultural product may go from the farmer to a processor. Later, it may go from a broker or wholesaler to a retailer, who will sell it to the consumer. Everyone involved in the process gets part of the sale price.

18-12. An on-farm roadside stand in California sells fresh cherries.

FOUR Ps

A manager must decide how marketing will be done. This often involves the four Ps: product, place, price, and promotion. Each is studied in planning what to produce. Never produce a product unless a market exists for it!

18-13. Appealing products get the attention of consumers. (Courtesy, Smith's Farm, Maine)

Product

The product is the item or service to be marketed. In agriculture, the product may be a plant part, an animal or animal products, by-products, or inputs. The product may be a service, such as plant or animal care. When selecting a product, consider its appearance and packaging. The products of agriculture are endless!

In selecting products, consider what people want. Managers learn about needs and conditions through market research. Market research gives strengths and weaknesses of competitors and areas of need. ***Competitors*** are others who provide the same or similar

18-14. A food company has created a tourist stop in Hawaii to promote its products.

products. Competition is an important part of free enterprise. Careful study of other producers tells if demand exists for a product.

One approach is to add value to existing products. This is often known as value-adding. A ***value-added product*** is one that has been made more usable for the consumer. The product is worth more. For example, a chicken begins as a live bird. Each step in processing adds value. First, the chicken becomes a whole carcass. Then the meat is cut into pieces. The pieces are deboned. The final product may be a breaded patty of white meat. Each step results in added value. The price per pound increases as work increases. Labor for the user of the product decreases.

In selecting the product, the manager considers the capability of the business. Can enough capital be obtained to do the project? Is the business near or far away from transportation? How does this product fit with other products from the company?

A niche market may be created when a new product fills a need never before identified. The product becomes something of a specialty and holds a unique place in the market.

18-15. A display promotes fruit (clementines, oranges, and grapes) on a street in LaMure, France.

Place

The second step in planning is to identify how and where the product will reach the consumer. "Place" includes both transportation and distribution of the product. For example, if a product is sold in urban areas but grown in rural areas, it must be moved to the city.

Sometimes, the product is sold directly to the consumer. This is ***direct marketing*** and is also a type of retail sales. A ***retail market*** is one that sells to the consumer. A ***wholesale market*** buys and resells to a retailer or another wholesaler. Wholesalers generally do not sell directly to the public.

18-16. This store sells fresh agricultural products direct to the consumer.

Both wholesalers and retailers may be independent or part of a chain. Independent businesses may have more power in selecting the products they sell. A local chain business, which is part of a larger corporation, often gives up some ability to make decisions.

Price

One difficult decision in marketing is pricing the product. The price must include all expenses and a profit. It must also be acceptable to the consumer. Often, price is set by figuring the cost of production, adding a profit, and comparing the prices of competing products.

The price may make a position for the product on the market. For example, a low price may indicate an economical item readily available to masses of people. A high price may suggest a luxury item intended to be purchased only rarely or by a select group.

Promotion

The next step in marketing is to get consumers to buy a product. ***Promotion*** is any activity taken to increase the public's awareness, interest, and purchasing of a product. Informing potential consumers about a product is important. Promotion is a creative activity limited only by the imagination of the manager.

Sometimes, a separate firm may be used to provide promotional services. These might include advertising, a public relations campaign, or coordination of special events. One promotion used in agriculture is face-to-face sales. For example, companies that produce pesticides may have salespeople who speak directly to consumers.

Promotion is an ongoing activity. Larger agricultural businesses have managers and departments committed solely to promotion of current and new products.

Career Profile

ADVERTISING SPECIALIST

A product has no value if no one will buy it. An advertising specialist develops ways to convince consumers to buy products.

The specialist may write advertisements for newspapers and magazines. An advertising specialist may make radio or television ads. Some advertising specialists set up displays at fairs or shows. The photo shows a display promoting agricultural research. (Courtesy, Agricultural Research Service, USDA)

Advertising specialists need a college degree in agribusiness, marketing, or communications. Learning continues on-the-job by working with a more experienced person. Advancement is based on assuming responsibility, being productive, and doing a good job. The ability to relate to people is very important.

18-17. Providing on-farm tours promotes agricultural products.

EVALUATION

In both management and marketing, evaluation is an ongoing function. Evaluation is a measure of the effectiveness or success of an activity. It is done both during and after a project or activity. Evaluation may be conducted either by people within the business or by a disinterested outside party. In either case, the goal of evaluation is to identify positives and negatives and to solve problems.

Evaluation is very important in agribusiness because it identifies and/or prevents problems before they can damage the company. Without evaluation, no one would know whether their work was doing what it should have done.

18-18. Auctions have been used to market hogs. (Courtesy, Mississippi State University)

REVIEWING

MAIN IDEAS

Management is deciding how a business is to be run. The strategies depend on the nature of the business. Management may be defined by functions, resources, and areas.

The five functions of management are planning, organizing, staffing, leading, and controlling. These five work with resources in finances, staff, inventory, and marketing.

The three areas of resource management relate to finances, staff, and inventory. Finance deals with capital, the assets of the business. Obtaining and using money are important duties.

Marketing is the process of moving the product to the consumer. Marketing planning should be done before production begins. Marketing commonly involves much planning. Marketing deals with identifying the product, selecting a place to sell the product, setting a price, and promoting the product.

Evaluation is the final step in marketing and management. It helps determine the success of any activity. Evaluation identifies areas to improve.

QUESTIONS

Answer the following questions. Use correct spelling and complete sentences.

1. What is management? Why is it important?
2. What are some skills a manager needs? Explain why.
3. Why do you think management can be explained so many different ways?
4. What is capital? How is it obtained?
5. What are the five functions of management?
6. What is marketing?
7. What is a consumer-driven market?
8. What is competition? Why should competitors be studied?
9. What is the difference between a wholesale and retail market?
10. How could a wholesale market impact consumer prices?
11. Why is evaluation important?

EVALUATING

Match the term with the correct definition. Write the letter by the term in the blank provided.

a. manager	d. management	g. capital
b. resource	e. inventory	h. promotion
c. accounting	f. collateral	

____ 1. Decision-making process in a business.

____ 2. Goods on hand.

____ 3. Assets of a business.

____ 4. What is promised a lender if a loan is not repaid.

____ 5. Any available material with value.

____ 6. System of recording financial transactions.

____ 7. Person responsible for making business decisions.

____ 8. An activity to increase public awareness of something.

EXPLORING

1. Invite a banker or other financial manager to speak to your class. Have the person bring and hand out loan application forms. Ask about sources of financing for agricultural businesses. Also, ask about strategies for managing your personal money. Write a summary describing the financing methods the banker discussed. In addition, write a letter thanking the speaker for his or her time.

2. Visit a marketing or public relations firm in your area. Tour the facilities to learn the different responsibilities or jobs in the business. Find out what agricultural products the business handles and what special techniques are used to promote these products. Prepare a written or oral report describing what you learned.

3. Visit an agricultural business in your area. Ask about the division of responsibilities and why those divisions were made. Follow the agricultural product as it moves through the business from beginning to end. Make a poster or bulletin board that describes the path the merchandise takes as it moves from the producer to the consumer.

19

WRITTEN AND ORAL COMMUNICATION

Have you heard someone say, "What do you mean?" Most everyone has a time when they fail to understand what another person is saying. We spend much of our time trying to exchange information. We need to be able to share information so that other people know what we are saying.

All of us can improve our communication skills. We can learn to speak and write better. Our abilities are never fully developed. Some people think they cannot improve. False! Everyone can continue improving how they go about sharing information.

Some people are very good with information. They quickly grasp what they read and hear. They are also quite good at sharing the information with other people. Communication abilities are learned skills. You will continue developing your skills with this chapter.

19-1. Communication involves many skills–speaking ability, eye contact, and personal appearance are positive in this photograph.

OBJECTIVES

This chapter is about communication. It explains what communication is and how each person can use the process to be more effective. The objectives of this chapter are:

1. Explain the communication process
2. Describe the important written means of communication
3. Identify the major parts of a business letter
4. Describe important oral communication skills
5. Explain how to prepare and make a public speech

TERMS

article
communication
entertaining speech
feedback
informative speech
letter
listening
medium
message
nonverbal communication
nonverbal cues
oral communication
persuasive speech
public speaking
receiver
report
resume
senses
source
verbal communication

http://www.FarmJournal.com

(This site has information on a wide range of subjects as well as links in agriculture.)

COMMUNICATION IS A PROCESS

Communication is the process of exchanging information. If information has not gone from one person to another, communication has not occurred. It is more than speaking and writing. Communication involves someone drawing meaning from what was spoken or written. How well the exchange occurs depends on the response of the person who got the information.

VERBAL AND NONVERBAL COMMUNICATION

All of us communicate each day. We use many ways to communicate. Some are verbal; others are nonverbal.

Verbal communication is the use of words to convey information. Words may be spoken or written. Words have meaning. Selecting and using the proper words is important. Communication is not very good if the wrong words are used. In fact, there may be no communication if the proper words are not used.

Nonverbal communication is the exchange of information without words. It influences verbal communication. The nonverbal cues influence the meaning of spoken words.

Nonverbal cues are the signals that people send that have meaning. People often do not know that they are sending these signals. Frowns, gestures, and posture are cues that tell much about how we feel. They accompany verbal signals and give meaning to them.

Some nonverbal cues are planned; others are not. Planned nonverbal cues result from a planned effort to communicate nonverbally. A good example is the dancer. Many body movements may be used to communicate feelings and emotions.

19-2. Communication is verbal and nonverbal. (Courtesy, Agricultural Research Service, USDA)

Unplanned cues result from our actions. Nodding or shaking our heads may happen without us knowing it. The same is true with how we hold our arms, how we stand, and the expressions on our faces. With conscious effort, we can improve nonverbal communication. We can smile more and have better posture. Eye contact is an important part of nonverbal communication in our culture. We can improve on eye contact by practicing looking into the eyes of another person when we are talking.

19-3. Sharing information about a tomato plant uses several senses.

THE SENSES

People get information through their senses. The ***senses*** are the ways people receive messages. They are sometimes known as receptors.

The five senses are sight, hearing, smell, taste, and touch. Most communication involves sight and hearing. Written words and symbols require sight. Spoken words require hearing. In communicating, these senses are used to exchange information. The person who wants to communicate needs to understand the communication process as related to the senses.

THE PROCESS

To communicate, someone must send a message to another person. The process involved may not be easy to see. But, it is there. The process has five parts. Four of the five must work for information to be exchanged. The fifth assesses how well the process has worked.

The five parts are:

1. Source—All communication begins with a ***source***. The source is known as the sender. This is where the message begins. Without a source, there would be no communication. The source initiates the message.

2. Message—The idea being shared is a ***message***. The source prepares the message for sending. How it is sent requires some thought. The message must be arranged to give meaning. The sender needs to know the person for whom the message is intended. Select a way to prepare the message that the receiver will understand. Messages may be in spoken words, written words, drawings, and other forms. Body movements and gestures are included in one-on-one communication. How the message is prepared is critical in the process. Can you get meaning from a message spoken to you in French? If you get a message in French and do not know French, the message has no meaning. Language and other ways of sending messages are known as codes. The receiver must be able to interpret the code.

Career Profile

TELEVISION JOURNALIST

A television journalist prepares and makes programs for broadcasting. As a popular means of communication, television is used to share information with masses of people. Some television journalists specialize in agriculture.

Education needs vary with the nature of the work. Most agricultural journalists have a degree in agricultural communications or a closely related area. Having a background in agriculture is important. Having good speaking skills is essential.

Most television journalists work at television stations. Some journalists work for government agencies, colleges, and agribusinesses. This photo shows a journalist preparing to interview a panel.

3. Medium—The ***medium*** is the channel that connects the sender with the receiver of the message. It links the two together. The medium selected depends on the nature of the message. In some cases, voice is used; in others, written words or symbols on paper are used to share information. Email and the World Wide Web are rapidly expanding as a medium. To use a medium, both the sender and receiver must use it. For example, a deaf person cannot understand spoken words. Sign language should be used.

4. Receiver—The ***receiver*** gets the message and draws meaning from it. The codes used in the medium are interpreted. Sometimes, receivers understand the message exactly. At other times, they do not. In communicating, always consider the receiver. Put yourself in their situation. What do you think they will understand?

5. Feedback—It is always a good idea for the sender to see how well he or she has communicated. This involves feedback. ***Feedback*** is the return channel from the receiver back to the sender. It helps the sender decide if the receiver drew the correct meaning from a message. If not, the sender repeats the message in a more appropriate form.

Why is this process important? It helps improve communication. Both the sender and receiver can use the process. This helps share information.

19-4. This conversation uses all five parts of the communication process.

WRITTEN COMMUNICATION

19-5. Written communication begins with organizing ideas.

Written communication involves using signs or symbols. The most important symbols are those known as letters. We use an alphabet with 26 letters. Letters are arranged to form words. The letters must be arranged properly for the spelling to be correct. In effect, words represent other things. An example is t-r-e-e. These four letters arranged in this order convey important information. If the letters are arranged as e-r-e-t, we do not get the same meaning.

Words are sequenced to form sentences. Several sentences may be formed into a paragraph. Paragraphs may be used to form a letter, report, or newspaper article. In this use, the word, letter, has a different meaning from a letter of the alphabet.

LETTERS

A ***letter*** is a written message. Most letters are words and sentences properly arranged to convey the intended meaning. Two kinds of letters are used: business and personal. Business letters tend to be more formal than personal letters. Business letters are used to apply for jobs, share information, and request assistance. Personal letters are written to friends and family members.

Most letters are written on paper and delivered by a courier or the Postal Service. Many people are now using email and the Internet to send letters. With either method of sending a written message, the sender needs to give careful attention to the message. Letters represent the sender when he or she cannot be present.

REPORTS

Reports are widely used. A ***report*** is a detailed written document. It may provide information about an experiment or be based on library research. Reports have titles, carefully written information, and are bound in some way. Sources of information used in preparing a report are listed.

Reports are prepared using an outline. Begin with a statement of the purpose of the report. Provide only information related to the purpose. If a report is prepared for a specific purpose, always follow the guidelines given.

ARTICLES

An ***article*** is factual writing about one topic. Articles are prepared for use in a newspaper or magazine. Many people can use the local newspaper to share information if they will write an article. Schools often have newspapers that use articles written by students. The articles may describe club events, student accomplishments, and FFA activities.

Before preparing an article, talk with the editor about what is needed. Write the article following an acceptable format. Submit it to the editor on time.

Most newspaper articles follow an upside-down pyramid style. This means that the most important information is presented in the first paragraph, known as the lead. Supporting information is presented in the paragraphs that follow.

RESUMES

A ***resume*** is a written summary of an individual's education, experience, and accomplishments. They are often needed when applying for a job. Resumes are sometimes known as personal data sheets or vitae.

BUSINESS LETTERS

A well-written business letter represents the sender positively. Letters should be typed or printed on paper. Use equipment that will give a good appearance. Use good quality paper. Most business letters are on white paper. The letter should be carefully folded and placed in a properly addressed envelope.

THE LETTER

Before beginning, decide the purpose of the letter and briefly outline how the content will be arranged. Keep it brief. Spell correctly. Use proper grammar and never use slang. Use terms the reader will understand. Read what you have written several times. Get another person to read the letter and offer suggestions. This will help you communicate with the receiver.

Always include the major parts of a business letter. These help organize the information. They also provide the receiver with your name and address. Without these, they cannot respond to you. Review the essential parts of a business letter in 19-6.

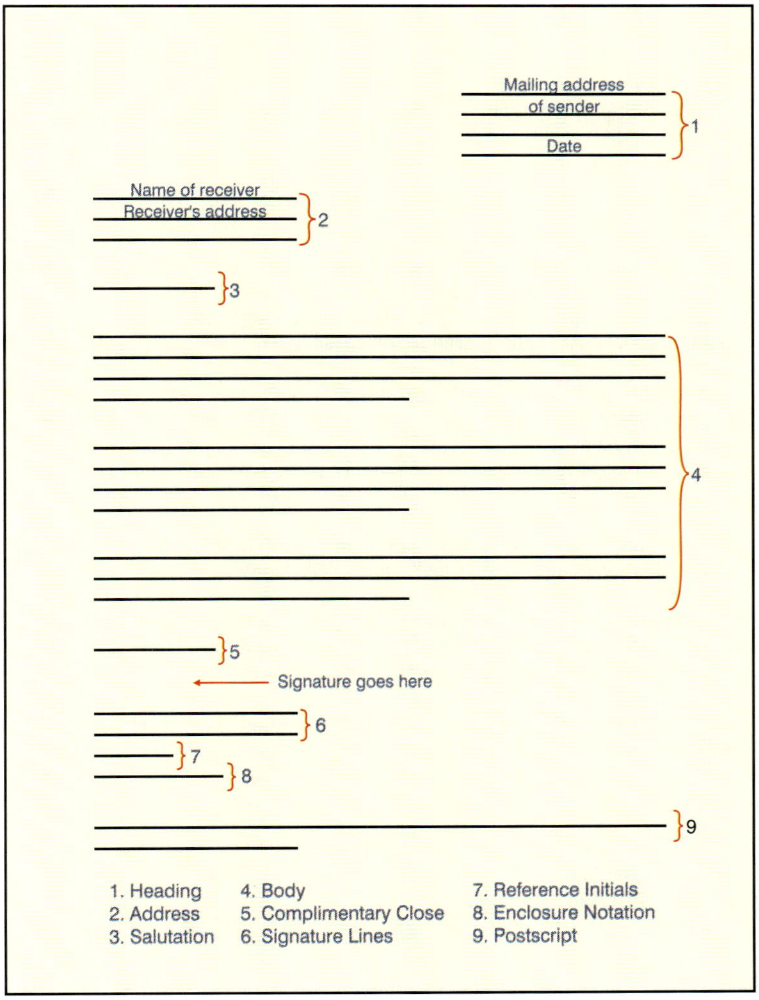

19-6. Parts of a business letter.

The essential parts of a business letter are:

- Heading—The sender's address and the date the letter was written are in the heading. In some cases, the sender's telephone number or email address will be included.
- Inside address—The address of the individual receiving the letter is the inside address. It should be correct and repeated on the envelope.

P. O. Box 512
Webster, OR 97333
January 22, 1999

Ms. Judy W. Jenkins, Manager
Alpharata Supply Company
P. O. Box 215
Webster, OR 97333

Dear Ms. Jenkins:

This is to invite you to be the guest speaker at the Webster FFA Chapter banquet on February 17, 1999. We would like for you to talk about what it takes to be successful in agribusiness. Your work as manager of Alpharata Supply Company makes you well qualified to be our speaker.

The banquet will be held in the Webster High School cafeteria beginning at 6:30 p.m. on Wednesday, February 17. We expect to have 136 parents and members present. We are allocating 14 minutes for your speech. Please let me know if you will have any special equipment needs. A podium will be at the head table for you to use.

Another part of our banquet will be presentation of awards. During that time, you will be presented the Honorary Chapter FFA Degree.

Since our banquet is only a few weeks away, I need to hear if you will be available in the next few days. You may write or, if you prefer, call me at 329-2464. I also have an email address if you would like to use it: jws@aol.com.

All of the FFA members are excited about the banquet. We certainly hope you will be our featured speaker.

Sincerely,

James William Smith
Webster FFA Chapter Secretary

19-7. Sample business letter.

- Salutation—The salutation is a greeting to the individual receiving the letter. It usually begins with "Dear" and includes the receiver's name. With adults, address them as "Ms." or "Mr."
- Body—The message being sent forms the body of a letter.
- Complimentary close—The complimentary close is a courteous expression that closes a letter. Examples include sincerely, yours truly, and sincerely yours.
- Signature—This is the sender's name. It is often typed four lines below the complimentary close. The sender signs his or her name in the space.

Computers and word processing equipment help prepare good letters. Use the spelling and grammar checking feature to improve the letter. Be sure the printer is in good condition. The copy sent should be neatly printed and easy to read.

THE ENVELOPE

An envelope is the wrapping that surrounds a letter. Select a standard size (usually number 10) envelope. Usually, the envelope should be the same kind and color of paper as the letter. Fold the letter in approximate thirds and insert in the envelope.

Be sure to properly write the address of the receiver and your return address on the envelope. Attach the necessary postage. If a courier is used, properly fill out the address label.

19-8. Sample no. 10 envelope.

RESUME

Resumes are most often used to apply for a job. You want an attractive resume that represents you well. A poorly prepared resume can result in your not getting a job.

CONTENTS

A resume should contain the needed information. It should not contain details that have little or no value or that are incorrect. The main items in a resume are:

- Your name and address (This may include telephone number and email address.)
- Date (This is the date the resume was prepared.)
- Career objective or goal
- Education (List certificates or degrees as well as schools attended. Emphasize the education and training that shows you are prepared for the job.)
- Work experience (Include both full-time and part-time jobs. Give the nature of the work and name of the employer.)
- Community or school activities (Include offices held, committee membership, and similar details.)
- Other (Include other items that reflect positively on you for the particular job.)

Do not include the names of references in a resume unless requested to do so. These can be provided later on a separate sheet of paper. It is also a good idea to omit photographs from a resume.

PREPARATION

Resumes are typed or printed on good quality paper. White paper is usually best. It is a good idea to use a word processor. This makes it easy to make corrections and prepare resumes for particular uses. Also, resumes can be saved on diskettes for future use. This means that you do not have to start over each time you need one.

A resume should have a good appearance. Be sure all information is accurate. All words should be spelled correctly. Use a printer that provides a neat copy.

Resume

Susan Lee Sloan
P. O. Box 580
Demorest, GA 30535
(800) 555-1111

Objective:	A position as a biotechnology technician with opportunities for increased responsibility and advancement
Education: **1997-1999**	North Georgia Tech, Clarkesville, Georgia Associate Degree in Horticulture Biotechnology Graduated May 12, 1999, with a 3.4 GPA
1993-1997	Central High School, Cornelia Georgia Graduated with honors, class rank was 12 out of 392
Experience: **1997-present**	The Orchard Golf Course, Turnerville, Georgia Part-time work included all areas of turf and landscape management
1995-1997	Frank's Lawn Maintenance Service, Cornelia, Georgia Part-time work included lawn care and landscape maintenance
College Activities: **1998-1999**	President, Horticulture Club President, Student Government Association Vice President, Scholarship Club Member, Landscape Design Team Member, College Ambassadors
1997-1998	Member, Horticulture Club Member, Landscape Design Team Representative, Student Government Association
High School Activities:	President, Honor Society President, CHS FFA Chapter Member, Debate Team Member, CHS Marching Band Delegate, State FFA Convention Member, Principal's Honor Club
Community Activities:	Member and Captain, Community Youth Tennis Team Active, Community Beautification Program Member, Sunday School Church
References:	Available Upon Request
Date Prepared:	May, 1999

19-9. Sample resume.

ORAL COMMUNICATION

Oral communication is using spoken words to share information. The sender selects the sounds to be used. The receiver must hear and interpret the sounds. This requires listening skills. This means that oral communication requires effort from both the speaker and listener or audience. Both must be active in the process. If not, communication will not occur.

19-10. A speaker must consider the audience. (Courtesy, National FFA Organization)

SPEAKING

Speaking involves communication with others on a one-to-one basis as well as before groups. In either case, it is important to know something about the receiver. Terms, speaking rates, language, and other methods should be selected that the listener (receiver) will understand.

Sharing information with our voices is important. We do it with the telephone as well as when we talk with people in person. Some people have jobs that require speaking to masses using the broadcast media. Radio and television are widely used.

It is best to use concrete words. These are words that communicate precisely. The listener has no doubt about what you have said.

Use short words. If the words are unfamiliar to the listener, offer a brief explanation. Use a few adjectives and adverbs to enhance meaning.

Use a good rate of speaking. Do not speak too fast. Some people will find it difficult to follow your message. Do not mumble or slur words.

Select a pitch that is easy for listeners. Pitch refers to the highness and lowness of your voice. Changing the pitch during speaking adds meaning to your message.

Smile when you speak. The smile translates to a "happy" voice. It also communicates feeling to the listener.

19-11. Good speaking and listening skills are needed for job success.

LISTENING

Listening is drawing meaning from oral communication. It involves the sense of hearing. Good listening requires concentration on what was said. Here are a few tips in listening:

- Prepare—Get ready to listen. Block other things out of your mind. Concentrate on what the speaker is saying.
- Understanding—Try to understand what the speaker is saying. Make interpretations of words and phrases. Watch the nonverbal cues from the speaker.
- Avoid bias—Have an open mind. Do not be so biased you do not allow a speaker's ideas to come through.
- Attention—Pay attention to what is going on. Do not read papers or talk with someone near you. If useful, take written notes on the main ideas.

19-12. Listening involves concentrating on what the speaker is saying.

- Location—Select a seat where you can hear. Sit near the speaker. Avoid being near noisy doors and windows. Do not sit where you can see outside. Sit where you will focus on the speaker.

AgriScience Connection

PREPARED PUBLIC SPEAKING

The best way to develop speaking skills is to practice. Young people usually have opportunities to learn while in school. Student activities help develop speaking abilities. Many organizations stress speaking before groups.

The National FFA Organization has a prepared public speaking career development event. Students begin by preparing and making speeches in their classes. A representative is selected to speak at a district event. The best speeches are chosen for the state or national levels. One speech is selected each year as the national winning prepared public speech.

PUBLIC SPEAKING

Public speaking is giving an oral presentation to a group. Speeches may be to civic clubs, church groups, and student assemblies. The sizes of groups will vary. Civic clubs may have 25 to 100 people. Conventions may have a few hundred to several thousand in the audience. Televised speeches are before audiences with millions of people.

19-13. Practice speaking before groups develops self-confidence.

KINDS OF SPEECHES

Speeches are of three types: informative, persuasive, and entertaining. Typically, people begin developing speaking skills with informative speeches.

An ***informative speech*** is one to give the audience new knowledge. Facts and dates are often used. Some people refer to informative speeches as technical reports.

A ***persuasive speech*** is to get the audience to believe something new. It may try to get people to take certain actions. Persuasive speeches include facts. The speaker also uses opinions or arguments to get people to act a certain way.

An ***entertaining speech*** is to get the audience to enjoy themselves. Jokes and amusing stories may be used. These speeches are perhaps the most difficult to give.

GIVING A SPEECH

In speaking, you want to do your best. You will want people to gain information from your efforts. As a speaker, your role is to share information—not to impress others with your speaking ability.

Following a few simple principles will help in giving a prepared speech.

19-14. Get information for a speech from good sources.

Getting Ready

Always be prepared! Giving a speech is easier when you have prepared thoroughly. Spend far more time getting ready than you will use in making the speech. Never give a speech without plenty of preparation.

Here are a few important areas in getting ready to give a speech:

- Topic—Select a topic that is interesting to you. Sometimes, you will be given a topic. If so, remember that all topics can be interesting—it is up to the speaker! Also, find out about the audience. The depth of information will depend on what your audience knows about the topic.
- Information—Often, preparation requires information. Use books, magazines, newspapers, the World Wide Web, and interviews with other people. Be sure all information is correct. Never give inaccurate information.
- Outline—Develop an outline for your speech. Include three main areas: introduction, body, and conclusion. Use the introduction to get the attention of the audience and show why the topic is important. Use the body to present the details of the topic. Speech bodies often have two to four main points. Use the conclusion to summarize and review the content of the speech.
- Development—Select a strategy for your speech. Write your ideas out well ahead of time. Go back over the ideas and polish them. Use questions, facts, and interesting stories to get the attention of the audience. Use a step-by-step process in unfolding your information. Avoid jokes and making statements that might embarrass another person.
- Visuals—With some speeches, visual aids can be used. Posters, charts, projected slides and transparencies, computer images, and real things are helpful. Be sure a visual contributes to your message. With large groups, be sure every one can see your visual. Good quality visuals make a speech better; poor visuals detract.

Speech Outline

TITLE: Waste Disposal

PURPOSE: To convey information about the proper disposal of solid waste materials

I. Introduction (facts about wastes made by people)

II. Body
  A. First main topic (wastes damage the environment)
    1. Subpoint number one (wastes are litter)
    2. Subpoint number two (wastes pollute water)
    3. Subpoint number three (wastes injure wildlife)
  B. Second main topic (reducing wastes)
    1. Subpoint number one (reducing throwaway materials)
    2. Subpoint number two (reuse materials)
    3. Subpoint number three (recycle)
  C. Third main topic (disposing of solid wastes)
    1. Subpoint number one (compost)
    2. Subpoint number two (use landfill)
    3. Subpoint number three (incinerate)

III. Conclusion (wastes can be reduced and properly disposed of)
  A. Summary of main points
  B. Action to take

19-15. Sample outline for a short speech.

- Practice—Repeatedly practice your speech. Get friends, teachers, and others to listen and give suggestions. Some speakers practice in front of a mirror. This helps them see what the audience will see. Another good strategy is to video tape your practice sessions. Study the tapes to see how you can improve. Get other people to offer their suggestions.

Making the Speech

You can give a good speech! On the day of your speech, arrive ahead of time. Get a feel of the place where you are speaking. Many people prefer a podium or speaker's stand. If you will use one, make the speaker's stand area comfortable. Arrange your visuals for ease of viewing. Speak in the room before the audience arrives to get a feel for how loud you will need to talk.

Practice with a microphone if one is used. Do not get too close to the microphone nor speak directly into it. Static and cracks in the sound are distracting to the audience. Most of the time, a microphone is 6 to 12 inches from the mouth. It is positioned at a 45-degree angle to your mouth so you will not blow into it.

When speaking before a group, stand on both feet. Do not lean on a table or speaker's stand. Look at the audience—eye contact is always important. Speak loudly, clearly, and distinctly. Use a pleasant voice. Use complete sentences, correct grammar, and pronounce words properly. Have your notes handy but refer to them no more than necessary. Be enthusiastic. Use variety in your voice. Occasional pauses will add emphasis. Use gestures but do not fidget. Place your hands at your side. Never put your hands in your pockets.

Always stay within the time allotted for your speech. During rehearsal, use a watch or clock to monitor the time. In giving the speech, stop on time and sit down. Afterward, thank the person who invited you to speak. Do so personally as well as send a letter of thanks.

19-16. With practice, speaking before a group gets easier.

REVIEWING

MAIN IDEAS

Communication is the exchange of information. It involves verbal as well as nonverbal methods. People receive information through their senses. Sight and hearing are most important. The best communication makes use of all senses.

Communication is a process that includes a source, message, medium, receiver, and feedback. Understanding the process helps in being a good communicator.

Written communication is using signs or symbols. The symbols are visual or graphic materials on paper or other material. Letters of the alphabet and numbers are used in written communication. Shapes, colors, and other means may be used. The most common written forms are letters, reports, articles, and resumes.

Oral communication is using spoken words to share information. It includes one-to-one communication as well as speaking before a group. Both speaking and listening skills are involved. Good preparation is an important part of giving a good speech.

QUESTIONS

Answer the following questions. Use correct spelling and complete sentences.

1. What is communication?
2. Distinguish between verbal and nonverbal communication.
3. What are the parts of the communication process?
4. What are the four most important forms of written communication?
5. What are the major parts of a business letter?
6. What is a resume? What are the major parts of a resume?
7. What is oral communication?
8. What are several suggestions to speak well?
9. What is listening? Why is it important in communication?
10. What is public speaking?
11. What should be done to ensure giving a good speech?
12. What should be considered in making a speech?

EVALUATING

Match the term with the correct definition. Write the letter of the term in the blank provided.

a. nonverbal communication
b. message
c. feedback
d. letter
e. listening
f. informative speech
g. receiver
h. medium

____ 1. Channel that connects the sender with the receiver.
____ 2. Exchanging information without words.
____ 3. Idea being communicated.
____ 4. Draws meaning from a message.
____ 5. Helps evaluate communication.
____ 6. A written message.
____ 7. Drawing meaning from oral communication.
____ 8. Speech that gives new knowledge to other people.

EXPLORING

1. Prepare a resume for yourself. Have your teacher offer suggestions on how to improve it. Save your work on a diskette so you can easily revise it in the future.

2. Write a sample business letter to a local business to ask about a job. Be sure to include all parts of the letter. Follow a good writing style. Get suggestions from your teacher on how to improve the letter.

3. Organize a public speaking event in your class. Have each person prepare a speech that meets guidelines of the National FFA prepared speaking career development event. Refer to information in the latest edition of the Career Development Events regulations.

20

THE FFA AND YOU

Success! Everyone wants to be successful. No one wants to fail. Sometimes, people wonder what they need to do to be successful. You have some idea of what is needed. You have already been successful in so many ways!

Education is essential in being successful. Knowledge and skill help us know how to go about work and daily living. We often need more from our education. We need to feel good about what we are learning. That is where the FFA comes in.

The FFA gives meaning to classes. Through the FFA, you can see the results of our work. Being active in the FFA offers opportunities for travel, fun, and meeting people. The FFA provides rewards for doing a good job.

20-1. The FFA helps members develop skills in many areas. This Washington member is proud of her winning agriscience water pollution project.

OBJECTIVES

This chapter describes the FFA and ways you can become involved. It covers how the FFA can help you be successful in your career and life. The objectives of this chapter are:

1. Explain the purpose and history of the FFA
2. Describe how to be an FFA member
3. Explain how the FFA is organized
4. Describe the activities offered by the FFA
5. Explain how to be a good FFA member

TERMS

- active FFA membership
- Career Development Events (CDE)
- Chapter FFA Degree
- charter
- FFA
- FFA advisor
- FFA motto
- Future Farmers of America
- Greenhand FFA Degree
- leader
- leadership
- official dress
- personal growth
- proficiency awards

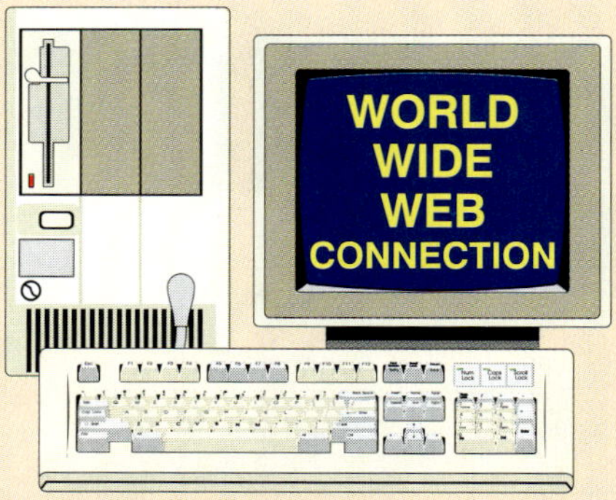

http://www.ffa.org

(This site has information on The National FFA Organization and links to other sources of agricultural information.)

PURPOSE AND HISTORY

Young people have so much potential! They have abilities that have not been developed. Too many young people do not know what they can do. Having opportunities in the FFA helps them realize their abilities.

PURPOSE

The purpose of the FFA is to help students develop to their full potential. It does so through agricultural education classes. The FFA is an integral part of the instruction. Many exciting activities are available.

The purpose of the FFA is to develop:

- Improved Agriculture
- Leadership
- Citizenship
- Patriotism
- Character
- Scholarship
- Cooperation
- Recreation
- Service
- Thrift

The FFA stresses three areas: leadership, personal growth, and career success. These areas are closely related.

20-2. FFA members planning a meeting.

20-3. An FFA member is serving in a leadership role.

Leadership

Leadership is the ability to influence other people to meet individual or group goals. Many skills are involved. It includes how we relate to other people.

People in leadership roles are leaders. A ***leader*** is a person who helps other people reach their goals. Effective leaders have traits that help them in their roles. With a little practice, these traits can be developed by anyone.

Overall, leaders are people with desirable traits in four areas:

- Personal Skills—These traits make it easy for other people to follow a leader. Examples are being responsible, being hard working, and being honest. We learn these traits from our early days as children. With effort, we can improve on them today.
- "How To" Skills—These traits help a leader share responsibilities. Examples are organizing a meeting, speaking to a group, and writing messages. These traits can be developed through the FFA and classes on leadership. Involving other people is essential to being a good leader.

20-4. Planning FFA events develops personal skills.

- "Thinking" Skills—These traits enable a leader to think and assess problems. Thinking something through is important before making a decision. Examples of "thinking" skills are analyzing situations, anticipating problems, and seeing opportunities.
- "People" Skills—These traits help a leader relate well with other people. They are not hard to learn. You can begin practicing them now! Examples include being trustworthy, respecting others, and having a positive attitude. Being a good communicator is an important "people" skill.

Personal Growth

Personal growth is developing skills to have a good life. It is leadership skills plus career skills. Social behavior and citizenship skills are included. Social behavior deals with our "manners." Manners help us respect other

AgriScience Connection

A GOOD HANDSHAKE

A good handshake helps people relate to each other. It is part of saying hello and good bye and in congratulating others. Handshakes are used to "seal" business deals. Many FFA events include handshakes.

Use the right hand in shaking. The hands should go together comfortably—palm to palm. Each should grasp the other person's hand. Do not squeeze too tightly nor have a "dishrag loose" hand. Shake but do not pump the other person's hand. Release the hands after a few seconds. Handshakes should not linger longer than the greeting.

A good handshake also has eye contact, a smile, and spoken words. Practice shaking hands with friends in you class. It will make it easier to do later.

people and be at ease around them. What we say to other people, how we dress, and how we go about eating our meals are examples of social behavior.

Citizenship deals with being a good member of our country. We obey laws, vote, and respect our government. We assume roles to make our communities good places to live.

Career Success

The FFA promotes career success in many ways. It recognizes excellence in agricultural education. It helps people develop skills to begin and advance in careers.

A related area is supervised agricultural experience (SAE). This is where students gain practical experience in many areas. Some have jobs. Others raise animals or grow plants. Some carry out research activities. The opportunities are endless!

The National FFA Organization has a program of Career Development Events. These activities relate to what is taught in agriculture classes. Many schools use Career Development Events as a part of the instruction.

20-5. "Learning by doing" is important in FFA and agriculture classes.

HISTORY

The FFA we know today evolved from the original organization. Students come first! Their needs, interests, and goals are most important. History helps us understand how the FFA became a dynamic organization.

Early Years

In the early 1920s, Future Farmers clubs were formed in Virginia. These clubs were for boys enrolled in agriculture classes in high school. The idea was very popular. Students gained from their participation. Teachers and leaders saw that these clubs were good.

Other states began Future Farmers clubs. Leaders saw a need for the states to share with each other. They saw the value of a national organization. An organizational meeting was held in 1928 in Kansas City, Missouri. Thirty-three delegates attended.

The organization was known as the ***Future Farmers of America.*** It was commonly called the FFA. It focused on farming and meeting the needs of rural students. Dues were only ten cents a year!

A Time of Growth

Many changes have occurred in the Future Farmers of America. New programs were developed for members. By 1934, all states except Rhode Island had FFA. Thousands of members were active in their local schools. A national convention was held each year in Kansas City, Missouri, for 70 years. In 1999, the National FFA Convention will be in Louisville, Kentucky.

The following are major events in the organization's history:

1928 Future Farmers of America was founded.

1939 National FFA Camp set up on land that formerly belonged to George Washington in Alexandria, Virginia. (The camp later became the National FFA Center and camping ended.)

1944 National FFA Foundation was formed to use funds from business and industry to support FFA activities.

1950 Public Law 740 was passed by Congress granting the FFA a federal charter.

1952 *The National Future Farmer* magazine was begun.

1965 Consolidation with the New Farmers of America (NFA) strengthened the FFA. (NFA was an organization similar to the FFA for African American students.)

1969 Female students were allowed to become members.

1971 National FFA Alumni Association was formed.

20-6. The National FFA Center was dedicated July 20, 1998, in Indianapolis, Indiana.

1988 Name of the organization was changed to National FFA Organization.

1989 Name of the *The National Future Farmer* magazine was changed to *New Horizons*.

1998 National FFA Center moved from Alexandria, Virginia, to Indianapolis, Indiana.

1999 Louisville, Kentucky, became site of National FFA Convention.

20-7. Leadership training serves beyond the FFA.

Today

Today, the letters "FFA" are used as the name for the Organization. "Future Farmers of America" is no longer the name. The delegates voted to change the name to the National FFA Organization (or FFA) in 1988.

The FFA has more than 450,000 members. More than 7,200 local high schools have FFA chapters. These are in all states plus the District of Columbia, Puerto Rico, and other island territories.

FFA activities reflect the broad scope of the agricultural industry. Students may be involved in any area. Agribusiness, horticulture, environmental science, and others are included. More activities now reflect the basic areas of science that are important in agriculture.

HOW TO BE A MEMBER

The National FFA Organization has four kinds of membership: active, alumni, collegiate, and honorary.

ACTIVE MEMBERSHIP

Active FFA membership is for students enrolled in agricultural education. These classes are offered in grades 7 to 12 in the schools. Membership can be continued beyond high school. This is usually until November 30 following the fourth National Convention after graduating from high school.

Students should be interested in the affairs of the FFA. Attending meetings is a membership requirement. National, state, and local dues are paid. All active members receive the *New Horizons* magazine. Members can participate in FFA activities and apply for awards.

New FFA members are known as Greenhands. This is because they have qualified for the ***Greenhand FFA Degree***. With further accomplishments, members earn the ***Chapter FFA Degree***. Outstanding individuals may receive the State FFA Degree from the state association. The American FFA Degree is the highest level of membership in the National FFA Organization.

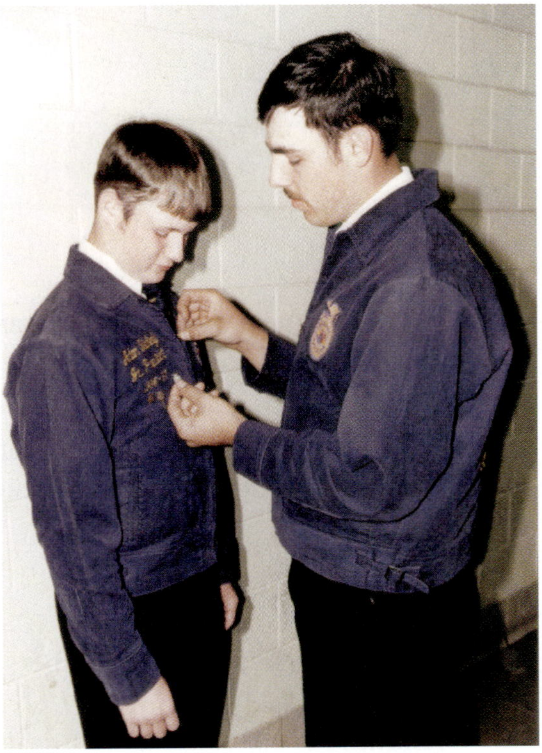

20-8. The Chapter FFA Degree pin is being awarded to a Greenhand.

Table 20-1. Greenhand and Chapter FFA Degree Requirements

The general requirements for being a Greenhand and earning the Chapter FFA Degree are summarized below. Refer to the Official Manual for complete details.

To Be a Greenhand:	To Have a Chapter FFA Degree:
1. Enroll in agriculture classes and plan an SAE	1. Have received the Greenhand FFA Degree
2. Learn and explain the FFA Creed, Motto, Salute, and FFA Mission	2. Have at least 180 hours of instruction in agriculture at or above the ninth grade level; have an SAE; and be enrolled in agriculture
3. Describe and explain the FFA emblem and colors	3. Participate in three functions in the chapter Program of Activities
4. Demonstrate the FFA Code of Ethics and proper use of the FFA jacket	4. Have earned and invested $150 or worked 45 hours after class
5. Know the history of the FFA, its constitution and bylaws, and Chapter Program of Activities	5. Have led a group discussion for 15 minutes
6. Own or have access to the Official FFA Manual and the FFA Student Handbook	6. Have demonstrated five parliamentary procedure abilities
7. Submit a written application for the Greenhand degree	7. Make progress toward individual FFA achievement
	8. Have a satisfactory scholastic record
	9. Submit an application for the Chapter FFA Degree

An agriculture teacher is known as an ***FFA advisor***. One duty of an advisor is to manage a local FFA chapter. The advisor sees that all members are qualified and organizes student leaders to manage the chapter.

OTHER MEMBERSHIP

The other categories of membership do not require enrollment in agricultural education.

Alumni membership is for former FFA members and others interested in the FFA Organization. The National FFA Alumni organization has offices with the National FFA. Alumni members pay dues.

Collegiate FFA membership is for students in college who are enrolled in agriculture in two- and four-year colleges. Collegiate chapters plan activities much as local FFA chapters. Collegiate members pay dues.

Honorary FFA membership is for adults who have been supportive of the FFA. Honorary members, who are voted on by FFA members, do not pay dues.

20-9. The FFA Official Manual gives details on many FFA activities.

Career Profile

FFA ADVISOR

Being an FFA Advisor is part of the role of an agriculture teacher. Advisors are so important to the success of a chapter. They work with FFA members in planning and carrying out a program of activities. They train teams for competitive events and help members develop leadership skills. Advisors counsel members and help them achieve awards.

FFA Advisors are trained as agriculture teachers. A part of the training is in FFA. This includes skills needed to run an FFA Chapter. College degrees in agricultural education are needed.

This photo shows one of the FFA Advisors at Franklin County High School, Georgia, helping the Sentinel set up a meeting room. (The owl is the symbol for the advisor. In the United States, owls are symbols for wisdom and knowledge.)

HOW THE FFA IS ORGANIZED

The FFA is organized on three levels: local, state, and national.

LOCAL CHAPTERS

Local secondary schools with agriculture classes may have an FFA chapter. This includes middle schools, high schools, and vocational centers. A local chapter receives a charter from the state association. A ***charter*** is an official act that recognizes a chapter. A ceremony is held. A framed certificate is presented to the school.

In local chapters, programs of activities are developed around the interests of the members. Members may be involved in judging livestock, landscaping projects, and fund raising. The FFA is flexible; members have many choices. There are activities in science, the environment, and food safety. Meetings are regularly held. Local chapters often have banquets, trips, and other events for members.

20-10. The Chapter Secretary keeps minutes of meetings.

A local chapter elects officers. Their role is to oversee the chapter. The officers include a president, vice president, secretary, treasurer, reporter, and sentinel. Some chapters have other officers, such as a chaplain. The agriculture teacher serves as the advisor.

STATE ASSOCIATIONS

State associations help local chapters and organize state-level events. A state advisor oversees FFA work in a state. The headquarters are usually in offices of the state department of education.

Some states are organized into regions, districts, or federations. These allow local chapters to participate before the state level.

State events include conventions, contests, and ceremonies. Local chapter members enter career development, leadership, and other events at the state level. Each state selects participants in these events to go to national competition.

A state association is chartered by the National FFA Organization. Each state has a team of officers. These are elected by delegates from local chapters at the state FFA convention. A state is allowed to send delegates to the National FFA Convention based on the number of FFA members in the state.

20-11. Delegates at a state convention.

NATIONAL ORGANIZATION

The National FFA Organization is set up to represent the interests of members. A National Constitution and Bylaws guide the organization. The Constitution can be amended by the delegates at the National FFA Convention. All members may review the Constitution in the *Official FFA Manual* (available from the National FFA Organization).

A board of six national officers oversees the FFA. One officer serves as president and one as secretary. Four are vice presidents. The officers are elected by delegates each year at the National FFA Convention. Officers represent all regions of the United States. They work closely with the National

FFA Advisor. The National FFA Advisor is the person in the federal government who is responsible for agricultural education in the United States.

A Board of Directors oversees the National FFA. This Board is made up of adults who understand agricultural education. Members of the Board are elected by professional groups, such as the state supervisors of agricultural education. Several members are appointed.

The National FFA Organization has a staff to carry out many activities. This includes all programs as well as preparing publications. Their offices are at the National FFA Center. Staff of the National FFA Foundation is also a part of the National FFA Center.

Delegates at the National FFA Convention vote on items about the FFA. They decide programs, membership, dues, and other areas. The student officers and Board of Directors work to carry out actions of the delegates.

The United States is divided into four FFA regions. Each region may have activities. Sometimes, state winners participate for region awards.

Table 20-2. FFA Regions

Western	Southern	Central	Eastern
Alaska	Alabama	Illinois	Connecticut
Arizona	Arkansas	Indiana	Delaware
California	Florida	Iowa	Maine
Colorado	Georgia	Kansas	Maryland
Guam	Louisiana	Kentucky	Massachusetts
Hawaii	Mississippi	Michigan	New Hampshire
Idaho	Puerto Rico	Minnesota	New Jersey
Montana	South Carolina	Missouri	New York
Nevada	Tennessee	Nebraska	North Carolina
New Mexico	Virgin Islands	North Dakota	Ohio
Oklahoma		South Dakota	Pennsylvania
Oregon		Wisconsin	Rhode Island
Texas			Vermont
Utah			Virginia
Washington			West Virginia
Wyoming			

ACTIVITIES

Opportunity! The FFA has many activities. The activities are in a wide range of areas for members with many interests. There is something for everyone in the FFA. A few of the opportunities are briefly described.

CAREER DEVELOPMENT EVENTS

The ***Career Development Events*** (CDE) allow members to show their skills. The areas are part of the classroom instruction in agricultural educa-

Table 20-3. Examples of FFA Career Development Events

Group or Team Events	Individual Events
Agricultural Mechanics	Extemporaneous Public Speaking
Agricultural Sales	Prepared Public Speaking
Dairy Cattle	
Dairy Foods	
Farm Business Management	
Floriculture	
Forestry	
Horse Evaluation and Selection	
Livestock	
Marketing Plan	
Meats Evaluation and Technology	
Nursery and Landscape	
Parliamentary Procedure	
Poultry	

tion. Some CDE are for individual members; others are for groups or teams. Most events focus on career skills. A few events focus on leadership skills.

CDE begins at the local chapter level. Students develop skills in classes related to careers in the agricultural industry. Many local chapters hold events to select the individual or team that will represent the chapter. These representatives go to district or state events. Usually, state winners go to national events.

20-12. FFA members often participate in fairs and shows.

PROFICIENCY AWARDS

Proficiency awards are for members who excel in skill development. It is the top achievement in an area. These awards are usually based on supervised agricultural experience (SAE). Proficiency awards are given at the local, state, and national levels. Sometimes, districts within a state may offer awards. Outstanding individuals are selected from each of the four FFA regions. One national proficiency award is given each year at the National FFA Convention.

Local FFA chapters receive application forms for proficiency awards. An FFA member who fills out an application form must have a record book that verifies what has been done. Local chapters send applications to the state association.

Table 20-4. Examples of FFA Proficiency Awards

Area	Entrepreneurship	Placement	Combined
Agricultural Communications			X
Agricultural Mechanical Technical Systems	X	X	
Agricultural Processing			X
Agricultural Sales/Service			X
Beef Production	X	X	
Cereal Grain Production			X
Dairy Production	X	X	
Diversified Crop Production	X	X	
Diversified Livestock Production	X	X	
Emerging Agricultural Technology			X
Environmental Science			X
Equine Science			X
Feed Grain Production	X	X	
Fiber Crop Production			X
Floriculture			X
Food Science and Technology			X
Forage Production			X
Forest Management			X
Fruit/Vegetable Production			X
Home and Community Development			X
Landscape Management			X
Nursery Operations			X
Oil Crop Production			X
Outdoor Recreation			X
Poultry Production			X
Sheep Production			X
Small Animal Care			X
Soil and Water Management			X
Specialty Animal Production			X
Swine Production	X	X	
Turf Grass Management			X
Wildlife Management			X

Complete details on proficiency awards are available from the National FFA Organization. The *Agricultural Proficiency Award Handbook* is available from the National FFA Center.

CHAPTER AWARDS AND ACTIVITIES

Local FFA chapters are recognized with the National Chapter Award program. These awards are for chapters that have most actively gone about the work of the FFA.

20-13. Local chapter members planning a program of activities.

Chapters are recognized for having quality programs. Emphasis is on member, chapter, and community development. Standards have been set for chapters to qualify for recognition. Officer teams and committees work hard to achieve these awards. They learn and have fun at the same time.

The Food for America Program is used by some chapters. This involves FFA members making presentations to elementary students about agriculture.

More details on chapter awards and activities are available from the National FFA Organization.

OTHER AWARDS

The National FFA has many other awards. These are for individuals and groups of members. In addition, chapters and states may have other awards. Three areas are included.

- Star Awards—Top individuals at the chapter, state, and national levels may receive "Star Awards." Chapters have the Star Greenhand Award for an outstanding new member. Chapters also have Star Farmer and Star Agribusiness Awards. States and the National level have Star Farmer and Star Agribusiness Awards. National winners must first receive the regional Star Award.
- Agriscience Awards—These awards go to students who make top achievements in science-based learning. Emphasis is on research in areas related to agriculture and the environment.
- Scholarships—Many scholarships for study beyond high school are available each year through the National FFA Organization.

OTHER ACTIVITIES

Many other activities are available to FFA members. These include state, national, and international events.

- Conventions—Members may participate in the state FFA convention and the National FFA Convention. Any FFA member who registers can attend the National Convention. These represent wonderful opportunities to meet people and polish leadership and career skills.
- International Travel—FFA members may travel to other countries. While there, they may live with families and learn about agriculture.
- Government Events—FFA members are often involved with legislative dinners and similar events. These events help members understand how laws are made. Sometimes, members may have internships with government agencies.
- Banquets—The FFA awards banquet is a regular event with most local chapters. Members are recognized for their hard work. Parents and others are made more aware of the FFA.
- Civic Events—FFA members are often involved in the activities of other organizations. In some areas, members may work with the Farm Bureau. In other communities, members may help the local Chamber of Commerce or other group.

The FFA Creed

I believe in the future of agriculture, with a faith born not of words but of deeds—achievements won by the present and past generations of agriculturists; in the promise of better days through better ways, even as the better things we now enjoy have come to us from the struggles of former years.

I believe that to live and work on a good farm, or to be engaged in other agricultural pursuits, is pleasant as well as challenging; for I know the joys and discomforts of agricultural life and hold an inborn fondness for those associations which, even in hours of discouragement, I cannot deny.

I believe in leadership from ourselves and respect from others. I believe in my own ability to work efficiently and think clearly, with such knowledge and skill as I can secure, and in the ability of progressive agriculturists to serve our own and the public interest in producing and marketing the product of our toil.

I believe in less dependence on begging and more power in bargaining; in the life abundant and enough honest wealth to help make it so—for others as well as myself; in less need for charity and more of it when needed; in being happy myself and playing square with those whose happiness depends upon me.

I believe that American agriculture can and will hold true to the best traditions of our national life and that I can exert an influence in my home and community which will stand solid for my part in that inspiring task.

Source: Official FFA Manual.

20-14. The National FFA Creed.

BEING A GOOD MEMBER

Being a good FFA member is more than paying dues. It means that you are involved. When involved, you learn more and feel better about yourself. Knowing about the organization is helpful.

A FEW DETAILS

The FFA has an impressive heritage. As a member, you benefit from what has been done to make it a top organization.

Motto

Twelve short words guide FFA members. These form the ***FFA motto***. The motto helps members in achieving their goals in life. These words form the motto:

Learning to do,
Doing to learn,
Earning to live,
Living to serve.

Colors

The official FFA colors are national blue and corn gold. The blue is a symbol of the blue field in the flag of the United States. The gold represents the matured fields of crops ready for harvest—a sign of success.

Emblem

The emblem represents the history, goals, and future of the FFA. A cross-section of an ear of corn contains an owl sitting on a plow with the sun rising in the background. An eagle rests on top of the corn.

20-15. The FFA Emblem. (Courtesy, National FFA Organization and used with written permission)

Each part of the emblem has meaning. Corn is an important crop grown in every state. The rising sun is about the future and the opportunities for FFA members. The plow represents the importance of work in being successful. The eagle is to remind FFA members of their nation and the freedoms we have. The owl is a symbol of knowledge and wisdom, which are needed for success in agriculture.

The FFA emblem is a registered trademark of the National FFA Organization. The *Official FFA Manual* describes how the emblem can be used.

Official Dress

Members are to dress appropriately for FFA events. The uniform worn by FFA members is ***official dress.*** The official dress provides identity and gives a well-known image.

A big part of official dress is the FFA jacket. Only members are to wear a jacket. Jackets should always be neat and clean. Jackets should have the proper emblem and lettering for you, your chapter, and state association. When worn, the jacket should be zipped to the top.

Official dress for male members is black slacks, white shirt, official FFA tie, black shoes and socks, and the jacket.

Official dress for female members is black skirt, white blouse, official FFA blue scarf, black shoes, and jacket. Black slacks are allowed for traveling and outdoor activities.

20-16. Members in official FFA dress.

Meeting Room Arrangement

FFA members find they have better meetings if the room is properly arranged. FFA symbols are a part of the arrangement. Each officer has a station. The symbol for the office is a part of the station.

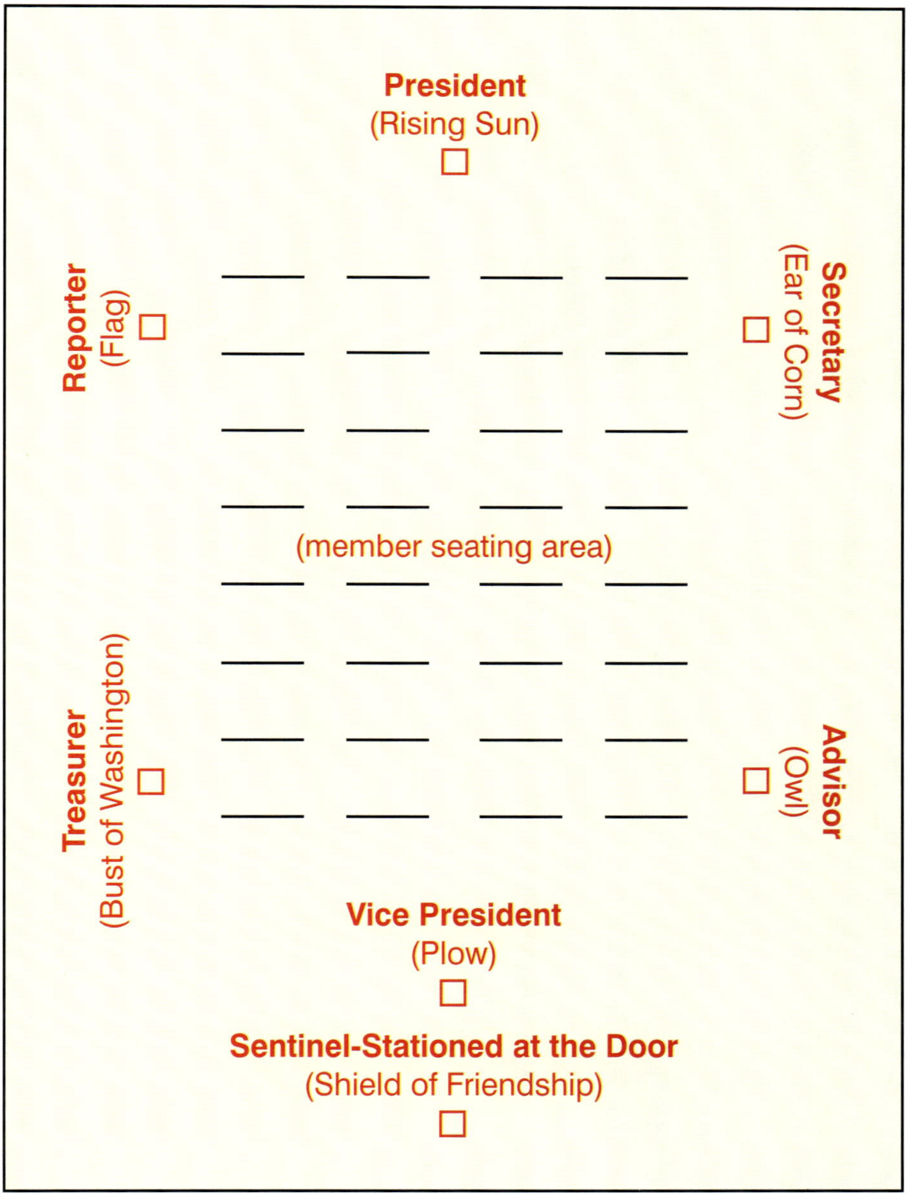

20-17. FFA meeting room arrangement.

20-18. A Chapter Sentinel is setting up a room for an FFA meeting.

20-19. Properly using a gavel in a meeting.

MEMBER RESPONSIBILITIES

Members get more out of the FFA when they are actively involved. Chapters are stronger when members take active parts. Each member is to uphold the ideals of the FFA and help it be an outstanding organization.

Look for ways to be involved. You will help your chapter and fellow members. Most important, you will be developing yourself. Here are a few tips:

- Learn about the FFA
- Participate in meetings and other activities
- Volunteer to serve on committees and be an officer
- Support chapter officers and other leaders
- Participate in leadership development activities
- Set up and carry out a good supervised agricultural experience program (SAE)
- Keep good records of your activities
- Take your studies seriously
- Respect the lives and rights of other people
- Avoid substances that abuse your body
- Practice good eating and rest habits
- Follow school and community regulations
- Strive to develop good human relations skills
- Tell your parents and others about the exciting opportunities in the FFA

REVIEWING

MAIN IDEAS

The FFA is the national organization for students enrolled in agricultural education. Students may be in grades 7 to 12, though most are in the high school grades. The purpose of the FFA is to help members develop their full potential in life.

Started in 1928 as the Future Farmers of America, the name was changed to FFA in 1988. This change better reflects the interests of the members. Membership requires enrollment in agricultural education and the payment of dues. All members receive the *New Horizons* magazine.

Many opportunities are available to FFA members. These include the Career Development Events and Proficiency Awards. Most of these are based on having supervised agricultural experience. Records should be carefully kept on SAE.

The FFA motto, colors, emblem, and official dress are much a part of each member's participation. The jacket is a well-known part of official dress. Only members are allowed to wear the jacket.

QUESTIONS

Answer the following questions. Use correct spelling and complete sentences.

1. What is the early history of the FFA?
2. What are the three main purposes of the FFA? Briefly explain each.
3. What are the four kinds of FFA membership?
4. What are major requirements for active FFA membership?
5. What are the three levels in the FFA organization?
6. How is the National FFA Organization administered?
7. What is a Career Development Event? Name two examples.
8. What is a Proficiency Award? Name two examples.
9. What is the motto of the FFA?
10. What are the official FFA colors?
11. What are the main items in the FFA emblem?
12. What is the official FFA dress for males? Females?
13. What are the responsibilities of FFA members? Name and explain two.

EVALUATING

Match the term with the correct definition. Write the letter of the term in the blank provided.

a. FFA
b. FFA motto
c. proficiency awards
d. FFA advisor
e. personal growth
f. Career Development Events
g. active FFA membership
h. charter

____ 1. Kind of FFA membership for students in secondary schools.
____ 2. Official designation of an FFA chapter.
____ 3. Three letters designating the name of the National FFA Organization.
____ 4. Developing skills to have a good life.
____ 5. FFA activities that allow members to demonstrate their skills.
____ 6. An agriculture teacher with responsibility for local FFA chapter.
____ 7. Twelve words that guide FFA members.
____ 8. Awards for FFA members who excel in particular areas.

EXPLORING

1. Invite a state FFA officer to be a resource person in your class. Ask him or her to explain important benefits of FFA membership.

2. Use the Internet and World Wide Web to learn more about the National FFA Organization. The web site address is: http://www.ffa.org

3. Study the *Official FFA Manual* and other materials that will help you set goals for your participation in the FFA. Prepare a brief written statement of what you would like to do. Share your goals with your teacher.

4. Join the FFA and get involved.

21

SAE: DEVELOPING CAREER GOALS AND SKILLS

Planning a trip is essential. Without planning, you will not know where you are going nor how to get there. Afterward, you will not know if you have gotten where you wanted to go. You will have wasted time and effort.

Life is like a trip. The kind of work we choose to do is an especially important area. We need to plan. We need to know where we are going and how to get there. We will not know all of the new opportunities that will come along. Being prepared helps take advantage of new opportunities.

Career planning helps develop the skills we need. It helps us to be successful. It is never too early to begin planning. Start now!

21-1. Completing high school is an important step in realizing career goals.

OBJECTIVES

This chapter describes career goal planning. It includes the use of supervised agricultural experience to help begin skill development. The objectives of this chapter are:

1. Explain goals and career ladders
2. Identify skills needed for career success
3. Explain supervised agricultural experience
4. List the benefits of supervised agricultural experience

TERMS

- career
- career ladder
- experience
- exploratory SAE
- goal
- goal setting
- major
- ownership SAE
- people skills
- placement SAE
- research and experimentation SAE
- self-employed
- training agreement
- training plan
- training station

http://www.stw.ed.gov

(This site has information on programs to aid the transition from school into gainful employment.)

CAREER PLANNING

A ***career*** is the general area of work a person follows. It is the direction of a person's life as related to work. It includes work-related activities. Careers may involve several jobs from entry as a young person until retirement.

Some people work for themselves. They have their own businesses, including farms. People who work for themselves are ***self-employed***. Many people work for other people, known as employers. Employers may be large corporations with thousands of workers or small businesses with only a few workers.

Work is a natural and integral part of life. All people are expected to have useful roles in our society. The kind of work we do and how well we do it is related to education. Planning is needed. Working hard to achieve our goals builds success.

21-2. An operator is using a ride-on water reel cranberry harvester. (Courtesy, Agricultural Research Service, USDA)

GOALS

People differ in what they want from a career. Some want fame and money. Others want to become a leader and provide service to other people. What do you want? Assess your interests, values, and abilities. Decide and set goals that will help you achieve what you want.

A ***goal*** is an end to be reached. It is something you want to achieve. Goals are best achieved when they are carefully set. We have to be realistic about our goals. Unrealistic goals may not be reached.

21-3. Insect cell cultures are being studied by entomologists. (Courtesy, Agricultural Research Service, USDA)

Goal Setting

Goal setting is making goals. It is describing what we want to accomplish in life. We have to look at what is important to us. Goals should be practical, challenging, and attainable. Of course, all goals require effort. Even the best goals are no good if we do not go about doing what we need to do to achieve them.

Different kinds of goals are set. These are based on the length of time to achieve them. A short-term goal is to be achieved in a year or less. Intermediate goals are to be achieved in 1 to 4 or 5 years. Long-term goals are to be achieved in 4 or 5 years. Often, short-term goals are steps toward intermediate and long-term goals.

Begin goal setting by writing down what you want to accomplish. Talk with your teacher or school counselor. Read materials. Assess your interests. Determine the general area in which

21-4. A teacher or counselor can help assess your career interests.

Goal Setting

Write your goals below. Follow each goal with a list of steps to follow and the ways and means for accomplishing each step. Set a deadline by which the ways and means are to be completed. Regularly evaluate how well you are doing in achieving your goals. Revise goals, steps, and ways and means as necessary.

Name ____________________ Date __________

Goal Number One: ____________________

Steps to Achieve Goal	Ways and Means for Steps	Date
1. ________	a. ________	______
	b. ________	______
2. ________	a. ________	______
	b. ________	______
3. ________	a. ________	______
	b. ________	______

Goal Number Two: ____________________

Steps to Achieve Goal	Ways and Means for Steps	Date
1. ________	a. ________	______
	b. ________	______
2. ________	a. ________	______
	b. ________	______
3. ________	a. ________	______
	b. ________	______

21-5. A handy form can be used in goal setting.

you would like to work—your career area. Identify an entry-level occupation in the career area. Learn the education and experience needed to begin. For example, a career in soil conservation might involve beginning as a soil technician. A college degree in agronomy or a closely related area would be needed.

Once made, goals often need to be revised. Times change. We may learn about new areas. Our interests may change. New goals may be needed. Revising goals is a normal activity. Actually, goal setting is a continuous process of modifying goals to meet new situations.

Ways and Means

Goals are of little benefit if we do not have any ideas about to how to achieve them. For each goal, ways and means of reaching it should be identified. These are the actions we need to take to reach the goal.

Progress comes about one step at a time. Just as goals may need to be revised, ways and means may also need to be revised. New opportunities may come along. These may make our goals and ways and means have new meaning. Modifications may be needed.

For example, a soil technician would include completing a college degree in the ways and means. A college would be selected that has the needed major. Financing the education is a part of ways and means.

Target Dates

Ways and means should have dates by which they are to be completed. We need to have deadlines. These help us wisely use resources, including our time. Sometimes, target dates may need to be revised.

Career Profile

SCHOOL COUNSELOR

School counselors help students in education, career, and personal areas. They usually have information on careers and schools that offer education. They can also help with personal problems. Counselors can give interest tests to help people better understand themselves.

Counselors in the schools are educators. They are usually trained as teachers with additional education in how to be a counselor. Counselors usually have college degrees in an area of education. Most have masters degrees. Many have advanced graduate degrees.

Job opportunities are with elementary, middle, and high schools.

21-6. Your counselor may have computer programs that will help in goal setting.

A soil technician can complete a degree in four years. This means that the target date for entering the career would be about four years after entering college.

CAREER LADDER

A ***career ladder*** is the upward movement of people in their careers. It comes about through advancement. People begin their careers in lower level jobs. They may move to higher level jobs, known as climbing a career ladder.

Moving up a career ladder is based on being productive in a job. This requires hard work and dedication to a career. New steps on the ladder bring more responsibility. Demands on the worker may be greater. Of course, advancement may result in better pay and other benefits.

Achieving career goals usually requires moving up the career ladder. Additional education and training are essential. In some cases, goals may be adjusted after beginning a career. People change jobs and the general direction of their careers. Some people change several times, others not at all.

Successful people often have to take risk in their careers. Progress often requires risk. Changing jobs involves risk. In doing so, never quit one job without having another job. Never change jobs until the advantages outweigh the disadvantages.

Moving to the top of a career ladder is not for everyone. People can be quite successful without pushing to get ahead. Moving to the top often places heavy demands on people. We have to decide if advancements are worth the effort required. Doing a job well is success in action!

CAREER SKILLS

Skills are required to carry out most jobs. The kinds of skills needed vary with the nature of the work. Entering a career requires education and training. Advancement requires keeping up to date and improving skills.

EDUCATION

Education can provide the preparation people need to enter a career. The kind of education needed varies with the nature of the work. High school classes, college courses, and other education should be selected to match career goals.

A high school diploma or higher level of education is needed for most jobs. In high school, select courses that will help you reach your career goals. Take the courses seriously. Make learning a high priority!

After high school, students may enroll in a community college, college, or university. A community college has programs that require two years or less to complete. Programs at colleges and universities require four years for a degree. Many universities offer advanced study at the graduate level.

21-7. Job skills in floral design are learned in horticulture classes.

Community colleges offer many opportunities for job-specific training. The programs often have close connections with employers. This helps in getting the first job. Community colleges are also known as junior colleges and technical schools. Many people earn a college degree. Most degrees are in specific majors. A ***major*** is a collection of courses in an area that helps meet degree requirements. Upon completing requirements of the major, a degree is granted. Most majors have courses that focus on specific areas of preparation. Fewer than 20 percent of today's jobs require a college degree. People without a college degree can be very successful. However, career advancement may require a degree.

PEOPLE SKILLS

People skills are the abilities that help people get along well together. These are very important in career success. Having good "people skills" helps one get along with other people and earn their respect.

Here are examples of important people skills:

- Being courteous
- Being honest
- Respecting other people
- Seeking suggestions from other people
- Having good communications skills
- Allowing other people to help make decisions
- Having a pleasant personality—with a smile on your face
- Helping other people feel good about themselves

Employers often have interesting experiences with people. Most employers say that good people skills are extremely important. People are more likely to be fired from lack of people skills than lack of job skills!

21-8. People skills include the ability to work together. (Courtesy, U.S. Fish and Wildlife Service)

EXPERIENCE

Experience is having personally done something. With careers, it is having worked in a particular job or location. Many agricultural jobs require experience.

Experience helps people know the nature of work. It helps us understand other people's situations. Experience on a farm is nearly essential if a person is going to be working with farmers.

Get experience by volunteering. Be eager to begin at a lower level. The best managers are people who began at the lower levels and worked their way up the career ladder. Of course, experience needs to be of good quality. It needs to relate to the chosen career area. Experience should be in a safe environment and with people who are honest and moral.

Students can begin getting experience while still in school. Some school programs involve various work activities. For example, high school agriculture classes often have supervised experience.

SUPERVISED EXPERIENCE

21-9. This student is measuring plant growth for a research laboratory. (Courtesy, Agricultural Research Service, USDA)

Supervised agricultural experience (SAE) is the application of class instruction in agriculture. A wide range of experiences can be had through SAE. Students can explore areas to help identify interests. Specific job skills can be developed. Some SAE allows students to earn money. SAE is an important part of many FFA activities.

HOW SAE WORKS

SAE is carefully planned and supervised. Planning helps define what is to be done. It prevents wasting time. The activities are carried out under the supervision of an adult. Each student keeps records on what has been done.

SAE is usually carried out after regular class hours at school. It may be on Saturdays, holidays, and during the summer. During school days, only a few hours may be given to SAE. Provisions of the Fair Labor Standards Act must be followed. These regulate the ages, hours, and other conditions related to the work of youth.

Safety is an important part of SAE. Always follow safe practices. Never take risks that might result in injury. Always use personal protection equipment, such as goggles and gloves, as needed.

KINDS OF SAE

SAE varies widely. Each student's SAE is based on the student's interests. Four kinds of SAE are typically used: exploratory, ownership, placement, and research/experimentation.

Exploratory

Exploratory SAE allows students to have a wide range of experiences. Beginning students use this type of SAE to help learn what they want to study in more depth. After exploring various areas, students can make better choices about education and careers.

Many areas can be explored. Some areas focus on careers; others focus on agriculture. All activities are carried out to get a good understanding of the area. The activities are not brief. Most require several hours or days to complete. Your teacher will help arrange experiences and supervise what you are doing. Here are a few examples:

21-10. Exploratory SAE can be in a school laboratory.

- Observing the work of a veterinarian.
- Observing an agricultural scientist working in a laboratory.
- Helping water plants in a greenhouse.

- Caring for small animals.
- Helping with crop research plots.
- Building bird feeders.
- Observing a forester cruise timber.

Ownership

Ownership SAE is owning and managing an enterprise in agriculture or a related area. Students own the materials and inputs. They assume the risk of ownership. Ownership SAE is sometimes known as entrepreneurship SAE.

Ownership involves getting the needed money to begin. All details must be planned. Careful management is needed to assure success. Records are kept on various phases of the enterprise. Regular assessment is made of progress.

With ownership SAE, students begin on a small scale and work to expand. An example is beef cattle production. A student might begin with one calf. Income from the calf or offspring might be used to increase the number of animals. In a few years, a small herd might be possible. Progress toward getting set up in the cattle business is important.

Ownership SAE is planned based on student needs and interests. Opportunities in the local area are also important. Examples of ownership programs are:

- Setting up a lawn care service.

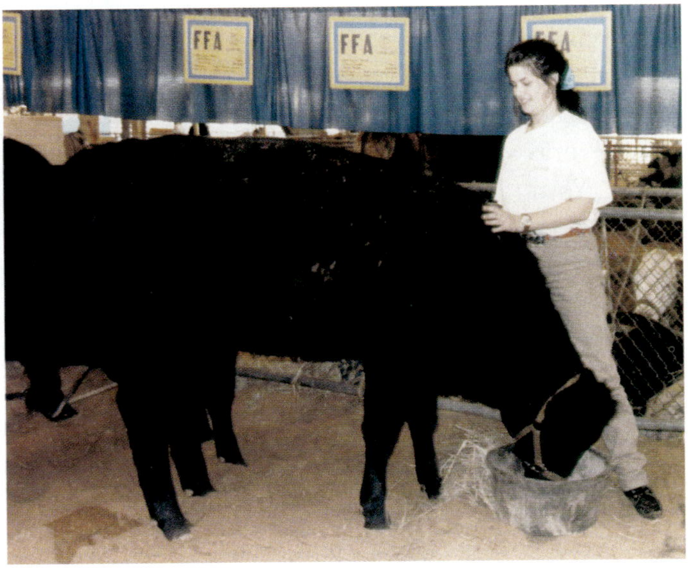

21-11. This student has an ownership SAE in livestock production.

- Raising and selling bedding plants.
- Raising sheep.
- Operating a horse shoeing business.
- Running a roadside fruit stand.
- Growing corn, cotton, wheat, or other crops.
- Raising and selling ornamental fish.
- Growing and selling fruit or vegetables.

Placement

Placement SAE is gaining work experience in agribusinesses, on farms, and in other ways. The student works for another person. Students may or may not be paid for their work. Usually, no pay is received for work in school laboratories. The products used or produced are not owned by the student. These belong to the employer. Placement SAE is sometimes known as cooperative work experience.

Placement SAE may be in any area of the agricultural industry. Farms, school farms, greenhouses, garden centers, camps, wildlife preserves, and retail stores are used for placement. Each local area varies. All cities and com-

AgriScience Connection

RECOGNITION FOR SAE

Students who excel with their SAE can gain recognition through the FFA. Many FFA activities are available to recognize top performance. Get the details on them from your agriculture teacher.

Each year, the National FFA Organization sponsors Career Development Events. These recognize members who have outstanding SAE. Awards are presented at the National FFA Convention.

Everyone likes to get a nice plaque. Those with the FFA emblem are special.

munities have placement opportunities. A quick review of the telephone directory may give many possibilities.

Examples of placement SAE include the following:

- Flower shop.
- Fish hatchery.
- Farm supplies store.
- Retail fruit and vegetable stand.
- Parts department of a farm equipment dealer.
- Biotechnology laboratory.
- Dairy farm.
- Veterinary clinic.
- Pet shop.
- Bee farm (apiary).
- Campground or nature area.

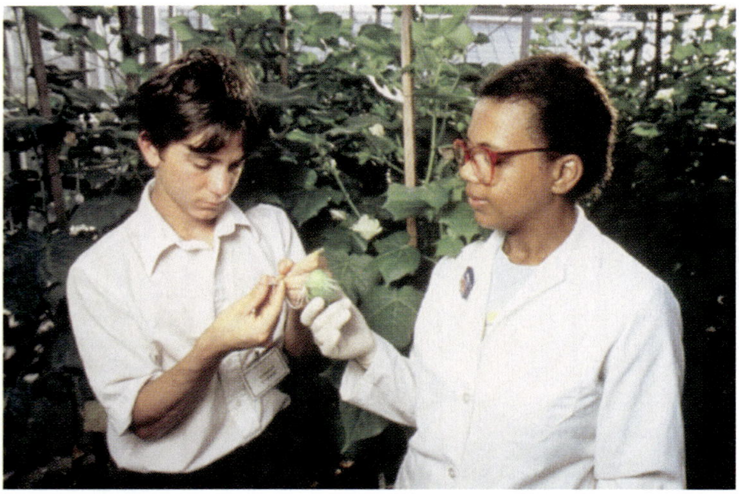

21-12. This student is receiving instruction in his placement SAE. (Courtesy, Agricultural Research Service, USDA)

Research and Experimentation

Research and experimentation SAE is science-based experience. It involves using science laboratory procedures to study a problem. The SAE may be carried out in a school laboratory or other location. Labs in colleges, businesses, and government agencies are often used. Scientists in these labs may direct the SAE. Students often get to use sophisticated equipment. They learn under the direction of a highly trained agricultural scientist.

The work may be to solve a problem in the community or investigate a new product. Some students choose environmental areas. They investigate sources of pollution and identify ways to prevent the pollution. In other cases, students work to develop new crops or methods of production. Some students are involved with biotechnology areas.

Here are a few examples:

- Working with an agronomist in breeding a new crop variety.
- Helping a food scientist develop new methods of keeping food fresh.
- Testing various media used in lab work.
- Collecting and testing water samples.
- Testing rates of fertilizer application.
- Developing methods of propagating horticultural plants.

Some students carry out research and experimentation SAE as projects in science. They combine agriculture and science to have an outstanding project. These may be a part of a science fair at the local school.

21-13. A research and experimentation SAE may involve lab work.

PLANNING

All SAE should be carefully planned. Every experience should have educational value. SAE should relate to what you are learning in your classes. If you are unsure about your goals, begin with an exploratory SAE. This will allow you to experience a variety of things. Some of what you learn will likely be of high interest to you.

Your teacher knows how to help in planning SAE. Begin with a discussion about what will be best for you. Be sure your parents understand the nature of SAE. Treat SAE as a time for learning.

Training Plan

A ***training plan*** is a form that lists the experiences to be gained in SAE. It lists the date each item is to be accomplished. Most training plans indicate when training will begin and end. The plan will give the name of the training station. A ***training station*** is the place where the SAE will be carried out. Training plans are most often used with placement SAE and research and experimentation SAE.

Training plans are developed jointly by the student and the teacher. The person who will supervise your work will also help. Most teachers provide a form to use. Consider the objective you have for the SAE. If your objective is to become a floral designer, for example, your experiences will be in a flower shop. You will do a wide range of activities associated with floral design. List these ahead of time. You want to be sure and get as much experience as possible. Check off all experiences as they are gained.

Training Agreement

A ***training agreement*** is a form that makes SAE official. It describes the conditions of an SAE. A training agreement shows the rate of pay and what each person is expected to do. Training agreements are signed by students, teachers, parents, and employers.

KEEPING RECORDS

Records are carefully kept on SAE. The training plan is a record of the experiences from an SAE. Additional records are needed of the hours worked, income earned, and other areas. Ownership SAE requires records of expense and income. These kinds of records help determine if the enterprise is profitable.

Records may be kept on paper or in a computer. Most schools use a standard record system. Records are needed in applying for FFA awards. Also, records are needed for tax purposes throughout life.

Name of Student ______________________ Teacher ________________

Occupational/Educational Objective ______________________________

Beginning Date __________________ Ending Date __________________

Training Station/Employer ______________________________________

Paid _____ Non-Paid _____ School Name __________________________

Experiences/Competencies	Date Accomplished	School-Related Instruction	Check When Done

21-14. Sample training plan form.

BENEFITS OF SAE

SAE has many benefits. Most of the benefits are to students. Schools, employers, and the community benefit from SAE. Parents and teachers also benefit from what students do in SAE.

The major benefits of SAE to students are:

- Helps in making career and education decisions.
- Develops self-confidence.
- Gives practical meaning to courses studied in school.
- Develops job skills.
- Promotes thinking skills.
- Applies record keeping skills.
- Promotes good money management.
- Teaches the work ethic.
- Helps develop the ability to assume responsibility.
- Gives practical experience in relating to other people.
- Helps make the transition from school to work.
- Helps achieve FFA awards.

Taking SAE seriously can provide even more benefits. Always do the best you can. Try to learn all that you can. Be cooperative and do what you say you will do. SAE helps give meaning to education and work.

21-15. SAE extends learning beyond the school. This shows a sensitive microphone being used to test for sounds in grain made by insects. (Courtesy, Agricultural Research Service, USDA)

REVIEWING

MAIN IDEAS

A career is the work a person follows throughout life. Some people work for themselves; others work for employers. Career success requires good planning. Begin with goals and ways and means of achieving the goals. Revise goals as new information is gained.

A career ladder is the sequence of jobs in a career. People begin in lower level jobs and advance. Good skills are needed to get ahead.

Career skills begin with education. Many people use high school to develop job skills and to prepare for additional education. In addition, "people skills" are needed. These deal with how people relate to each other.

Experience is important in getting a job. A good way for students to get experience is with SAE. SAE provides the opportunity to apply skills learned at school in areas related to career or education. SAE is carefully planned. Four types are used: exploratory, ownership, placement, and research and experimentation. Many benefits are gained from SAE.

QUESTIONS

Answer the following questions. Use correct spelling and complete sentences.

1. What is a goal?
2. What are the kinds of goals based on length of time?
3. What are ways and means?
4. What is a career ladder?
5. What are three important people skills?
6. What is SAE?
7. What kinds of SAE programs are used?
8. What records are kept of SAE experiences?
9. What is a training plan?
10. What are three benefits of SAE to students?

EVALUATING

Match the term with the correct definition. Write the letter of the term in the blank provided.

a. career	d. major	g. people skills
b. goal setting	e. experience	h. ownership
c. career ladder	f. training station	

____ 1. Type of SAE that involves owning and managing an enterprise.

____ 2. Collection of college courses that lead to a degree.

____ 3. Upward movement in a career.

____ 4. General area of work in a person's life.

____ 5. Process of setting goals.

____ 6. Abilities that help people get along well together.

____ 7. The place where SAE is carried out.

____ 8. Having personally done something.

EXPLORING

1. Interview a senior student who has had supervised agricultural experience. Determine the kind of SAE, career or education objective, and how it was carried out. Get suggestions that will help as you plan an SAE.

2. Invite an employer or personnel officer to serve as a resource person in class. Have the person discuss the importance of people skills. Ask about the major problems people have in their work. Also, ask how these problems are handled.

3. Go through the goal-setting process for yourself. Make a list of short-term, intermediate, and long-term goals. List possible ways and means of reaching the goals. Set dates by which you wish to achieve each goal or ways and means.

22

PERSONAL SKILLS

We interact with other people. We do so in our families, schools, and jobs. How we go about this builds good relationships. Paying careful attention to other people raises our personal well-being.

Other people are important. They have useful lives and opportunities. They have families who love and need them. All people are worthy of our respect. No human life should be treated poorly.

Personal skills help us in dealing with other people. These skills help us know what to do and when to do it. They help people around us rest at ease and not feel threatened. Personal skills help us avoid being embarrassed. They help us get along with other people.

22-1. Good personal skills help people in many ways. (Courtesy, National FFA Organization)

OBJECTIVES

This chapter covers important areas in relating to other people. The objectives of this chapter are:

1. Explain the role of personality in making a first impression
2. Describe the importance of personal appearance
3. Apply social etiquette standards
4. Describe how to achieve personal excellence

TERMS

accessory
dress
etiquette
extrovert
fabric
grooming
health
introvert
personal hygiene
personality
posture
protocol
wellness
Western style

http://www.balancenet.org

(This site has information on education and work, including getting the first job.)

MAKING A GOOD IMPRESSION

Success is 90 percent personality! Some people may disagree. They may feel that personality is important but not 90 percent of success.

All agree that the first impression we make is very important. Good impressions are essential where people interact with each other. Long-term success includes many factors. The human personality factor is always important.

PERSONALITY

Personality is the visible part of a person. It is what you see when you first meet another person. Personality includes both visual and spoken qualities. What we say and how we say it are parts of the "first impression." It is how people respond to each other. The personality of another person is how that individual responds to you. Likewise, your personality is how you respond to other people and are viewed by them.

Types

An individual's personality is made up of a wide range of human emotions. Some are complex. Carl Jung, a psychologist, tried to simplify personality by placing people in two groups: introverts or extroverts. Looking at what Jung had to say probably helps us understand some aspects of human behavior.

22-2. Personality types vary and influence relationships. (Courtesy, National FFA Organization)

An ***introvert*** is a person who focuses on himself or herself. They are not particularly interested in other people and events. A person who is shy is often said to be an introvert.

An ***extrovert*** is a person who focuses on the outer world. They are interested in other people and events. They make friends easily and are sociable. Extroverts are interested in what is going on around them.

Balance

People have both introvert and extrovert tendencies. These are usually kept in balance. Sometimes, one will dominate. That will give a person a distinct personality.

We can strive to improve our personalities. Introverts can take the initiative to reach out to other people. They can develop more extrovert traits. Ex-

Career Profile

FOOD SCIENTIST

A food scientist studies foods and develops ways to assure we have good food to eat. The methods used help preserve nutrients needed for good health.

Food scientists have college degrees in food science or a closely related area, such as biology. All have college degrees. Most have masters degrees. More are getting doctoral degrees in food science.

Food scientists work with large food companies, government agencies, and universities. The photo shows a food scientist sampling lemon juice to test for natural acid and sugar levels. (Courtesy, U.S. Department of Agriculture) troverts who are "stuck on themselves" and overbearing may lack sincerity. They can work to show a sincere interest in others.

PERSONALITY IS LEARNED

An individual's personality largely results from the environment in which they grow up. People are what they have learned to be! As people have new experiences, their personality changes.

We can work to improve our personalities. Studying and practicing the desired qualities are helpful. For example, we can practice smiling and looking into another person's eyes when introduced.

The first few words spoken to another person set the tone of the relationship. Saying the rights things and using good grammar are important. We can practice and improve on these. Friends can work together to help each other.

PERSONAL APPEARANCE

First impressions of other people are based on their grooming, dress, posture, and general health. These are associated with personality. An individual's personality is reflected in their choice of clothing, their grooming, and the way they carry their bodies.

22-3. Personal appearance is important in our success.

CLOTHING

What we wear says much about us. Our clothes project an image to other people. If this is hard to believe, look at the people around you. Does what they are wearing convey an impression to you?

The kind of clothing we wear is our ***dress***. How we dress is important. Everyday school clothing differs from

22-4. Selecting the best tie may require the opinion of another person.

what is worn to the prom. Dress is a part of the first impression we get of people. Wearing the proper clothing helps in making a good first impression and in being successful.

Clothing is intended to protect and enhance the human body. Proper dress is needed to get a job and advance in it. Each year, new styles are introduced. These are fads that last only a few months or years. Always be careful in wearing the latest fashions. Some new fashions reflect extremes in dress. They may not be appropriate in the world of work!

Clothing should be clean, pressed, and free of holes. Dirty clothing is not professional. Some work results in clothing getting dirty. It is expected. Properly launder dirty clothing. Repair any minor holes or rips before they get too large. After wearing clothing that does not need to be cleaned, properly hang it on a hanger.

Style

The ***Western style*** of clothing is worn in the United States. It includes suits, shirts and blouses, jackets, shoes, and pants and slacks. This is different from the western wear of cowboys—boots, hats, and jeans. Western style dominates in North America and Europe. It is the clothing worn in the business world and is often contrasted with African and Eastern clothing.

New fashions come along each year. Fashion refers to the design and fabric of clothing. Clothing fashion designers try to create demand each year for new fashions. They vary colors, lengths, and other features.

Wear appropriate clothing that projects the kind of person you want to be. Dressing conservatively is best (traditional) rather than in the latest fads. In dressing for work, being conservative is far better. Many employers do not appreciate the latest fads. Because clothing costs money, we need to be sure to make wise buying decisions.

22-5. Clothing projects us in a certain way. (Courtesy, National Cotton Council of America)

Fit

Clothing should fit properly. Know the sizes you wear. Teenagers are often growing fast. Their sizes may change often. Try on clothing at the store before buying it. Note the sizes that fit best. People who know proper fit should help you decide what to get. Stores have sales people who can help.

Ideally, clothing should drape over the body. Clothing should be neat and comfortable. Sometimes, clothing needs to be altered. Taking-up or letting-out clothing should be done with care.

Pant legs, sleeves, and other parts of clothing should be the proper length. Neck and waist sizes are important with some clothing.

22-6. Get the assistance of a qualified person in fitting new clothing. (Courtesy, Smith & Byars Clothing)

22-7. A label on a shirt identifies the cloth as sea island cotton—a long staple (fiber) cotton known for comfort.

Fabrics

Fabric is the material used in making clothing. Cotton, silk, and wool are popular natural fibers. Nylon and rayon are widely used synthetic materials. Some fabrics are blends of two or more fabrics, such as a cotton and polyester. The kind of fabric influences how the clothing will wear and should be cleaned.

In selecting fabric, consider the care needed to keep the clothing looking good. Cotton may require ironing to remove the wrinkles. Using permanently pressed cotton reduces the need for ironing. Some fabric can be washed. Other fabric must be sent to a dry cleaner.

When the clothing will be worn is important in fabric selection. Cotton is very durable and often used for work clothes. Silk is a fragile fabric that has especially good qualities. Wool is warm and durable.

Color

Clothing color communicates formality and informality. Navy blue, dark gray, and black are good color choices for professional attire. Light colors are more suitable for casual clothing. Colors, fabrics, and patterns should coordinate. Avoid bright, neon colors in clothing for work.

Accessories

Accessories include ties, scarves, belts, shoes, and jewelry. These items complete a clothing outfit.

Ties and scarves should be color coordinated with the rest of the outfit. Always properly wear them. Wearing ties and scarves incorrectly detracts from a professional appearance.

Belts and shoes are often coordinated. Often, both are the same color and finish of leather. Keep shoes cleaned and polished.

Jewelry can complement an outfit, especially for females. Always select conservative jewelry and wear it with moderation. Dangling jewelry detracts from appearance. Never wear jewelry when using machinery or chemicals. If in doubt about the appropriateness of jewelry, do not wear it.

GROOMING

Grooming is the neatness of an individual's personal appearance. It includes hair, nails, and skin. Grooming includes how your hair is styled and the care and cleanliness of your nails and skin.

Hair

Hair is a top priority in grooming. It should be clean, cut, and fixed in a conservative style. Never use a fad hair style in the work setting. This includes the cut as well as the color. Style hair so it is out of the face and neat. For safety, tie long hair back when using power tools or machinery.

AgriScience Connection

HAIRDRESSING

Hair is an easily changed feature of the human body. It can be cut, colored, set, and changed in other ways. All people want good hair.

Caring for hair involves brushing and shampooing your hair and eating right. Brushing removes tangles and spreads scalp oil throughout the hair. Shampooing cleans the hair. With oily hair, shampooing may be needed each day. Dry hair should be shampooed once or twice a week. Eating right assures the body of having nutrients to keep hair healthy.

Avoid using harsh hair treatments. Excessive heat from dryers or curlers can damage the hair. Some hair care products injure the hair. Use products that do not contain damaging chemicals. Select the right shampoo for your hair.

22-8. Brushing helps keep the hair in good condition.

Other body hair may need to be removed. Males shave to remove facial hair. Facial hair is not a benefit for young males in their career. Females may shave or use a hair remover on their legs and underarms. Some of these practices vary with traditions.

Personal Hygiene

Personal hygiene involves keeping your body clean and free of odor. Regular baths are needed. Bathing is an important part of personal hygiene. Use soap and water to wash the body. Rinse all soap off. Some people prefer soaps that are gentle; others prefer deodorant soaps.

22-9. A variety of products are available for use in having good personal hygiene.

Body odor may be a problem. Use deodorant to help prevent body odor due to perspiration. Select a brand that is appropriate for you. Do not use one that contains ingredients to which you are allergic. Never wait for someone to tell you that you have body odor. Prevent it from the start! Use deodorant following each bath.

Teeth and breath are included in personal hygiene. Having clean teeth is a boost to your personality—you can smile! Regular brushing also prevents tooth decay. Regular dental checkups are needed. Sometimes, cavities develop and need to be filled. Some people need braces to position their teeth properly. However, having braces is expensive. Be sure it is a high priority for you before beginning the investment. Some people use mouthwash to help control bad breath. Spicy foods add to bad breath. Regularly brushing your teeth helps prevent bad breath.

Keeping your nails neatly trimmed adds to personal appearance. Painting the nails is an option if done conservatively. Brightly colored or gaudy nails may be unprofessional.

Skin

The skin should be kept clean. Young people sometimes are concerned about skin blemishes. Concern is natural. We all want a good appearance. Help prevent skin problems by washing the skin, eating properly, and getting enough rest. Major problems should be treated by a dermatologist.

The popularity of tattoos has increased in recent years. Most tattoos are permanent. Some can be removed only with very expensive procedures. Tattoos have a negative impact in the professional world. Who wants to have a design permanently injected under their skin?

22-10. Brushing your teeth helps prevent decay and adds to your appearance.

22-11. Use sun screen to protect the skin from ultraviolet sun rays. Use a screen with an SPF of at least 15.

POSTURE

Posture is how we hold and position our bodies. It includes how the body is carried by our skeleton and how we move.

Stand erect. Hold the head up and shoulders back. Do not slump. When standing up, place weight equally on both feet. Avoid propping or leaning on chairs, tables, and walls.

When walking, move with enthusiasm. Moving slowly may signify low job productivity. Success depends on showing that we are energetic and capable.

HEALTH

Health is physical and mental condition. Healthy people are happy and enjoy life. They are better workers and give a more favorable impression to other people. Fortunately, we know much about how to have good health.

Wellness is taking steps to have good health. We can choose some things that keep us well. These can become part of our life style. Here are a few activities that promote wellness:

- Do not use drugs and tobacco products.
- Get regular exercise.
- Get adequate rest—six to eight hours of sleep each night.
- Eat a nutritious breakfast and other meals.
- Maintain the weight that is right for body type.

22-12. Using an exercise bike promotes wellness.

Some schools and communities have wellness programs. These typically focus on fitness. The goal is to have people who are in good health and physically fit. Participating in wellness programs helps prevent disease because the body is fit. People can practice wellness in their homes—exercise properly, eat right, avoid harmful substances, and get adequate rest.

ETIQUETTE

Etiquette is a system of social behavior. It helps people get along with each other. Etiquette is sometimes known as "manners."

22-13. A sense of humor helps in getting along with people.

The rules of etiquette cover many areas of our lives. How to dress for an event, greet another person, and set a table for a meal are parts of etiquette. Telephone, email, and driving etiquette are included. Over time, etiquette rules change. The rules also vary with the culture and status of the people involved.

Protocol is a special area of etiquette that deals with rules for events attended by government officials. It is used to assure that officials receive proper respect based on their level of position.

Relating with other people requires appropriate etiquette. Four areas are included here.

COURTESY

A courteous person follows etiquette rules in daily living. This includes conversations and other areas of interacting with people.

Conversations

Etiquette begins with what we say to other people. Our culture emphasizes being polite. Important words in polite conversation are "please," "may I," "thank you," "I'm sorry," and "excuse me." Make these words part of your daily vocabulary. This will help you win people over. They will appreciate your being polite. Slang and profanity are not a part of a polite conversation.

People who keep up with current events are better able to carry on conversations with people. Know the other person's interests. Talk about things to which they can relate. Avoid controversial subjects. Always allow other people to express their thoughts. Do not interrupt them when they are talking.

Speaking to people is also important. Happily saying "good morning" or "how are you today" helps you get along with people.

Actions

Etiquette includes many areas. Holding a door and allowing others to pass through it is a courteous act. Allowing others to go before you in ordering food, selecting a place to sit, and using a vending machine are other examples. This does not mean that we let other people take advantage of us. It means that we achieve our goals while boosting other people.

MAKING INTRODUCTIONS

An introduction is used to get acquainted with a person you do not know. You may need to introduce yourself to another person or introduce two people who do not know each other. Knowing how to do this will make it go smoothly. Everyone will feel good about the encounter. With introductions, always look the other person in the eye. Say "hello" or "good morning."

One way to meet someone you do not know is to say "Hello, my name is ___ ___. It is good to meet you." The other person will likely give their name. It is customary for people to shake hands following an introduction.

22-14. A good handshake is part of an introduction.

If you are introducing people to each other, say "Susan, I would like for you to meet my friend (or parent) Bill Jenkins. Bill this is Susan O'Neal." In making introductions, be sure to carefully pronounce the names of people. Always make everyone feel comfortable. Be confident.

EATING

Eating etiquette varies with the formality of the

meal. Many meals are now eaten in fast food places. Formal etiquette rules do not apply in these restaurants. Restaurants with menus and table service are more formal. Those with white clothes on the tables are even more formal, especially if the restaurant offers fine dining.

Using a napkin is always important. This is true in a fast food or fine dining restaurant. Place the napkin on your lap. Occasionally, use it to wipe your mouth if you feel that any food may be on your lips. Never blow your nose on a napkin. After your meal, place the napkin beside your plate. In a fast food restaurant, put it and other trash in the proper container.

With formal meals, each place where a person will sit may have several eating utensils. Plates, forks, knives, spoons,

22-15. Using a table napkin during meals is a part of etiquette.

22-16. A formal place setting for one person. (The napkin could be beside or under the forks, in the plate, or at the top of the bread plate.)

glasses, and other pieces may be present. These form a table setting. A napkin is also a part of the table setting.

You will need practice to know which piece of tableware to use and how to hold it. As a rule, begin using tableware from the outer edge of the setting and move inward as the meal progresses. Casually observe someone who knows about table etiquette. Before a banquet or other formal meal, use a reference to read about etiquette or get a demonstration from your teacher.

SHAKING HANDS

A handshake is an important part of culture in the Western world. A good handshake makes a strong first impression. Practice shaking hands with your friends. This will help you to be more confident and relaxed with other people.

A good handshake is firm and friendly. Extend your right hand with the palm open to fit into the palm of the other person's hand. Lightly grasp the other person's hand and make a slight upward and downward movement for a few seconds. Release the hand. Do not squeeze too tightly. Do not hold the other person's hand too long. While shaking hands, look the other person in the eye. Smile and offer a greeting, such as "hello" or "nice to meet you."

PERSONAL EXCELLENCE

You want to be the very best you can possibly be! To be less is a waste of your talents. Excellence comes about when we use our abilities properly. Sometimes, we need to be reminded of what is needed.

EXCELLENCE IN OUR WORK

Excellence in our work goes a long way in measuring success. Liking our work helps us achieve excellence. Beginners have a lot to learn. You can learn! With effort, you can achieve whatever you wish in your career.

A few guidelines to personal excellence in work are:

- Have a positive attitude. Do not complain.
- Be enthusiastic. Use your energy in a positive way toward your work.
- Get along with coworkers. Give a little extra to be a good member of the work team.

- Be flexible. Take on new tasks with enthusiasm.
- Be dependable. Always be at work on time and do what you say you will do. Live up to your word! Do not miss work except in rare situations. If you must miss, inform your superior ahead of time.
- Follow instructions. Do what the boss says with a smile and enthusiasm.
- Use resources wisely. Do not waste materials. Do not waste your time nor the time of other workers.
- Be physically fit. Get plenty of rest. Do not go to work tired.
- Do the work correctly. Be accurate. Take pride in what you do.
- Never complain. Do not complain at work nor to other people after work. Employers want loyal workers. Never criticize your boss.

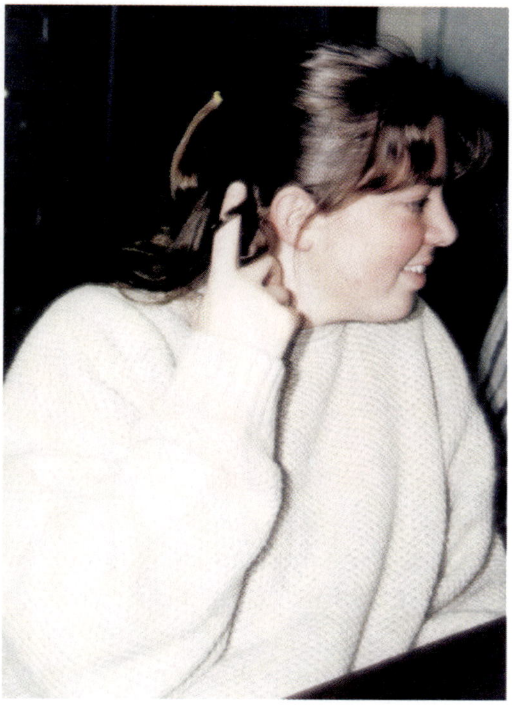

22-17. Avoid "playing" with your hair when around other people.

EXCELLENCE WITH OTHER PEOPLE

Other people are important. We need to respect our families, friends, and coworkers. All people should be respected.

A few guidelines for personal excellence with our family and friends are:

- Follow the rules of etiquette. Being courteous will build pride and happiness.
- Respect the wishes and rights of others.
- Help make life better. Do a little extra.
- Remember people on special days. Send a birthday card or a special greeting.
- Be responsible. Do not be reckless or endanger other people.
- Resolve conflicts before they grow. Differences exist among people. Keep them to a minimum.
- Keep your room and living area neat. Do a little extra. Put your clothes, books, and other belongings where they should be.

22-18. Strive to get along with all people. (Courtesy, National FFA Organization)

- Use money wisely. Families never have enough money. Avoid wasting resources.

EXCELLENCE WITH CITIZENSHIP

All of us in the United States live in communities or neighborhoods, cities or towns, and states. The number of people is increasing each year. Getting along with each other is essential. Taking pride in where we live is important.

A few guidelines for personal excellence in citizenship are:

- Keep informed. Know what is going on in government and community organizations. Read the local newspaper.
- Register to vote and vote. When you become 18 years of age, you will be eligible to vote. Always know as much as possible about the candidates and vote for the one who will provide the best leadership in the position.
- Drive responsibly. Getting a license to operate a motor vehicle is important to young people. Driving is a big responsibility. Care is needed to prevent injury to other people and destruction to property.
- Follow the laws. Always obey the law. Expect to be punished if you violate a law.
- Pay taxes. We all get benefits from our government. We should help provide money for these benefits.
- Take pride in where you live. Do not litter. Keep your area neat and in good order.

REVIEWING

MAIN IDEAS

Success often depends on how other people see us. Our personality includes both visual and spoken qualities. Some people are introverts; others are extroverts. If we do not like our personality, we can do something about it. Remember, personality is learned!

Personal appearance is an area we can do a lot about. The clothing we wear, our grooming, and our posture make up our personal appearance. We are born with certain attributes that cannot be changed. We build on them!

Etiquette is a system of behavior that allows people to relate to each other. It helps people know what to do in various situations. Courtesy is important. Introducing people and using good manners when eating are skills that can be learned.

Personal excellence can be achieved in our work, relationships with other people, and as a citizen. Each of these helps people lead self-fulfilling lives.

QUESTIONS

Answer the following questions. Use correct spelling and complete sentences.

1. What is personality?
2. How do people with introvert and extrovert personalities differ?
3. How do people get their personality?
4. What are the major considerations in clothing?
5. What is grooming? What important areas are included?
6. What can a person do to promote wellness?
7. Why is etiquette important?
8. What important words help people have polite conversations?
9. What is one way of introducing yourself to another person?
10. What is a good handshake?
11. What are three guidelines for personal excellence in work? With other people? With citizenship?

EVALUATING

Match the term with the correct definition. Write the letter of the term in the blank provided.

a. personality
b. introvert
c. extrovert
d. dress
e. fabric
f. personal hygiene
g. grooming
h. wellness
i. etiquette
j. posture

____ 1. Taking steps to have good health.
____ 2. How we hold and position our bodies.
____ 3. The visible part of another person.
____ 4. A personality type that focuses on the outer world.
____ 5. A personality type that focuses inward on the individual.
____ 6. The kind of clothing we wear.
____ 7. The material used to make clothing.
____ 8. System of social behavior.
____ 9. Having a clean body free of odor.
____10. Neatness of personal appearance.

EXPLORING

1. Organize your class to have a style show. Get a local clothing store to help by instructing students in how to model clothing and to provide sample garments to be modeled. Have clothing for different activities, such as working in an office, going to a prom, and attending church.

2. Role play various areas of personal development. Areas to include are making introductions, shaking hands, and greeting each other.

3. Select a leader in the local community. Invite the person to serve as a resource person in class and discuss personal skills needed for success.

23

LEADERSHIP SKILLS

Do people do what we want them to do? Many times people do things that we do not understand or wished they had not done. Be smart! Become a leader and help others take the best actions.

Leaders develop the skills they use. They are not born with leadership skills. Most leaders work hard at continuing to improve their abilities as leaders. Just look in a bookstore at the available material to help adults develop as leaders!

Agriculture classes, the FFA, and supervised experience are excellent places to begin developing leadership skills. You can be a leader! It begins with a positive attitude toward developing yourself.

23-1. Leaders work hard to develop and improve their skills. (Courtesy, National FFA Organization)

OBJECTIVES

This chapter introduces important areas in leadership development. With practice of these areas, you can enhance your abilities. The objectives of this chapter are:

1. Describe leadership
2. List attributes of leaders
3. Explain styles of leadership
4. Describe team building in leadership
5. Explain how to organize, plan, and conduct a meeting
6. Explain how minutes are kept

TERMS

adjournment
attribute
autocratic leader
call to order
democratic leader
follower
influence
laissez-faire leader
leadership style
meeting
meeting program
minutes
new business
old business
order of business
presider
team building

http://www.whitehouse.gov

(This site provides access to the Executive Branch of the Federal Government.)

LEADERSHIP

Leadership is helping people reach goals. The goals may be individual goals or group goals. To lead, there must be followers. A ***follower*** is an individual who goes about reaching goals under the direction of a leader. Several individuals may work together as a group of followers.

A person cannot be a leader without followers. Care is needed to help followers work together. Leaders help people set goals and ways of reaching the goals. Leaders must carefully work with followers. Leaders help followers understand problems. They encourage followers to move ahead. Leaders recognize followers for their accomplishments.

Followers and leaders need to be equal. A leader does not set himself or herself up as "better than" the followers. The leader always treats followers with respect.

23-2. Leaders feel at ease in front of a group.

LEADERSHIP CONCEPTS

Leadership is not easy to understand. Leaders vary in how they relate to followers. Many people have studied leadership. They have tried to describe what is involved. Insight is growing in the nature of leadership.

Four areas important to leaders are:

- *Influence*—***Influence*** is encouraging an individual or group to take action. The leader does not tell people what to do. Most people do not want to be told what to do nor how to do it. A leader offers advice and experiences. A leader helps other people see what will work.

Leaders provide the best influence by providing information and allowing people to make their own decisions. A caring and sharing relationship exists between good leaders and followers.

- *Process*—People do not become leaders by virtue of the position they hold. People are leaders because of how followers view them. Leaders understand the process of working with people. The process includes encouraging people and being respected by the group. As a process, leaders help people find answers. Leaders do not have the best answers. Helping other people is the major attribute of leadership as a process.

  Leaders must be trusted. People will not follow a leader they do not trust. Honesty and sincerity are important.

- *Relationship*—A leader and followers must have a good relationship. Disagreements must be settled without disrupting the work. Each should encourage the other to do their best. Further, each must be considerate of the other. They must care about each other's needs and interests.

  The relationship between leaders and followers must involve respect. Without respect, reaching goals is difficult. Respect involves trust. Followers must feel that the leader is honest and can be trusted.

23-3. How followers respond to a leader determines the effectiveness of the leader.

- *Service*—A leader provides a service to a group. Leaders must often sacrifice their own wishes for the welfare of the group. Leaders fulfill many different roles in a group. They must often work longer and harder than followers.

ROLES VARY

The roles of people vary. A person might be a leader in one situation and a follower in another. Roles may swap depending on the situation. The most effective groups allow role changes.

Knowledge and experience are important in being a leader. As situations change, group leaders may change because of the knowledge and experience of its members. A person needs to have information to be an effective leader. They need to understand their followers as well as the problem or goal being achieved. Experience helps a leader know how to act and what happens as the result of making certain decisions.

LEADER ATTRIBUTES

An ***attribute*** is a trait of an individual. Most people have many traits that make them leaders and followers. Some leadership attributes can be learned. This makes it important to study leadership attributes.

Several leadership attributes are listed. As you review the list, assess attributes you have and attributes you need to develop. Knowing those you need to develop will help you improve your leadership skills.

Table 23-1. Attributes of an Effective Leader

- Hard Working
- Trustworthy
- Knows Strengths and Weaknesses
- Self-confident
- Assumes Responsibility
- Helps People Understand Each Other
- Flexible
- Knows How to Follow Directions
- Can Plan and Conduct Meetings
- Can Speak Effectively
- Can Lead Discussions
- Knows How to Organize Events
- Can Think Logically
- Can Anticipate and Solve Problems
- Can Visualize Changes
- Can Make Decisions
- Understands the Needs of People
- Accepts Differences of Opinion
- Always Has a Positive Attitude

LEADERSHIP STYLES

Leadership style is how leaders go about their roles. Leaders vary in their approach to motivating other people to action. Leaders have been placed in three groups based on their leadership style. Leaders vary in the styles they use depending on the nature of the group being led.

The three leadership styles are:

- Autocratic style—An ***autocratic leader*** makes decisions and tells people what to do. The people being led may provide input. This style may work best with people who need to be told what to do. It is dictatorial. In a democracy, we feel people should have a role in making decisions.
- Laissez-faire style—A ***laissez-faire leader*** allows members to make decisions. The leader does not interfere with the group of followers. Sometimes, followers are in a better position to make a decision than the leader. They may have information the leader does not have.

Career Profile

PUBLIC OFFICIAL

Public officials are in leadership roles. They may be elected or appointed to the positions they hold. Some positions involve agriculture, the environment, and related areas.

Jimmy Carter is a high school leader who became President of the United States. In high school, he was a member and officer in the local FFA chapter. He gained experience as a leader. Through college and military service he continued to develop leadership skills. He expanded his leadership skills as Governor of the State of Georgia. He went on to be elected President followed by a distinguished career as a civic leader.

FFA leadership training was certainly beneficial to President Carter.

23-4. Leaders use a leadership style appropriate for the group.

- Democratic style—A ***democratic leader*** allows followers to help make decisions. The leader participates in decision-making. Most leaders in the United States try to use the democratic style of leadership. We want people to be involved!

Other ways of classifying leadership style could be used. Situational leadership involves selecting a style based on the ability of the followers in certain situations. Leaders need to be flexible. Look at the people who are being led. How can they best achieve their goals?

TEAM BUILDING

Team building is getting people to work together as a group. All members of the group share some common interests and goals. The commonality may be in one specific area or a broader area of interest. Groups get far more done when they "think as one." One role of a leader is to form a group into a team.

23-5. Team building is important to group sports.

An example of where team action is needed is with a fund-raising event for a club. If the group decides to sell fruit, much more will be sold if all members help with selling. All members need to work together. To get team action, all members should be involved in making the decision about the fund raiser. Sharing in decision making helps get shared action later.

TEAM CHARACTERISTICS

Success with a group depends on the leader and how the members have bonded as a team. Members have to want to be a team. They have to help each other achieve. They have to work together unselfishly.

Three characteristics of a good team are:

- Unity—This means the members of a group feel "connected." All feel a part of the same thing. Each member has a role. Members know they are expected to perform their role.
- Relationships—Members need to have good relationships with each other. They need to feel comfortable. They need to share information that builds overall performance. Good relationships thrive on reward—not big rewards but regular and frequent encouragement of each other.
- Working together—With a team, members work together to achieve a goal. Differences among individuals are put aside. All follow the procedures that have been set. People get far more done working together and sharing in a common effort.

BUILDING A TEAM

Team building involves three major phases: forming, norming, and performing. These phases usually occur in sequence but are continual.

- Forming—Forming is getting people together as a team. People act as members of a team rather than as individuals. This comes about when the members share in deciding their mission. The individuals must trust each other.
- Norming—Norming is getting people to work as one unit. The team is cohesive. Members think for the good of the team and not just of themselves. Every member must feel welcome. Members share a commitment to each other. Decisions are made with participation from all members. In norming, the members must put the group first.
- Performing—In performing, the team is working together to achieve its goals. The members are closely attached to each other. They are a productive group. Each member reinforces the work of the other members. Performing

23-6. Team building involves important group work.

includes assessing how well the group is doing. Adjustments are made to improve the group.

The best teams have a shared vision. All members of the team have a similar mental image of the future. They all have common goals to reach. The role of a leader is to see each person as a valuable, contributing team member.

MEETINGS

Meetings are popular events in our society. A ***meeting*** is a group of people who gather for a shared purpose. Members of the group vary. Some are highly interested in the meeting. Others are not actively involved. A good leader finds a way to get all who attend involved.

Good meetings have value to members. Everyone who is a member of an organization should attend its meetings. People are more likely to go if the program is good.

Meetings range from small groups to large conventions. Most local clubs have small meetings of fewer than 100 people. Large conventions may have several thousand people attend. A professional planner is likely in charge of a convention.

23-7. An officer team planning a meeting.

WHY MEETINGS ARE HELD

Meetings are held to conduct business or to share information. Having a good club or FFA chapter requires well-planned meetings. Members get involved and are motivated to do more in a good meeting. Leaders can use meetings to build teams among members.

The main reasons meetings are held are to:

- Provide an educational program, including guest speakers
- Carry out the affairs of an organization

23-8. The people in this meeting are learning about biotechnology.

- Recognize members for outstanding work
- Announce future events
- Organize members into work groups
- Give members a place to develop leadership skills

PLANNING A MEETING

Good meetings require planning. This helps meet the needs of members and promote attendance. The group will be stronger. The members will feel they are getting more benefits from their membership.

Some organizations use a committee to plan meetings. This committee should meet before a meeting is to be held. Always plan well in advance. This allows more details to be worked out. The meeting will be better organized.

Here are some things to do in planning a meeting:

- Determine the overall purpose of the meeting
- Identify business to be acted on

AgriScience Connection

LEARNING HOW TO PRESIDE

Learning how to be a presider comes with practice. Just as other skills, the ability to preside over a meeting is learned. Take advantage of opportunities in your school and community to develop these skills.

Study materials on being a presider. Use books, video tapes, and other materials. Many of these are available in the local public library or the library of your school. Observe people who are presiders. Note their style and how they conduct themselves. Volunteer to serve in a leadership and presiding role in the FFA chapter or other organization.

23-9. The site for a meeting has been arranged ahead of time.

- Organize an educational program
- Set a date and time for the meeting
- Select a convenient location for the meeting
- Assign responsibilities for the meeting to different members
- Develop an order of business for the meeting

ORDER OF BUSINESS

An ***order of business*** is the plan that lists the events or items in a meeting. Events are listed in the sequence in which they are to occur. The names of individuals responsible for each item are also listed. Orders of business are sometimes known as agendas or programs.

A written order of business should be prepared well in advance of the meeting. Carefully review it to see that all details are covered. Most orders of business are no more than a page long. Large conventions may have a program several pages long.

A standard format is used for most meetings. The name of the group, date, time, and place of the meeting are listed at the top. Items on the order of business are listed in sequence. The name of the person responsible for each item is listed next to the item.

The common items to include and their sequence are:

- Call to order—The ***call to order*** is the announcement that the meeting is now beginning. The name of the presider is listed. A welcome may be given. The purpose of the meeting is often announced. In some meetings, the order of business may be approved by the members. Often, the same person will be in charge of the entire meeting. The person may be the group's president or other leader. Each new item is announced and the person responsible is introduced to the group.

- Action on minutes—The minutes of the last meeting are read and approved. Sometimes, written copies may be handed out and the minutes are not read. Corrections to the minutes should be made before they are approved.
- Reports of committee chairs and officers—This is the time in a meeting when people report on the work the committee is doing.
- Old business—***Old business*** refers to items remaining from the previous meeting of the group. These items should be discussed and voted on before new business.

Order of Business
Meeting of FFA Chapter
Clinton High School
March 18, 1999
10:00 a.m.
Agriculture Classroom

Call to Order: President—Margie Oswego
Welcome
Announcements

Minutes of Previous Meeting: Secretary—James Gonzales

Reports:
Treasurer—Angelica Loviza
Social Committee—Jason Sloan

Old Business:
Completing plans for making tour of horticulture industry

New Business:
Selecting Delegates for the State Convention
Planning escargot sale

Program:
Introduction—Jay Smith
"Agriculture in Our State"—Honorabale Rick Perry, Commission of Agriculture

Adjournment: President—Margie Oswego

23-10. A sample order of business.

- New business—***New business*** refers to items being brought before the group for the first time. The group may vote on each item of new business.
- Program—The ***meeting program*** is the informative or educational part of a meeting. It should be designed to meet the needs of members.
- Announcements—Announcements of future events are made following the program and before adjourning.
- Adjournment—***Adjournment*** is the time when meetings end. Most meetings have a set time to end; others end when all items in the order of business have been covered. Having a set time usually moves meetings along. Adjournment is typically handled by the same person that called the meeting to order.

PRESIDING OVER A MEETING

A ***presider*** is the person who runs a meeting. They call the meeting to order on time. Presiders follow the order of business and keep the program moving along. This includes announcing the item to be covered and introducing the person who will cover an item.

Usually, the presider is the president or chair of the group. They should have abilities in serving as a presider. They are always friendly, fair, and honest.

23-11. A presider may use a gavel.

Several desirable traits of a presider are:

- Takes the responsibility of presiding seriously
- Plans for the meeting well ahead of time
- Arrives early to arrange the meeting room
- Has a professional appearance
- Begins and ends meetings on time
- Uses good communication skills, including speaking clearly and loud enough to be heard
- Follows the rules of the organization
- Is courteous and fair to all members

23-12. After a meeting, thank people for attending.

- Allows any members who wish to speak on issues to do so but never relinquishes control of the meeting
- Uses parliamentary procedure, as appropriate
- Has a sense of humor and graciously corrects errors
- Returns the meeting room to the way it was before the meeting was held
- Follows up after a meeting to thank people for their efforts in making it successful

Decisions by groups are often made using parliamentary procedure. Most groups adopt the procedures in *Robert's Rules of Order.* This book describes how parliamentary procedure is used. It gives suggestions for presiders.

MINUTES

Minutes are the official written records of a meeting. The information is usually recorded by the secretary during a meeting. The order of business provides an outline for the minutes. Following the meeting, the secretary edits the notes into the proper form. The minutes are acted on at the next meeting.

The minutes may be recorded as notes during the meeting on paper or with a computer. If a computer is used, be sure to regularly save the information in a file. This assures that you will have the minutes in case of power or computer failure. A failure could result in the minutes being lost. Once lost, it is nearly impossible to come up with accurate minutes. Computer-kept minutes should be printed out for distribution to members and for a permanent record.

Most organizations have a minutes book. Sometimes, this book is known as a secretary's book. This is a book with the minutes of the organization over a long time. It serves as the official record of decisions made by the members in their meetings. Minute books should be kept in a safe place.

Minutes of a Regular Meeting

____

(Organization Name)

____

(Date)

Call to Order: The meeting was called to order at _____ (time) by the president, ____________________________ (name).

Roll: ________ (number of members present)

Minutes: Minutes of the previous meeting were read and approved.

Treasurer's Report: The treasurer reported a balance on hand of $_________ in the checking account and $ ________ in savings.

Committee Reports: ____________________________ (name) gave the following for the Recreation Committee: (include a brief statement summarizing the report).

Program: Example—Jane Smith presented slides on Australia.

Old Business: Example—Members were reminded to bring toys to the next meeting for the "Toys for Tots" campaign.

New Business: Example—A motion was made by Jill Olson that the organization sell Christmas wreaths. Sam Carter seconded the motion, and the motion carried.

Adjourn: The meeating Adjourned at _________ (time).

______________________ ______________________

Signed: President Signed: Secretary

23-13. Sample format for minutes.

REVIEWING

MAIN IDEAS

Leadership is helping people achieve group goals. A leader must have followers to lead. Important concepts in being a leader include influence, process, relationship, and service. Leaders have traits that allow them to function. Fortunately, all people can develop these traits.

Leadership style is how a leader goes about the role of leading. Styles range from the leader being a dictator to the leader allowing the members to do about whatever they want. Democratic leadership is between the two extremes.

Team building is an important part of leading a successful group. People on a team must have unity, good relationships, and be willing to work together. Teams usually go through three steps: forming, norming, and performing.

Leaders are often responsible for running meetings. Most meetings are held to conduct business or have an informative program. Careful planning is needed for a meeting to be a success. An order of business should be prepared. The presider should take the responsibility seriously. Official written records are kept of most meetings. These are known as minutes.

QUESTIONS

Answer the following questions. Use correct spelling and complete sentences.

1. What is needed for a person to be a leader?
2. What are the four concepts of leadership important to leaders? Briefly explain each.
3. What are the attributes of a leader? (List four.)
4. What are three leadership styles? Briefly explain each.
5. What is team building? Why is it important?
6. What are three characteristics of a good team?
7. Why are meetings held?
8. Why is planning important to the success of a meeting? What should be done in planning a meeting?
9. What is an order of business? Why are they prepared?
10. What are minutes?

EVALUATING

Match the term with the correct definition. Write the letter of the term in the blank provided.

a. order of business
b. minutes
c. follower
d. democratic leader
e. team building
f. meeting
g. presider
h. adjournment

____ 1. Time when a meeting ends.
____ 2. Plan or sequence of events in a meeting.
____ 3. Person who runs a meeting.
____ 4. Official written records of a meeting.
____ 5. An individual who achieves goals under the direction of a leader.
____ 6. A leadership style that allows followers to help make decisions.
____ 7. A group of people who have gathered for a shared purpose.
____ 8. Getting people to work together as a group.

EXPLORING

1. Attend a meeting of local government officials as an observer. Groups include county commissioners, town boards, boards of education, and planning commission members. Note the sequence of the meeting and how items are handled. Prepare a written report of your observations. Include suggestions on how the meeting could have been run more effectively.

2. Arrange a workshop or class session on parliamentary procedure. Use *Robert's Rules of Order* as the main reference. (The latest edition of this book is available from Scott, Foresman and Company.) Develop skills in parliamentary procedure by conducting practice business meetings. Have members take turns making motions, serving as presiders, and handling items of business.

3. Attend a parliamentary procedure contest. Note how members of the teams make motions and handle business. Note how the presiding officer goes about his or her duties.

GLOSSARY

Note: The number in parenthesis following the definition refers to the page in the text where the term is first defined.

A

Accessories—clothing items to complete an outfit for business or personal dress; examples include ties, scarves, and jewelry **(466)**.

Accident—an event or condition that occurs unintentionally **(93)**.

Accounting—a system of recording financial transactions **(381)**.

Active FFA membership—students in grades 7 to 12 enrolled in agricultural education who have paid membership dues to join the FFA **(421)**.

Adjournment—the time when a meeting ends **(492)**.

Agribusiness—nonfarm work or activities in the agricultural industry **(3)**.

Agricultural education—education in and about agriculture **(21)**.

Agricultural experiment station—research units associated with land-grant colleges **(31)**.

Agricultural industry—all activities to meet the needs of people for food, clothing, and shelter **(3)**.

Agricultural mechanics—using mechanical devices to do agricultural jobs **(191)**.

Agricultural technology—technology used in agriculture; science-based inventions used in agriculture **(39)**.

Agriculture—growing crops and raising animals to meet the needs of people for food, clothing, and shelter **(5)**.

Agriculture policy—laws and government actions that influence agriculture **(12)**.

Agriscience—use of science in producing food, clothing, and shelter materials **(113)**.

Agronomy—study of plants used for crops **(115)**.

Air—mixture of gases that surrounds the earth **(171)**.

Animal—an organism that obtains food; a member of the Animal Kingdom **(116)**.

Animal by-product—a secondary product from the slaughter of animals, such as hides, hair, and hooves **(244)**.

Animal science—study of animals used for food and other purposes, primarily the study of livestock **(117)**.

Animal well being—caring for an animal so all its needs are met and it does not suffer **(269)**.

Annual—a plant that completes its life cycle in one growing season **(278)**.

Applied research—investigations to solve specific questions with results that may be used immediately **(221)**.

Appropriate technology—technology people can use **(39)**.

Aquaculture—farming in water; culturing fish, molluscs, crustaceans, and aquatic plants **(7)**.

Article—factual writing about one topic; frequently published in magazines, journals, and newspapers **(398)**.

Artificial insemination—using implements to place sperm in the reproductive tract of a female **(267)**.

Attribute—the trait of an individual **(483)**.

Audit—an examination of the financial records by a person who is trained to do so **(381)**.

Auger—a device made of a steel shaft wrapped with screw threads that is used to bore holes **(202)**.

Autocratic leader—a leader who makes decisions and tells people what to do **(484)**.

B

Balanced ration—a ration that contains all of the nutrients an animal needs in correct proportions **(266)**.

Basic research—investigations that attempt to answer scientific questions but do not have an immediate practical application **(221)**.

Bedrock—parent material of soil **(315)**.

Beef—meat from cattle; cattle used to produce meat **(241)**.

Biennial—a plant that completes its life cycle in two growing seasons **(279)**.

Binomial nomenclature—a system of naming organisms based on scientific similarities and differences **(277)**.

Biotechnology—using science to change organisms or their environment or to get products from organisms **(131)**.

Botany—study of plants **(115)**.

Brace—the device for holding and turning an auger **(202)**.

Breed—animals within a species that have been bred for specific qualities that are distinctive and consistent **(268)**.

Business—a person or group that produces and/or sells goods and services **(81)**.

C

CGIAR—the Consultative Group on International Agricultural Research; promotes sustainable agriculture in developing countries **(33)**.

Caliper—a tool used for measuring thicknesses or diameters **(198)**.

Call to order—statements or procedures by a leader to announce that a meeting is about to begin **(490)**.

Canning—a method of food preservation that involves heating food materials and containers and sealing the container **(343)**.

Capital—available resources of a farm, agribusiness, or other entity, including money, equipment, animals, and raw materials **(380)**.

Carcass—muscles and bones of a slaughtered animal **(242)**.

Career—general area of work a person follows **(441)**.

Career Development Event (CDE)—official FFA activities that allow members to show their career and agricultural skills **(427)**.

Career ladder—upward movement of a person in a career through progressive advancements **(445)**.

Castration—removing the male testicles **(271)**.

Chapter FFA Degree—the second level of active FFA membership **(421)**.

Charter—an official act that recognizes a state FFA association or local chapter **(424)**.

Chemistry—study of the makeup of materials (matter) and reactions by materials **(118)**.

Cheval—meat from horses **(241)**.

Chisel—a wedge-shaped cutting tool for wood, metal, and other materials **(200)**.

Chromosome—threadlike structure in a genome that contains genetic material and protein **(139)**.

Circuit—complete path for the flow of electrons **(214)**.

Claw hammer—tool used for driving nails into wood **(202)**.

Clay—smallest mineral particles in soil **(311)**.

Cloning—making two or more organisms out of one using tissues **(132)**.

Codex Alimentarius—a group that sets standards for food products in international trade **(163)**.

Cold processing—refrigeration; used to store perishable food products at low temperatures but above freezing **(342)**.

Collateral—what is promised to a lender if a loan is not repaid **(381)**.

Commercial agriculture—producing plants and animals to sell **(15)**.

Commodity—an agricultural product such as grain or fiber **(163)**.

Communication—process of exchanging information **(393)**.

Competition—rivalry between two or more businesses that produce the same or similar products from which consumers will choose **(87)**.

Compliance—fulfilling all the legal regulations associated with a business **(382)**.

Compression stroke—occurs when a piston compresses the fuel and air mixture in a cylinder of an internal combustion engine **(212)**.

Concentrate—a feed ingredient high in energy and low in fiber **(266)**.

Conclusion—statement of the value of a hypothesis based on experimentation **(231)**.

Conductor—material that transmits electricity **(215)**.

Conifer—evergreen tree that usually produces seeds in cones **(69)**.

Conservation—wise use of resources **(183)**.

Consumer—a person, business, or agency that uses goods and services **(85)**.

Consumer-driven market—a market in which the consumer makes decisions about buying and selling that have an influence on the overall market **(383)**.

Control—individual or group that does not receive the experimental variable or treatment **(229)**.

Cooperative—an association to provide services to members **(85)**.

Corporation—association of members for doing business; a legal entity with the same freedoms as people have in doing business **(84)**.

Culture—characteristics of people arising from their customs; the way people live as passed from one generation to the next **(155)**.

Current electricity—flowing electrons in a circuit **(214)**.

Curriculum—classes and learning experiences that students have in school **(27)**.

Custom—long established way of doing something **(155)**.

Cuttings—sections of plant leaves, stems, or other parts that grow into a plant; a method of cloning **(133)**.

D

Data—information collected in doing research **(228)**.

Debeaking—removing the tip of a bird's beak **(272)**.

Debt—what is owed to another **(380)**.

Deciduous—tree or other plant that sheds its leaves in the winter **(69)**.

Dehorning—removing the horns on an animal **(272)**.

Dehydration—drying; the nearly complete removal of water from a food product **(344)**.

Demand—amount of something that is available, especially the amount people will buy at a given price **(76)**.

Democratic leader—a leader who allows and encourages followers to make decisions and seeks to involve all people in the process **(485)**.

Developed country—a country that is industrialized; jobs are available so people can work and earn money **(153)**.

Developing country—a country that is not industrialized; many people have low incomes **(153)**.

Development—creating something new, such as a useful product or service; also known as R&D **(31)**.

Dicot—a plant with two seed leaves (abbreviation of dicotyledon) **(281)**.

Direct marketing—products sold by a producer directly to consumers **(386)**.

Dividend—profit made by a corporation that is paid to owners of stock **(85)**.

Docking—removing the tail, especially with young lambs **(271)**.

Domesticated—an organism that has been tamed, confined, and bred for human use **(64)**.

Dress—the kind of clothing people wear **(463)**.

E

Earplug—a device that fits in the ear to protect it from injurious sounds **(99)**.

Economic system—how people go about doing business, including how things are created, owned, and exchanged **(78)**.

Economics—the study of the system by which people meet their needs, with emphasis on financial areas **(75)**.

Ecosystem—all of the parts of an organism's environment that influence how it lives and grows **(121)**.

Electricity—the flow of electrons in a conductor **(214)**.

Endangered species—organism species threatened with becoming extinct **(175)**.

Engine systems—components that are essential for an engine to operate **(211)**.

Entertaining speech—an oral presentation that evokes emotion, such as pleasure or laughter, from the audience **(407)**.

Entomology—the study of insects **(117)**.

Entrepreneurship—creating goods or services to meet unique needs **(88)**.

Environment—surroundings of an organism **(121)**.

Environmental science—study of the environment **(121)**.

Erosion—loss of soil by the action of natural forces, such as water and wind **(183)**.

Etiquette—a system or procedure of social behavior **(471)**.

Exhaust stroke—piston forces burned fuel out of the cylinder of an internal combustion engine **(213)**.

Experience—personally benefiting from doing something **(447)**.

Experiment—any method for testing a hypothesis **(226)**.

Experimental variable—condition or variable manipulated in an experiment **(229)**.

Exploratory SAE—a type of SAE used by beginning students in agricultural education to learn about different areas and interests **(449)**.

Export—a product that leaves a country **(57)**.

Extinction—the complete disappearance of a species from the earth **(174)**.

Extrovert—a person who focuses on the outer world and has high interest in other people **(462)**.

F

FDA (Food and Drug Administration)—an agency of the U.S. government responsible for assuring quality in selected products **(163)**.

FFA—the organization for students enrolled in secondary agricultural education **(30)**.

FFA advisor—an agriculture teacher who oversees an FFA chapter **(422)**.

FFA motto—a twelve-word statement that guides FFA members in achieving worthy goals in life **(433)**.

Fabric—the material used in making clothing **(466)**.

Fabrication—cutting or shaping products into forms that are more convenient **(336)**.

Farm—a place where farming occurs **(6)**.

Farming—using land and other resources to grow crops and raise animals **(6)**.

Farrowing—the birth of piglets **(249)**.

Feedback—return channel from the receiver back to the sender in the communication process **(396)**.

Feeder animal—young animal that is fed for growth and finish to produce meat **(241)**.

Feral animal—a domesticated animal that has returned to the wild **(172)**.

Fertilizer—a material added to soil to increase nutrients for plant growth **(320)**.

Field plot—small area that is used for growing experimental crops or testing new methods **(31)**.

Flower—the first reproductive part of a plant **(287)**.

Fluid milk—liquid milk products **(64)**.

Follower—an individual who goes about reaching goals under the direction of a leader **(481)**.

Food additive—any substance added to food to increase taste appeal, improve consistency, aid in preservation, or meet another need **(350)**.

Food Guide Pyramid—a graphic representation of human nutrient needs and the foods that will provide for the needs **(60)**.

Food poisoning—an illness caused by eating spoiled or contaminated food **(334)**.

Food preservation—treating food to keep it from spoiling **(49)**.

Food processing—steps used to prepare raw food for consumption **(335)**.

Forestry—the production and manufacturing of trees into products **(8)**.

Fossil fuel—material from formerly living organisms that is used to provide energy **(178)**.

Free enterprise—way of doing business that allows people to own and operate businesses with a minimum of government control **(80)**.

Freezing—exposing foods and other materials to temperatures below the freezing point of water; a method of food preservation **(340)**.

Fruit—formed following a flower from the walls of the fertilized ovary **(289)**.

Future Farmers of America—an organization for students in agriculture classes founded in 1928; today it is known as the FFA **(419)**.

G

GATT (General Agreement on Tariffs and Trade)—agreement among countries that apply regulations uniformly on international trade **(164)**.

Game—species of wildlife that are hunted for food and sport **(174)**.

Gene—the unit of heredity on a chromosome **(140)**.

Gene mapping—process of locating and identifying genes by trait **(140)**.

Gene splicing—joining DNA from one organism with the DNA in another **(141)**.

Genetic code—how the information in the genes of an organism is arranged **(140)**.

Genetic engineering—artificially changing the genetic makeup of an organism **(136)**.

Genome—structure in a cell that contains heredity material **(139)**.

Germination—process of a plant growing from a seed **(291)**.

Gestation—period of development of young in its mother's body **(268)**.

Global economy—economic situation resulting from the trade between countries **(161)**.

Global positioning system—a satellite-based approach to locate exact positions on the earth **(366)**.

Goal—an end to be reached, such as a career goal **(441)**.

Goal setting—making goals; the process of describing desired life accomplishments **(442)**.

Grading—sorting products by size, kind, and quality **(335)**.

Grafting—placing a section of one plant onto another plant so that the section grows **(134)**.

Greenhand FFA Degree—the first level of active FFA membership **(421)**.

Grooming—the neatness of an individual's personal appearance **(467)**.

Ground fault circuit interrupter (GFCI)—device used on electrical wires to protect from shock **(104)**.

Growth implant—small pellet or other material containing hormones placed in an animal to promote growth **(136)**.

H

Habitat—the physical area where an organism lives **(176)**.

Hammer—tool made for driving or pounding **(202)**.

Hand tool—small powerless tool used to do mechanical jobs **(195)**.

Hardwood tree—deciduous trees with fine grain wood **(69)**.

Harvester—a mechanical device that gathers or picks crop products **(44)**.

Hatch Act—a Federal act passed in 1887 to set up a system of agricultural experiment stations **(24)**.

Hazard—a situation where risk of injury or loss is present; danger **(93)**.

Health—physical or mental condition of the body **(470)**.

Herbicide—a substance used to control plant pests **(47)**.

Horizon—visible layers in a soil profile **(314)**.

Hormone—a substance produced by an organism that has a specific effect on its growth or other behavior **(136)**.

Hormone residue—traces of hormones in food and other materials **(163)**.

Horticulture—production and use of plants for personal appeal or food **(115)**.

Hunger—discomfort caused by a need for food **(150)**.

Hypothesis—a proposition about relationships between variables that will be tested using research methods; what the scientist believes will happen **(226)**.

I

Import—products brought into a country from another country **(58)**.

Influence—encouraging an individual or group to take action **(481)**.

Informative speech—an oral presentation that gives the audience new knowledge **(407)**.

Insecticide—substance used to control insect pests **(48)**.

Insulator—material that is not a good conductor of electricity; used to protect conductors **(215)**.

Intake stroke—bringing fuel and air mixture into an internal combustion engine **(212)**.

International trade—buying and selling of goods by two or more nations or businesses within the nations **(57)**.

Introvert—a person who focuses on himself or herself and does not show high interest in other people **(462)**.

Invention—a new device or product; new ways of doing work **(42)**.

Inventory—goods on hand; merchandise or production materials in possession of a manufacturer, retail source, or other entity **(379)**.

Irradiation—exposing food to radiant energy to kill spoilage organisms **(344)**.

Issue—question or problem that has more than one possible answer **(142)**.

L

Lactation—process of milk production by female mammals **(242)**.

Laissez-faire leader—a leader who allows members to make decisions without interference **(484)**.

Lamb—meat from a young sheep; a young sheep **(241)**.

Leader—a person who helps other people achieve goals **(416)**.

Leadership—the ability to influence other people to meet individual or group goals **(416)**.

Leadership style—how a leader goes about fulfilling his or her role **(484)**.

Letter—a written message; two main types are used: personal and business **(397)**.

Life cycle—length of time from beginning to death of an organism; from germination to death of a plant; from birth or hatching to death of an animal **(278)**.

Lime—material applied to land to lower the acidity of soil (raise the pH) **(318)**.

Linen—cloth material made from fibers of the flax plant **(68)**.

Listening—drawing meaning from oral communication **(405)**.

Livestock—large animals raised on farms or ranches for food or other uses **(245)**.

Loam—soil with equal amounts of sand and silt and a smaller amount of clay **(312)**.

Loan—money or other capital that has been borrowed **(380)**.

M

Major—collection of courses in an area that helps meet degree requirements from a college **(446)**.

Major elements—nutrients needed in largest quantity by plants **(319)**.

Malnutrition—result of food deficiencies in an organism **(149)**.

Marketing—getting what people want to them in the desired form **(15)**.

Management—the way an agribusiness, farm, or other enterprise is run; doing things to help an agribusiness achieve its goals **(375)**.

Manager—a person who makes decisions and uses resources to reach goals **(375)**.

Mass selection—a method of crop improvement that involves saving seed from the best plants **(46)**.

Materials—articles or supplies used in constructing **(191)**.

Material Safety Data Sheet (MSDS)—printed information with chemical products that describes how to safely use and store the product **(107)**.

Measuring device—a tool used in making measurements **(196)**.

Meat—muscle tissue of an animal **(241)**.

Mechanical technology—using power or other devices that increase force **(43)**.

Medium—the channel that connects the sender with the receiver in the communication process **(396)**.

Meeting—a group of people who gather for a shared purpose **(487)**.

Meeting program—informative or educational part of a meeting **(492)**.

Message—the idea being shared in the communication process **(395)**.

Mineral—a material on the earth that has never lived **(177)**.

Minor element—nutrients needed by plants in small quantities **(319)**.

Minutes—official written records of a meeting **(493)**.

Money—medium of exchange; anything exchanged for goods or services **(77)**.

Monogastric—a digestive system, such as that in swine, consisting of a stomach with one compartment **(264)**.

Monopoly—business condition when there is no competition; one source controls products and prices **(87)**.

Monocot—a plant that has one seed leaf (abbreviation for monocotyledon) **(281)**.

Morrill Act—a Federal law passed in 1862 to set up a system of land-grant schools to teach agriculture and related areas **(23)**.

Mutton—meat from a sheep that is more that one year old **(241)**.

N

Natural fiber—fiber made from a plant or animal source **(67)**.

Natural resources—all of the things found in nature, including living organisms, minerals, soil, water, and air **(9)**.

New business—items being brought before a group in a meeting for the first time **(492)**.

Nonrenewable natural resource—a resource that is not replaced once it is used **(181)**.

Nonverbal communication—the exchange of information without words **(393)**.

Nonverbal cues—signals that people send that have meaning **(393)**.

Nutrient—any substance required for life **(265)**.

Nutrition facts panel—a label on food products that provides nutrition information **(66)**.

Nutritional deficiency—condition that develops from a lack of proper nutrients **(149)**.

O

Official dress—uniform worn by members at FFA events **(434)**.

Old business—items on an order of business remaining from a previous meeting **(491)**.

Operator's manual—written document with information on safe operation and maintenance of equipment **(102)**.

Oral communication—using spoken words to share information **(404)**.

Order of business—a plan that lists events or items in a meeting **(490)**.

Organic matter—any product from living organisms; decaying plant and animal waste **(309)**.

Organism—a living thing **(114)**.

Ornamental horticulture—producing plants for their beauty; includes foliage and flower plants as well as landscaping and landscape maintenance **(8)**.

Ownership SAE—type of SAE in which the student owns and manages a production agriculture enterprise or business venture **(450)**.

P

Packaging—placing food in an appropriate container for food safety and consumer use **(336)**.

Parasite—organism that lives in or on another organism, such as a round worm in cattle **(49)**.

Particle gun—device used to send (shoot) DNA into a cell during gene splicing **(141)**.

Particulate mask—mask that covers the nose and mouth to remove dust particles from air that is inhaled **(100)**.

Partnership—business owned by two or more people **(83)**.

Parturition—birth process **(269)**.

Pathogen—an organism that causes disease **(338)**.

People skills—abilities that help people get along well together **(447)**.

Perennial—a plant that needs three or more years to complete its life cycle **(279)**.

Perishable food—food that does not keep well; food susceptible to spoilage **(333)**.

Personal growth—developing skills to have a good life **(417)**.

Personal hygiene—cleanliness of body; being free of odor and dirt **(468)**.

Personal protective equipment (PPE)—devices to protect people from injury **(96)**.

Personality—the visible part of a person, especially as related to spoken qualities **(461)**.

Persuasive speech—an oral presentation designed to get the audience to do something **(407)**.

Pest—anything that causes injury to animals, plants, or property **(47)**.

Pesticide—a substance that is used to control pests **(47)**.

Pesticide residue—traces of pesticides found on materials, including food products **(163)**.

pH—a scale for measuring the acidity or basicity of soil **(317)**.

Phloem—tissue that carries plant products, such as glucose, from production sites to other parts **(284)**.

Photosynthesis—process in plants that converts water and carbon dioxide to glucose sugar and oxygen **(286)**.

Physics—the study of the physical nature of objects, including heat, light, electricity, and mechanics **(120)**.

Placement SAE—type of SAE in which the student gains experience working for another person, business, or farm **(451)**.

Plantain—a fruit similar to the banana **(152)**.

Plant—an organism that makes its own food; a member of the Plant Kingdom **(114)**.

Planter—a mechanical device that places seed or vegetative reproductive parts of plants in the soil **(44)**.

Pleasure animal—an animal kept for fun or companionship, such as a dog **(244)**.

Pliers—a wrench made with two levers for grasping or cutting bolts, wire, and other materials **(204)**.

Plow—a tool that loosens soil and may shape the soil into rows or other forms **(43)**.

Polled animal—cattle that naturally do not have horns **(246)**.

Popular literature—books, newspapers, magazines, and other publications for the general public **(223)**.

Pork—meat from swine **(241)**.

Posture—how a person holds and positions his or her body **(470)**.

Poultry—domesticated birds, such as chickens and turkeys **(242)**.

Poverty—condition when people do not have enough money to meet basic needs for food, clothing, and housing **(58)**.

Power stroke—occurs as a piston in an internal combustion engine is forced backward by the heat produced when fuel burns **(213)**.

Power tool—any tool with power from a motor or engine **(205)**.

Precision farming—a form of variable rate technology that combines information and technology to manage crops **(365)**.

Preservation—keeping resources so that they are not consumed **(185)**.

Preserving—treatments used to keep food from spoiling **(336)**.

Presider—person who runs a meeting **(492)**.

Price—amount of money exchanged in buying or selling a product or service **(76)**.

Private ownership—when people own property **(80)**.

Proactive—taking preventative actions before problems arise **(358)**.

Proficiency award—recognition from the FFA for students who excel with their SAE **(428)**.

Promotion—any activity to increase the public's awareness, interest, and purchasing of a product **(387)**.

Property—anything that has value that can be exchanged **(80)**.

Proprietary R&D—new technology developed and owned by a private business **(363)**.

Protocol—area of etiquette that deals with rules for events attended by government, business, and other officials **(471)**.

Prototype—a test model of a new product which serves as a pattern for future production **(34)**.

Public speaking—giving a formal oral presentation to a group **(407)**.

Q

Quality of life—the environment in which we live, including food, housing, health, air, water, and other conditions **(57)**.

R

R&D—the combination of research and development to create new technology **(362)**.

Ration—feed given an animal in a 24-hour period **(266)**.

Receiver—the individual who gets a message and tries to draw meaning from it **(396)**.

Recombinant DNA—new DNA formed by genetic engineering **(141)**.

Recycling—reusing materials from a product to make another product **(183)**.

Remote sensing—collecting information about something from a distance **(368)**.

Renewable natural resource—a resource, such as water, that can be restored after it is used **(180)**.

Report—detailed written document or oral presentation **(398)**.

Research—using systematic methods to answer questions **(31)**.

Research and experimentation SAE—type of SAE with a science base that involves laboratory or field investigations **(452)**.

Resource—any material with value **(378)**.

Respirator—device that covers the nose and mouth or the full face that purifies air before it is inhaled **(100)**.

Resume—written summary of an individual's education, experience, and accomplishments **(398)**.

Retail market—an enterprise that sells to consumers **(386)**.

Risk—possibility of loss **(81)**.

Root—vegetative part of plant that grows underground **(282)**.

Roughage—feed ingredient high in fiber, such as grasses and legumes **(266)**.

Ruminant—a digestive system, such as that in cattle, with the stomach made into compartments **(265)**.

S

Safe—being free from harm and danger **(93)**.

Safety—preventing injury or loss **(95)**.

Safety glasses—glasses worn to protect the eyes **(97)**.

Safety testing—using proposed new products on a small scale to determine if they cause any safety problems **(363)**.

Sand—small particles of rock; largest particles in soil **(311)**.

Saw—a tool with sharp teeth for cutting materials **(199)**.

Science—knowledge of the world based on careful observation of objects and events **(113)**.

Scientific fact—information or results that can be shown at any time **(232)**.

Scientific law—a broad sweeping statement about phenomena based on scientific facts **(232)**.

Scientific literature—journals and other published materials in technical language for use by scientists **(224)**.

Scientific method—a step-by-step process for solving scientific problems **(221)**.

Screwdriver—a tool with a wood or plastic handle on a long metal shaft, with one end of the shaft shaped to fit screws **(205)**.

Secondary agricultural education—instruction in agriculture offered in middle and high schools **(27)**.

Seed—reproductive structure formed in a plant ovary that becomes a new plant **(291)**.

Self-employed—a person who works for himself or herself **(441)**.

Self-sufficient farming—producing crops and animals for personal consumption; not commercial **(14)**.

Service animal—animal that helps people in some way, such as a guard dog **(244)**.

Silt—medium-sized mineral particles in soil **(311)**.

Site-specific farming—using practices based on the specific needs of a location; also known as variable rate technology **(52)**.

Slope—steepness of land; measured as feet of rise to run **(321)**.

Smith-Hughes Act—a Federal law passed in 1917 which provided for agricultural education and other vocational education in public schools **(24)**.

Smith-Lever Act—a Federal law passed in 1914 which set up the Cooperative Extension Service **(24)**.

Softwood tree—evergreen trees with a coarse grain wood **(69)**.

Soil—outer layer of the earth's surface that supports plant growth and other life **(169)**.

Soil conservation—using practices that protect the soil from loss or damage **(183)**.

Soil profile—a vertical cross section of soil **(314)**.

Soil science—the study of soil **(118)**.

Soil structure—the way soil particles arrange themselves; known as aggregates or peds **(313)**.

Soil texture—size of particles in soil **(311)**.

Soil triangle—graphical representation of soil textures based on amount of sand, silt, and clay **(312)**.

Sole proprietorship—business owned by one person **(82)**.

Source—the sender or initiator of a message **(395)**.

Spoilage—food ruined by bacteria, fungi, or foreign material **(334)**.

Squanto—a Native American who taught colonists in North America about agriculture **(21)**.

Square—tool for getting angles and marking materials **(199)**.

Staff—employees; the people who work for a farm, business, agency, or other employer **(379)**.

Stem—vegetative part of a plant that supports leaves, buds, and other organs **(283)**.

Sub-Saharan Africa—area of African continent south of the Sahara desert **(152)**.

Subsoil—layer of soil beneath the topsoil **(315)**.

Suburban farming—using small areas of land in residential and business areas of cities or areas surrounding cities to produce crops and animals **(7)**.

Supervised agricultural experience—application of instruction in agriculture through real-world experiences; experiences are planned by the teacher, student, and others; four types: exploratory, placement, ownership, and research and experimentation; also known as SAE **(29)**.

Supply—amount of something that is available **(75)**.

Sustainability—using resources in a way that guarantees continued opportunity to use them **(370)**.

Sustainable agriculture—using practices that maintain the ability to grow crops and raise livestock **(125)**.

Sustainable resource use—using resources so that they last a long time **(182)**.

System—a thing or process of several parts that work together for one function or purpose **(355)**.

T

Tariff—a tax placed on imports or exports **(164)**.

Team building—getting people to work together as a group **(485)**.

Technology—use of inventions in working and living; application of science to improve productivity **(39)**.

Technology system—combining ideas and machines to achieve a purpose such as agricultural production or processing **(356)**.

Theory—statement of a principle that explains potential solutions to a problem **(232)**.

Tissue culture—a kind of cloning that uses cells or small clusters of cells from a parent to produce a new living thing **(132)**.

Tool—an implement used to do a mechanical job **(191)**.

Topsoil—the top few inches of soil high in organic matter that supports plant growth **(315)**.

Toxin—a chemical by-product that sickens or kills other organisms **(334)**.

Trade balance—difference between the imports and exports of a country **(161)**.

Trade barrier—government policies that slow or stop international trade **(162)**.

Training agreement—a written document that describes the terms of SAE that is signed by the student, parent(s), and teacher as well as the employer if one is involved **(454)**.

Training plan—a written document that lists the experiences to be gained in an SAE **(454)**.

Training station—the place where SAE experiences are carried out **(454)**.

Transgenic—an organism that has been modified by having a gene from another organism transferred into it **(46)**.

Transgenic organism—the product organism resulting from altering the genetic material into an organism **(137)**.

U

Udder—milk-producing organ containing the mammary glands **(247)**.

V

Value-added product—a product that has been made more usable for the consumer, such as by processing **(385)**.

Variable rate technology—varying cropping practices in a field based on conditions in small segments of the field **(364)**.

Vascular system—made of xylem and phloem in plants; transports fluids within organisms **(284)**.

Vegetarian—a person who eats foods from plant origin and excludes meat from his or her diet **(66)**.

Verbal communication—use of words to convey meaning **(393)**.

W

WTO (World Trade Organization)—an organization that promotes international trade and well being **(164)**.

Water conservation—using practices that prevent water loss and protect water supplies **(184)**.

Wellness—positive actions to have good health; freedom of disease **(470)**.

Western style—the style of clothing worn in the United States, as contrasted with clothing of Africa and areas of Asia **(464)**.

Wholesale market—an enterprise in the marketing process that buys and resells products to retailers or other wholesalers **(386)**.

Wholesome food—nutritious food that contributes to good health and well being **(333)**.

Wildlife—living things that have not been domesticated **(172)**.

Winter annual—a plant that completes its life cycle, which includes growing, in the winter **(278)**.

Wrench—a tool for grasping and turning bolts, screws, and nuts **(203)**.

X

Xylem—tissue that carries water and minerals from root hairs throughout the plant **(284)**.

Z

Zoology—the study of animals **(117)**.

BIBLIOGRAPHY

Alexandratos, Nikos. *World Agriculture: Towards 2010, An FAO Study.* New York: John Wiley & Sons, Inc., 1995.

Alpha Tau Talk, *Newsletter of the Penn State Chapter of Phi Delta Kappa.* University Park: Pennsylvania State University, 1994.

Biondo, Ronald J., and Jasper S. Lee. *Introduction to Plant and Soil Science and Technology.* Danville, Illinois: Interstate Publishers, Inc., 1997.

Biondo, Ronald J., and Charles B. Schroeder. *Introduction to Landscaping: Design, Construction, and Maintenance,* 2nd ed. Danville, Illinois: Interstate Publishers, Inc., 2000.

Brown, Lester R. *State of the World.* New York: W. W. Norton & Company, 1997.

Drache, Hiram M. *History of U.S. Agriculture and Its Relevance to Today.* Danville, Illinois: Interstate Publishers, Inc., 1996.

Ecosystems. Paramus, New Jersey: Globe Fearon Educational Publisher, 1995.

Holland, I. I., and G. L. Rolfe. *Forests and Forestry,* 5th ed. Danville, Illinois: Interstate Publishers, Inc., 1997.

Hunter, Sharon, Marshall Stewart, Brenda Scheil, Robert Terry, Jr., and Steven D. Fraze. *Developing Leadership and Personal Skills.* Danville, Illinois: Interstate Publishers, Inc., 1997.

International Food Information Council Foundation. [Online]. http://ificinfo.health.org/

Janick, Jules, and James E. Simon, ed. *New Crops.* New York: John Wiley & Sons, Inc., 1993.

Lee, Jasper S., Jim Hutter, Rick Rudd, Lyle E. Westrom, Austin M. Bull, Chris Mohr, and Jody Pollok. *Introduction to Livestock and Compantion Animals,* 2nd ed. Danville, Illinois: Interstate Publishers, Inc., 2000.

Lee, Jasper S., James G. Leising, and David E. Lawver. *AgriMarketing Technology.* Danville, Illinois: Interstate Publishers, Inc., 1994.

Lee, Jasper S., and Diana L. Turner. *Introduction to World AgriScience and Technology,* 2nd ed. Danville, Illinois: Interstate Publishers, Inc., 1997.

Maynard, Donald N., and George J. Hochmuth. *Handbook for Vegetable Growers.* New York: John Wiley & Sons, Inc., 1997.

Newman, Michael E., and Walter J. Wills. *Agribusiness Management and Entrepreneurship,* 3rd ed. Danville, Illinois: Interstate Publishers, Inc., 1994.

Official FFA Manual. Alexandria, Virginia: National FFA Organization, 1999.

Osborne, Edward W. *Biological Science Applications in Agriculture.* Danville, Illinois: Interstate Publishers, Inc., 1994.

Phipps, Lloyd J., and Glen M. Miller. *AgriScience Mechanics.* Danville, Illinois: Interstate Publishers, Inc., 1998.

Porter, Lynn, Jasper S. Lee, Diana L. Turner, and J. Malcolm Hillan. *Interstate's Environmental Science and Technology.* Danville, Illinois: Interstate Publishers, Inc., 1997.

Schroeder, Charles B., Eddie Dean Seagle, Lorie M. Felton, John M. Ruter, William Terry Kelley, and Gerard Krewer. *Introduction to Horticulture: Science and Technology,* 3rd ed. Danville, Illinois: Interstate Publishers, Inc., 2000.

Smith, Robert Leo. *Elements of Ecology.* New York: HarperCollins Publishers, Inc., 1992.

United Nations. [Online]. http://www.un.org/

INDEX

D

E

F

G

H

I

J

L